Edexcel

Chemistry for A2

Graham Hill
Andrew Hunt

Hachette UK's policy is to use papers that are natural, renewable and recyclable products and made from wood grown in sustainable forests. The logging and manufacturing processes are expected to conform to the environmental regulations of the country of origin.

Orders: please contact Bookpoint Ltd, 130 Milton Park, Abingdon, Oxon OX14 4SB. Telephone: (44) 01235 827720. Fax: (44) 01235 400454. Lines are open 9.00 – 5.00, Monday to Saturday, with a 24-hour message answering service. Visit our website at www.hoddereducation.co.uk

© Graham Hill, Andrew Hunt 2009
First published in 2009 by
Hodder Education,
An Hachette UK Company
338 Euston Road
London NW1 3BH

Impression number 5 4 3
Year 2013 2012 2011

All rights reserved. Apart from any use permitted under UK copyright law, no part of this publication may be reproduced or transmitted in any form or by any means, electronic or mechanical, including photocopying and recording, or held within any information storage and retrieval system, without permission in writing from the publisher or under licence from the Copyright Licensing Agency Limited. Further details of such licences (for reprographic reproduction) may be obtained from the Copyright Licensing Agency Limited, Saffron House, 6-10 Kirby Street, London EC1N 8TS.

Cover photo © Alfred Pasieka/Science Photo Library
Illustrations by Ken Vail Graphic Design, Cambridge and Barking Dog Art
Typeset in Goudy 10.5pt by Ken Vail Graphic Design, Cambridge
Indexing by Indexing Specialists (UK) Ltd
Printed in Dubai

A catalogue record for this title is available from the British Library.

ISBN: 978 0340 959305
The Publishers would like to thank the following for permission to reproduce copyright material:

Photo credits: p.1 Photodisc; **p.2** Maximilian Stock/Science Photo Library; **p.4** Martyn f. Chillmaid/Science Photo Library; **p.16** Clive Freeman/Science Photo Library; **p.20** *tl, tr* Oxford Scientific Films/Photo Library, *bl* Corbis, *br* Tony Craddock/Science Photo Library; **p21** *tr* Carlos Bach/Science Photo Library, *bl, br* Charles D. Winters/Science Photo Library; **p.25** Volker Steger/Science Photo Library; **p.27** Charles D. Winters/Science Photo Library; **p.30** *l* Sam Ogden/Science Photo Library, *r* Maximilian Stock/Science Photo Library; **p.31** Robert Gugliemo/Science Photo Library; **p.33** Corbis; **p.38** *l* Steve Gschmeissner/Science Photo Library, *r* Tek Images/Science Photo Library; **p.41** David Campione/Science Photo Library; **p.42** Ria Novosti/Science Photo Library; **p.45** Charles D. Winters/Science Photo Library; **p.47** *l* Maximilian Stock/Science Photo Library, *tr* Paul Rapson/Science Photo Library, br Andrew Lambert/Science Photo Library; **p.51** *t* Andrew Lambert/Science Photo Library, *b* Getty Images; **p.55** Getty Images; **p.57** *tl* Patrick Landmann/Science Photo Library, *tr* European Space Agency Multimedia Gallery, *bl* Science Photo Library; **p.58** Metropolitan Museum of Art, Art Resources/Scala Florence; **p.61** Charles D. Winters; **p.63** Bob Gibbons/Science Photo Library; **p.64** Corbis; **p.68** *tr* Maximilian Stock/Science Photo Library, *bl* Andrew Lambert/Science Photo Library; **p.72** *t* Bjanka Kadic/Science Photo library, *(all)* Andrew Lambert; **p.74** *t* Martin Fowler/Alamy, *c* BSIP, Chassenet/Science Photo Library, *b* Junior Bildarchiv/Alamy; **p.76** Jim Dowdalls/Alamy; **p.77** Mauro Fermariello/Science Photo Library; **p.79** *t* Steve Taylor/Science Photo Library, *b* Geoff Kidd/Science Photo Library; **p.80** Pascal Goetscheluck/Science Photo Library; **p.82** Roger Scruton; **p.86** Steve Gschmeissner/Science Photo Library; **p.87** Shelia Terry/Science Photo Library; **p.92** *(both)* Andrew Lambert/Science Photo Library; **p.96** *t* Andrew Syred/Science Photo Library, *b* Anthony Cooper/Science Photo Library; **p.97** Susan Wilkinson; **p.98** Maximilian Stock/Science Photo Library; **p.101** BSIP, Chassenet/Science Photo Library; **p.104** Chris Sattleburger; **p.106** *tr* BSIP, Chassenet/Science Photo Library, *br* Cordelia Molloy/Science Photo Library; **p.107** Corbis; **p.108** Corbis; **p.110** *tr* Gareth Price; *bl* Mauro Fermariello/Science Photo Library; **p.112** Mauro Fermariello/Science Photo Library; **p.115** *(both)* Philippe Psaila/Science Photo Library; **p.116** Gareth Price; **p.119** Corbis; **p.121** Sinclair Stammers/Science Photo Library; **p.123** Jerry Mason/Science Photo Library; **p.124** James Holmes/Science Photo Library; **p.125** Michael Donne/Science Photo Library; **p.127** Jurgen Scriba/Science Photo Library; **p.131** Photodisc; **p.133** David Wentraub/Science Photo Library; **p.134** NASA/Science Photo Library; **p.137** *(both)* Andrew Lambert/Science Photo Library; **p.139** Roger Scruton; **p.145** Roger Scruton; **p.146** Andrew Lambert/Science Photo Library; **p.149** Roger Scruton; **p.150** *t* Mike Devlin/Science Photo Library, *b* Colin Palmer/Alamy; **p.153** Martin Bond/Science Photo Library; **p.154** Courtesy of Honda UK; **p.155** *t* Courtesy of Honda UK, *b* Jim Varney/Science Photo Library; **p.158** Alistair Laming/Alamy; **p.159** Andrew Lambert/Science Photo Library; **p.163** Istock; **p.164** *t* Andrew Lambert/Science Photo Library, *(all)* Martyn f. Chillmaid/Science Photo Library; **p.165** Ian Hooton/Science Photo Library; **p.167** Biosym Technologies/Science Photo Library; **p.170** Philippa Unwins/Science Photo Library; **p.171** Andrew Lambert/Science Photo Library; **p.174** Tate Picture Library; **p.180** *t* AJ Photo/Science Photo Library, *b* Rex Features; **p.185** Andrew Lambert/Science Photo Library; **p.186** Mireille Vautier/Alamy; **p.193** Martyn f. Chillmaid; **p.196** Advertising Archives; **p.198** *l* Saturn Stills/Science Photo Library, *r* Istock; **p.200** Roger Scruton; **p.205** t Colin Underhill/Alamy, *b* Roger Scruton; **p.213** Science Photo Library; **p.214** Kenneth Eward/Science Photo Library; **p.215** *l* Roger Scruton, *r* Jeremy Burgess/Science Photo Library; **p.216** Images Etc. Ltd/Alamy; **p.217** James Holmes/Science Photo Library; **p.218** Getty Images; **p.220** Sally & Richard Greenhill/Alamy; **p.221** Advertising Archives; **p.222** Corbis; **p.223** Eye of Science/Science Photo Library; **p.225** *t* James Bell/Science Photo Library, *b* Istock; **p.234** Colin Cuthbert/Science Photo Library; **p.239** PA Photos; **p.240** Leonard Lessin/Science Photo Library; **p.241** Geoff Tompkinson/Science Photo Library; **p.242** Tim Evans/Science Photo Library; **p.243** Peter Menzel/Science Photo Library.
b = bottom, *c* = centre, *l* = left, *r* = right, *t* = top

Acknowledgements: Every effort has been made to trace all copyright holders, but if any have been inadvertently overlooked the Publishers will be pleased to make the necessary arrangements at the first opportunity.

Although every effort has been made to ensure that website addresses are correct at time of going to press, Hodder Education cannot be held responsible for the content of any website mentioned in this book or associated resources. It is sometimes possible to find a relocated web page by typing in the address of the home page for a website in the URL window of your browser.

Edexcel
Chemistry
for A2
Graham Hill
Andrew Hunt

A Note for Teachers

Edexcel Chemistry for A2 Network disc

The *Edexcel Chemistry for A2 Network disc* which accompanies the Student's Book and Dynamic Learning Student website provides a complete bank of resources for teachers and technicians following the Edexcel specification. Powered by Dynamic Learning, the Network disc contains the same interactive version of the Student's Book that is available on the Dynamic Learning Student's website, plus every resource that teachers might wish to use in activities, discussions, practical work and assessment.

These additional resources on the Network disc include:

- a synoptic Topic overview for each of the fifteen topics showing how the text and resources cover and follow the Edexcel specification, including coverage of 'How Science Works'. These synopses also indicate how the resources on the Network disc can be used to create a teaching programme and lesson plans
- Introductory PowerPoints for each topic
- Practical worksheets for students covering the entire course
- Teacher's and technician's notes for all practicals showing the intentions of each practical, a suggested approach to it, health and safety considerations, the materials and apparatus required by each group and answers to the questions on the worksheets
- additional Weblinks
- additional Activities involving data analysis, application and evaluation
- 3D Rotatable models of molecules of interest
- answers to all Review questions in the Student's Book, as well as answers to all Extension questions available to students from the Dynamic Learning Student website and to questions in the additional Activities on the Network disc
- Interactive objective tests for each topic with answers.

All these resources are launched from interactive pages of the Student's Book on the Network disc. Teachers can also search for resources by resource type or key words to allow greater flexibility in using the resource material.

A Lesson Builder allows teachers to build lessons by dragging and dropping resources they want to use into a lesson group that can be saved and launched later from a single screen.

The tools provided also enable teachers to import their own resources and weblinks into the lesson group and to populate a VLE at the click of a button.

Risk assessment

As a service to users, a risk assessment for this text and associated resources has been carried out by CLEAPSS and is available on request to the Publishers. However, the Publishers accept no legal responsibility on any issue arising from this risk assessment; whilst every effort has been made to check the instructions for practical work in this book and associated resources, it is still the duty and legal obligation of schools to carry out their own risk assessment.

Contents

Introduction

Welcome to *Edexcel Chemistry for A2*. This book covers everything in the Edexcel specification with 9 topics for Unit 4 and 6 topics for Unit 5. Test yourself questions throughout the book will help you to think about what you are studying while the Activities give you the chance to apply what you have learned in a range of modern contexts. At the end of each topic you will find exam-style Review questions to help you check your progress.

Student support for this book can be found on the **Dynamic Learning Student** website, which contains an interactive copy of the book. Students can access a range of free digital resources by visiting **www.dynamic-learning-student.co.uk** and using the code printed on the inside front cover of this book to gain access to relevant resources. These free digital resources include:

- Data tables, for use when answering questions
- Tutorials, which work through selected problems and concepts using a voiceover and animated diagrams
- Practical guidance to support experimental and investigative skills
- Weblinks that provide access to relevant social, environmental and economic contexts and help with the more demanding concepts and practical techniques.

All diagrams and photographs can be launched and enlarged directly from the pages. There are also Learning outcomes available at the beginning of every topic, and answer files to all Test yourself and Activity questions. Extension questions, covering some ideas in greater depth, are available at the end of each topic. The Student Online icon, shown on the right, indicates where a resource such as a Data sheet, Tutorial, Practical guidance or Extension questions is provided on the Dynamic Learning Student website. All other resources are linked to interactive areas on the page or associated with particular pages in the resources menu. In addition, all these resources can be saved to your local hard drive.

DL www

With the powerful Search tool, key words can be found in an instant, leading you to the relevant page or alternatively to resources associated with each key word.

You will need to refer to the Edexcel Data booklet when answering some of the questions in this book. This will help you to become familiar with the booklet. This is important because you will need to use the booklet to find information when answering some questions in the A2 examinations. You can download the Data booklet from the Edexcel website (www.edexcel.com/quals/gce/gce08/chemistry/Pages/default.aspx). The booklet includes the version of the periodic table that you use in the examinations.

Acknowledgements

We would like to acknowledge the suggestions from Ian Davis, Neil Dixon and Tim Joliffe – teachers who commented on our initial plans. The team at Hodder Education, led by Katie Mackenzie-Stuart, has made an extremely valuable contribution to the development of the book and the Network disc and website resources. In particular, we would like to thank Anne Trevillion, the project manager, Anne Wanjie, Deborah Sanderson, Anne Russell and Tony Clappison for their skilful work on both the print and electronic resources.

Graham Hill and Andrew Hunt
April 2009

Unit 4

General Principles of Chemistry I – Rates, Equilibria and Further Organic Chemistry

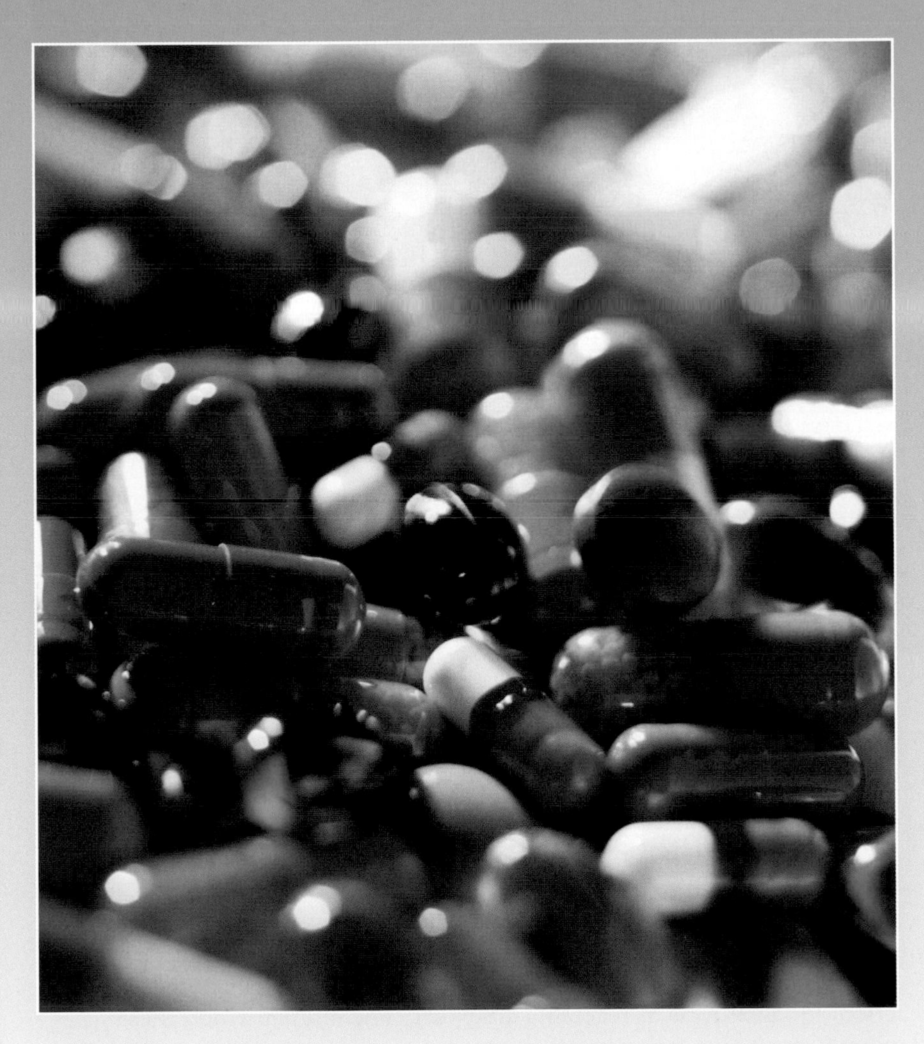

1 How fast? – rates
2 How far? – entropy
3 Equilibria
4 Application of rates and equilibrium ideas
5 Acid–base equilibria
6 Stereochemistry
7 Carbonyl compounds
8 Carboxylic acids and their derivatives
9 Spectroscopy and chromatography

1 How fast? – rates

Reaction kinetics is the study of the rates of chemical reactions. Several factors influence the rate of chemical change including the concentration of the reactants, the surface area of solids, the temperature of the reaction mixture and the presence of a catalyst. Chemists have found that they can learn much more about reactions by studying these effects quantitatively. They can then set up models to simulate the data and make predictions about the impact of changing the conditions.

Chemists apply these models to drug design and to the formulation of medicines to make sure that patients receive treatments which are effective for some time without causing harmful side-effects. The models can also account for the damage arising from pollutants in the atmosphere and help chemists to suggest ways for reducing or preventing the problems.

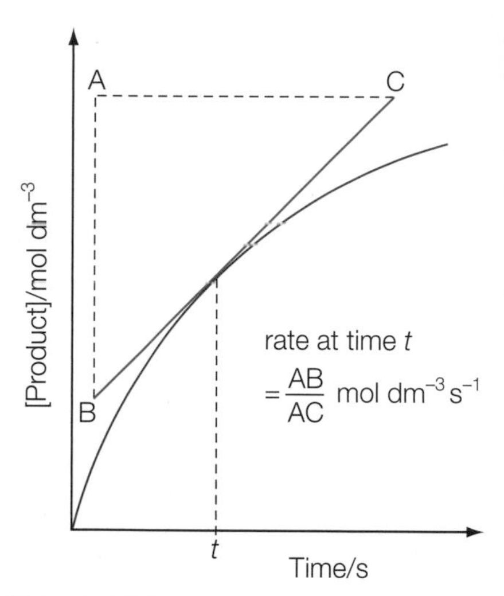

Figure 1.2▲
A concentration–time graph for the formation of a product. The rate of formation of product at time t is the gradient (or slope) of the curve at this point.

Figure 1.1▲
Understanding the factors which determine the rate and direction of chemical change is essential to the design of productive, safe and profitable chemical processes.

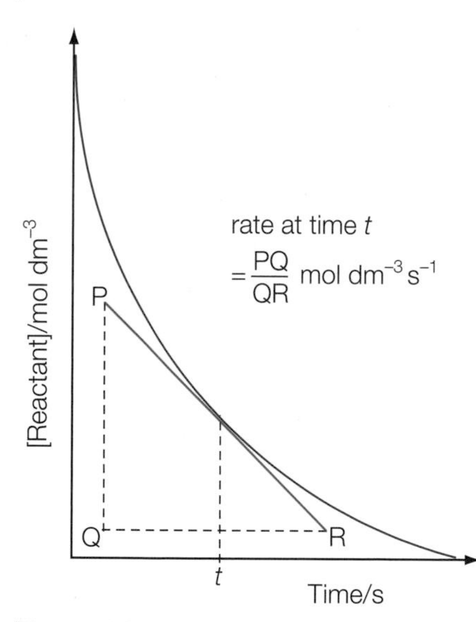

Figure 1.3▲
A concentration–time graph for the disappearance of a reactant. The rate of loss of reactant at time t is the gradient (or slope) of the curve at this point.

Definition

The **rate of reaction** measures the rate of formation of a product or the rate of removal of a reactant.

$$\text{Rate} = \frac{\text{change in measured property}}{\text{time}}$$

1.1 Measuring reaction rates

Balanced chemical equations tell us nothing about how quickly the reactions occur. In order to get this information, chemists have to do experiments to measure the rates of reactions under various conditions.

The amounts of the reactants and products change during any chemical reaction. Products form as reactants disappear. The rates at which these changes happen give a measure of the rate of reaction.

Chemists define the rate of reaction as the change in concentration of a product, or a reactant, divided by the time for the change. Usually the rate is not constant but varies as the reaction proceeds. The first step in analysing the results of an experiment is to plot a concentration–time graph. The gradient (or slope) of the graph at any point gives a measure of the rate of reaction at that time (Figures 1.2 and 1.3).

Test yourself

1 Each of the following factors can change the rate of a reaction. Give an example of a reaction to illustrate each one.
  a) the concentration of reactants in solution
  b) the pressure of gaseous reactants
  c) the surface area of a solid
  d) the temperature
  e) the presence of a catalyst.
2 How does collision theory account for the effects of altering each of the factors a) to e) in question 1?
3 In a study of the hydrolysis of an ester the concentration of the ester fell from 0.55 mol dm^{-3} to 0.42 mol dm^{-3} in 15 seconds. What was the average rate of reaction in that period?
4 The gaseous oxide N_2O_5 decomposes to NO_2 gas and oxygen.
  a) Write a balanced equation for the reaction.
  b) If the rate of disappearance of N_2O_5 is 3.5×10^{-4} mol dm^{-3} s^{-1}, what is the rate of formation of NO_2?

Practical methods

Ideally chemists look for methods for measuring reaction rates which do not interfere with the reaction mixture as in Figures 1.4–1.6. Sometimes however, it is necessary to withdraw samples of the reaction mixture at regular intervals and analyse the concentration of a reactant or product by titration as illustrated in Figure 1.7.

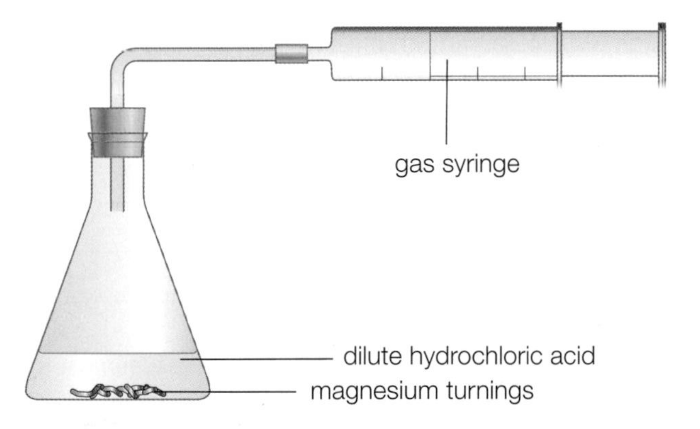

Figure 1.4▲
Following the course of a reaction with time by collecting and measuring the volume of a gas formed.

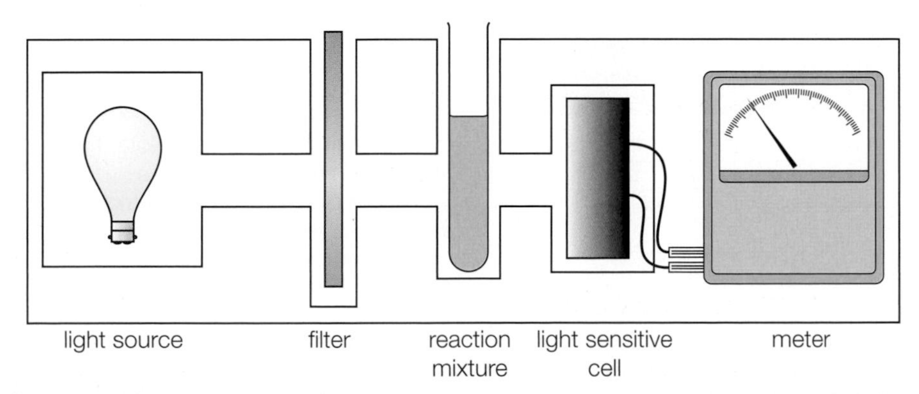

Figure 1.5▲
Using a colorimeter to follow the formation of a coloured product or the removal of a coloured reactant.

Activity

Investigating the effect of concentration on the rate of a reaction

Bromine oxidises methanoic acid in aqueous solution to carbon dioxide. The reaction is catalysed by hydrogen ions.

$Br_2(aq) + HCOOH(aq) \rightarrow 2Br^-(aq) + 2H^+(aq) + CO_2(g)$

The reaction can be followed using a colorimeter.

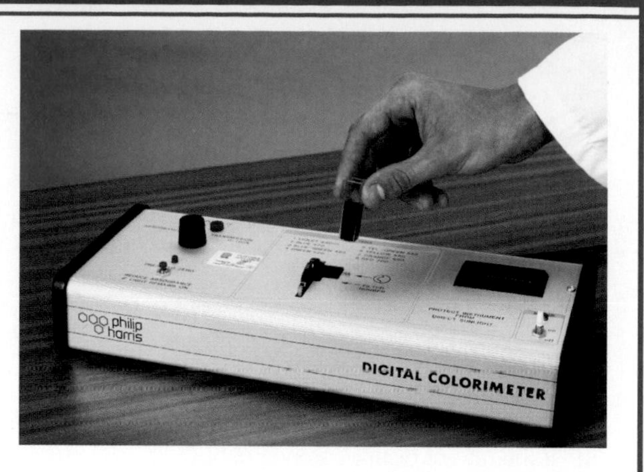

Figure 1.6▶
A colorimeter can be used to follow the changes in concentration of coloured chemicals during a reaction.

Table 1.1 shows some typical results. The concentration of methanoic acid was kept constant throughout the experiment by having it present in large excess.

Time/s	Concentration of bromine/10^{-3} mol dm^{-3}
0	10.0
10	9.0
30	8.1
90	7.3
120	6.6
180	5.3
240	4.4
360	2.8
480	2.0
600	1.3

Table 1.1▲
Results of an experiment to investigate the rate of reaction of bromine with methanoic acid. Note that the bromine concentrations are multiplied by 1000. The actual bromine concentration at 90 seconds, for example, was 0.0073 mol dm^{-3}.

1. Explain why it is possible to follow the rate of this reaction using a colorimeter.
2. Suggest a suitable chemical to use as the catalyst for the reaction.
3. Explain the purpose of adding a large excess of methanoic acid.
4. Plot a graph of bromine concentration against time using the results in Table 1.1.
5. Draw tangents to the graph and measure the gradient to obtain values for the rate of reaction at two points during the experiment. Take values at 100 s and 500 s. (Remember when calculating the gradients that the bromine concentrations are 1000 times less than the numbers in the table.)
6. Table 1.2 shows values for the reaction rate obtained by drawing gradients at other times on the concentration–time graph. Plot a graph of rate against concentration using the two values you have found in answering **5** and the values in Table 1.2.

Time/s	Concentration of bromine/10^{-3} mol dm^{-3}	Rate of reaction from gradients to the concentration–time graph/10^{-5} mol dm^{-3} s^{-1}
50	8.3	2.9
200	5.0	1.7
300	3.5	1.2
400	2.5	0.8

Table 1.2◀
Rate values obtained by finding gradients of tangents to the concentration–time graph. Note that the rate of reaction values are multiplied by 100 000. The actual rate at 300 seconds, for example, was 1.2×10^{-5} mol dm^{-3} s^{-1}.

7. How does the bromine concentration change with time?
8. How does the rate of reaction change with time?
9. How does the rate of reaction depend on the bromine concentration?

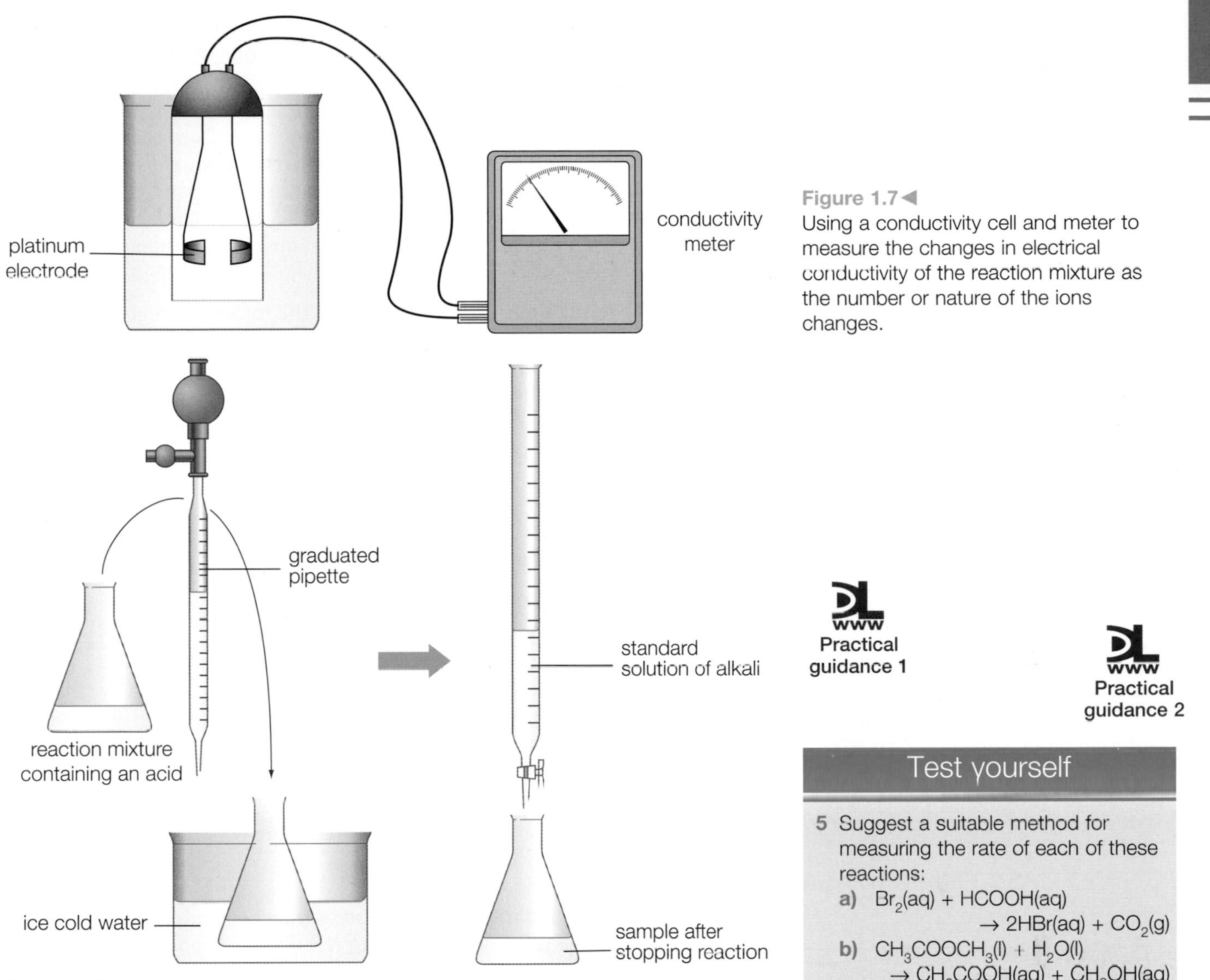

Figure 1.7◀
Using a conductivity cell and meter to measure the changes in electrical conductivity of the reaction mixture as the number or nature of the ions changes.

Figure 1.8▲
Following the course of a reaction catalysed by an acid by removing measured samples of the mixture at intervals, stopping the reaction by running the sample into an alkali and then determining the concentration of one reactant or product by titration. Further samples are taken at regular intervals.

Practical guidance 1

Practical guidance 2

Test yourself

5 Suggest a suitable method for measuring the rate of each of these reactions:

a) $Br_2(aq) + HCOOH(aq) \rightarrow 2HBr(aq) + CO_2(g)$

b) $CH_3COOCH_3(l) + H_2O(l) \rightarrow CH_3COOH(aq) + CH_3OH(aq)$

c) $C_4H_9Br(l) + H_2O(l) \rightarrow C_4H_9OH(l) + H^+(aq) + Br^-(aq)$

d) $MgCO_3(s) + 2HCl(aq) \rightarrow MgCl_2(aq) + CO_2(g) + H_2O(l)$

1.2 Rate equations

Chemists have found that they can summarise the results of investigating the rate of reaction in the form of a rate equation. A rate equation shows how changes in the concentrations of reactants affect the rate of a reaction.

Take the example of a general reaction for which x mol A react with y mol B to form products:

$$xA + yB \rightarrow \text{products}$$

The equation which describes how the rate varies with the concentrations of the reactants takes this form:

$$\text{rate} = k[A]^n[B]^m$$

where [A] and [B] represent the concentrations of the reactants in moles per cubic decimetre (litre).

The powers n and m are the reaction orders. The reaction above is order n with respect to A and order m with respect to B. The overall order is $(n + m)$.

Note

A rate equation cannot be deduced from the balanced equation: it has to be found by experiment. In this general example the values of n and m in the rate equation may or may not be the same as the values of x and y in the balanced equation for the reaction.

Definition

Writing the formula of a chemical in square brackets is the usual shorthand for **concentration in mol dm^{-3}**. For example: [A] represents the concentration of A in mol dm^{-3}.

The rate constant, k, is only constant for a particular temperature. The value of k varies with temperature (see Section 1.3). The units of the rate constant depend on the overall order of the reaction.

Overall order	Units of the rate constant
Zero	$mol\,dm^{-3}\,s^{-1}$
First	s^{-1}
Second	$dm^3\,mol^{-1}\,s^{-1}$

Table 1.3▲

Worked example

The decomposition of ethanal to methane and carbon monoxide is second order with respect to ethanal. When the concentration of ethanal in the gas phase is $0.20\,mol\,dm^{-3}$, the rate of reaction is $0.080\,mol\,dm^{-3}\,s^{-1}$ at a certain temperature. What is the value of the rate constant at this temperature?

Notes on the method

Start by writing out the rate equation based on the information given. There is no need to write the equation for the reaction because the rate equation cannot be deduced from the balanced chemical equation.

Substitute values in the rate equation, including the units as well as the values. Then rearrange the equation to find the value of k. Check the units are as expected for a second order reaction.

Answer

The rate equation: rate = k[ethanal]^2

Substituting: $0.080\,mol\,dm^{-3}\,s^{-1} = k \times (0.20\,mol\,dm^{-3})^2$

Rearranging: $k = \dfrac{0.080\,mol\,dm^{-3}\,s^{-1}}{(0.20\,mol\,dm^{-3})^2}$

Hence: $k = 2.0\,dm^3\,mol^{-1}\,s^{-1}$

Definition

In a **rate equation** such as 'rate = $k[A]^m[B]^n$', the constant k is the **rate constant**. The powers n and m are the **orders** of the reaction with respect to the reactants A and B that appear in this equation. The **overall order of the reaction** is $(n + m)$.

Test yourself

6 The rate of decomposition of di(benzenecarbonyl) peroxide is first order with respect to the peroxide. Calculate the rate constant for the reaction at 107 °C if the rate of decomposition of the peroxide at this temperature is $7.4 \times 10^{-6}\,mol\,dm^{-3}\,s^{-1}$ when the concentration of peroxide is $0.02\,mol\,dm^{-3}$.

7 The hydrolysis of the ester methyl ethanoate in alkali is first order with respect to both the ester and hydroxide ions. The rate of reaction is $0.000\,69\,mol\,dm^{-3}\,s^{-1}$, at a given temperature, when the ester concentration is $0.05\,mol\,dm^{-3}$ and the hydroxide ion concentration is $0.10\,mol\,dm^{-3}$. Write the rate equation for the reaction and calculate the rate constant.

First order reactions

A reaction is first order with respect to a reactant if the rate of reaction is proportional to the concentration of that reactant. The concentration term for this reactant is raised to the power one in the rate equation.

$$\text{Rate} = k[X]^1 = k[X]$$

This means that doubling the concentration of the chemical X leads to a doubling of the rate of reaction. The rate of reaction is proportional to the concentration of the reactant. This means that a plot of rate against concentration gives a straight line passing through the origin (Figure 1.9).

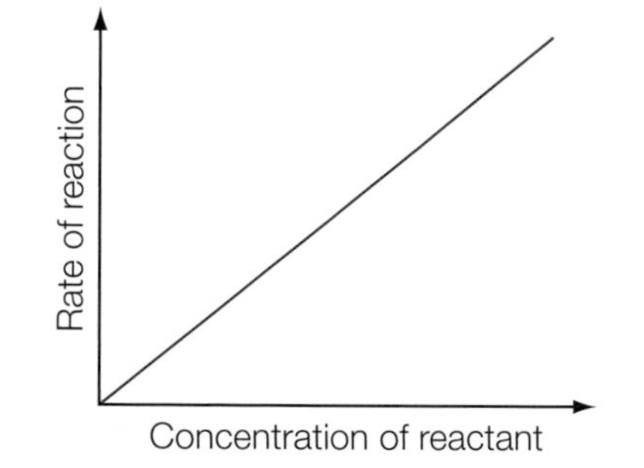

Figure 1.9▲
The variation of reaction rate with concentration for a first order reaction. The graph is a straight line through the origin showing that the rate is proportional to the concentration of the reactant.

Definition

The **half-life** of a reaction is the time for the concentration of one of the reactants to fall by half.

One of the easier ways to spot a first order reaction is to plot a concentration–time graph and then study the time taken for the concentration to fall by half (Figure 1.10). This is the half-life, $t_{½}$. It can be shown mathematically from the rate equation that for a first order reaction:

$$t_{½} = \frac{0.69}{k}$$

where k is the rate constant.

This shows that, at a constant temperature, the half-life of a first order reaction is the same wherever it is measured on a concentration–time graph.

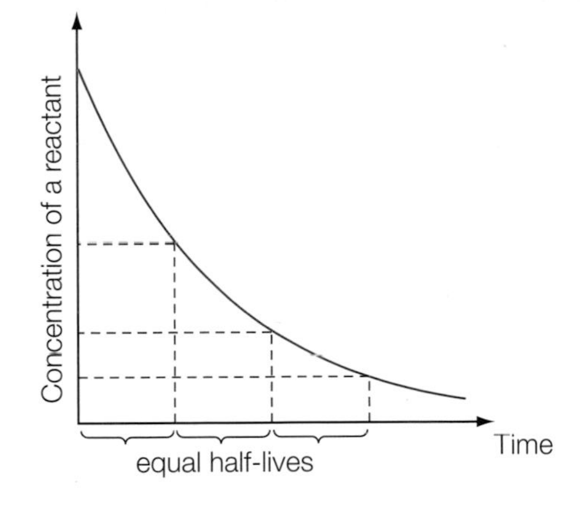

Figure 1.10◀
The variation of concentration of a reactant plotted against time for a first order reaction. The half-life for a first order reaction is a constant so it is the same wherever it is read off the curve. It is independent of the initial concentration.

Second order reactions

A reaction is second order with respect to a reactant if the rate of reaction is proportional to the concentration of that reactant squared. This means that the concentration term for this reactant is raised to the power two in the rate equation. At its simplest the rate equation for a second order reaction takes this form:

$$\text{rate} = k[\text{reactant}]^2$$

This means that doubling the concentration of X increases the rate by a factor of four.

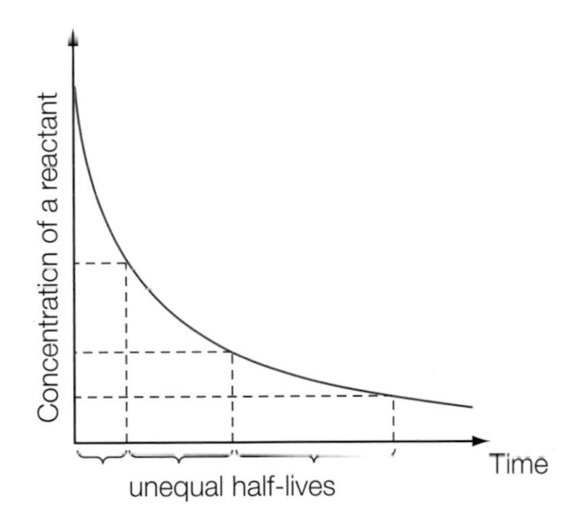

Figure 1.11▲
The variation of concentration of a reactant plotted against time for a second order reaction. The half-life for a second order reaction is not a constant. The time for the concentration to fall from c to $c/2$ is half the time for the concentration to fall from $c/2$ to $c/4$. The half-life is inversely proportional to the starting concentration.

The variation of rate with concentration for a second order reaction can be found, as before, by drawing tangents to the curve of the concentration–time graph (Figure 1.11). However, a rate–concentration graph is not a straight line for a second order reaction but instead a curve, as shown in Figure 1.12.

Test yourself

8 Refer to your answers to the activity in Section 1.1.
  a) From your rate–concentration graph, what is the order of the reaction of the reaction of bromine with methanoic acid with respect to bromine?
  b) i) Determine three values for half-lives for the reaction from your concentration–time graph.
     ii) Are your values consistent with your answer to a)?

9 Explain how the rate constant can be found from a rate–concentration graph such as in Figure 1.9.

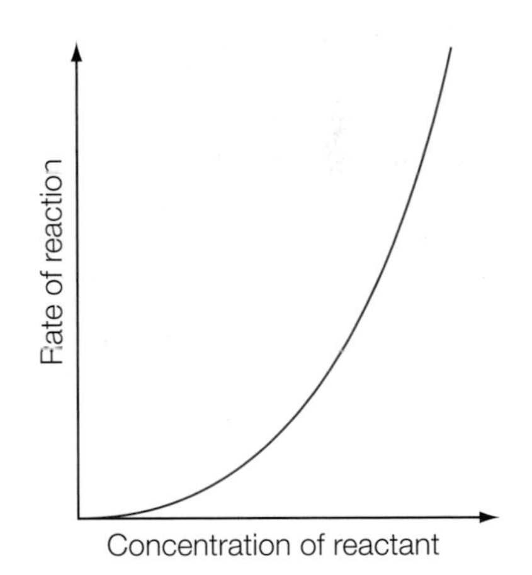

Figure 1.12▲
A concentration–time graph for a second order reaction.

Test yourself

10 The rate of reaction of 1-bromopropane with hydroxide ions is first order with respect to the halogenoalkane and first order with respect to hydroxide ions.
  a) Write out the rate equation for the reaction.
  b) What is the overall order of reaction?
  c) What are the units of the rate constant?

Note

It can be shown mathematically that for a second order reaction, with the rate equation in the form rate = $k[A]^2$, a plot of $1/[A]_t$ against time is a straight line, where $[A]_t$ is the concentration at time t. The slope of the line = k.

Zero order reactions

Note

Any term raised to the power zero equals 1. So $[\text{reactant}]^0 = 1$.

At first sight is seems odd that there can be zero order reactions. A reaction is zero order with respect to a reactant if the rate of reaction is unaffected by changes in the concentration of that reactant. Chemists have found a way to account for zero order reactions in terms of the mechanisms of these reactions (see Section 1.4).

In a rate equation for a zero order reaction, the concentration term for the reactant is raised to the power zero:

$$\text{rate} = k[\text{reactant}]^0 = k \text{ (a constant)}$$

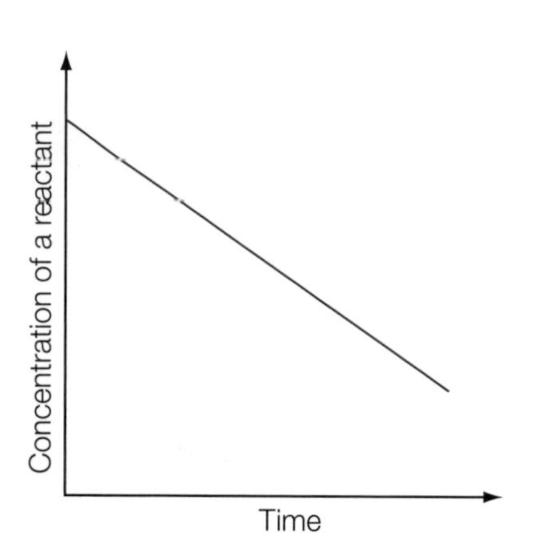

Figure 1.13▲
The variation of concentration of a reactant plotted against time for a zero order reaction. The gradient of this graph measures the rate of reaction. The gradient is a constant so the rate stays the same even though the concentration of the reactant is falling.

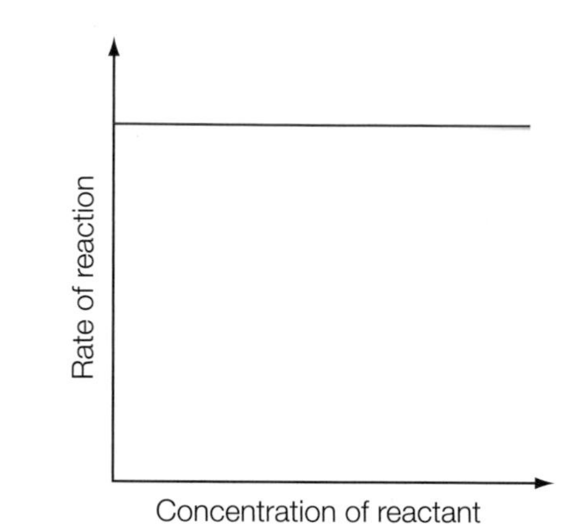

Figure 1.14▲
The variation of reaction rate with concentration for a zero order reaction.

Test yourself

11 Ammonia gas decomposes to nitrogen and hydrogen in the presence of a hot platinum wire. Experiments show that the reaction continues at a constant rate until all the ammonia has disappeared.
  a) Sketch a concentration–time graph for the reaction.
  b) Write both the balanced chemical equation and the rate equation for this reaction.

The nitration of methylbenzene is an example where the conditions can be such that the reaction is zero order with respect to the aromatic compound. The methylbenzene reacts with nitronium ions formed from nitric acid (see Section 12.8). The reaction creating the NO_2^+ ions is relatively slow but as soon as the ions form they react with methylbenzene. As a result the rate is not affected by the methylbenzene concentration.

The initial-rate method

The most general method for determining reaction orders is the initial-rate method. The method is based on finding the rate immediately after the start of a reaction. This is the one point when all the concentrations are known.

The investigator makes up a series of mixtures in which all the initial concentrations are the same except one. A suitable method is used to measure the change of concentration with time for each mixture (see Section 1.1). The results are used to plot concentration–time graphs. The initial rate for each mixture is then found by drawing tangents to the curve at the start and calculating their gradients.

Worked example

The initial-rate method was used to study the reaction:

$$BrO_3^-(aq) + 5Br^-(aq) + 6H^+(aq) \rightarrow 3Br_2(aq) + 3H_2O(l)$$

The initial rate was calculated from four graphs plotted to show how the concentration of BrO_3^-(aq) varied with time for different initial concentrations of reactants with the results shown in Table 1.4.
What is:

a) the rate equation for the reaction
b) the value of the rate constant?

Experiment	Initial concentration of BrO_3^-/mol dm^{-3}	Initial concentration of Br^-/mol dm^{-3}	Initial concentration of H^+/mol dm^{-3}	Initial rate of reaction/mol dm^{-3} s^{-1}
1	0.1	0.10	0.10	1.2×10^{-3}
2	0.2	0.10	0.10	2.4×10^{-3}
3	0.1	0.30	0.10	3.6×10^{-3}
4	0.2	0.10	0.20	9.6×10^{-3}

Table 1.4▲

Notes on the method

Recall that the rate equation cannot be worked out from the balanced equation for the reaction.

First study the experiments in which the concentration of BrO_3^- varies but the concentration of the other two reactants stays the same. How does doubling the concentration of BrO_3^- affect the rate?

Then study in turn the experiments in which the concentrations of first Br^- and then H^+ vary while the concentrations of the other two reactants stay the same. How does doubling or tripling the concentration of a reactant affect the rate?

Substitute values for any one experiment in the rate equation to find the value of the rate constant, k. Take care with the units.

Tutorial

Answer

From experiments 1 and 2: doubling $[BrO_3^-]_{initial}$ increases the rate by a factor of 2. So rate $\propto [BrO_3^-]^1$.

From experiments 1 and 3: tripling $[Br^-]_{initial}$ triples the rate. So rate $\propto [Br^-]^1$

From experiments 2 and 4: doubling $[H^+]_{initial}$ increases the rate by a factor of 4 (2^2). So rate $\propto [H^+]^2$.

The reaction is first order with respect to BrO_3^- and Br^- but second order with respect to H^+.

The rate equation is: rate $= k\,[BrO_3^-][Br^-][H^+]^2$

Rearranging this equation, and substituting values from experiment 4:

$$k = \frac{\text{rate}}{[BrO_3^-][Br^-][H^+]^2}$$

$$= \frac{9.6 \times 10^{-3}\,\text{mol dm}^{-3}\,\text{s}^{-1}}{0.2\,\text{mol dm}^{-3} \times 0.1\,\text{mol dm}^{-3} \times (0.2\,\text{mol dm}^{-3})^2}$$

$$k = 12.0\,\text{dm}^9\,\text{mol}^{-3}\,\text{s}^{-1}$$

Test yourself

12 This data refers to the reaction of the halogenoalkane 1-bromobutane (here represented as RBr) with hydroxide ions. The results are shown in Table 1.5.

a) Deduce the rate equation for the reaction.

b) Calculate the value of the rate constant.

Experiment	[RBr]/mol dm^{-3}	$[OH^-]$/mol dm^{-3}	Rate of reaction/mol dm^{-3} s^{-1}
1	0.020	0.020	1.36
2	0.010	0.020	0.68
3	0.010	0.005	0.17

Table 1.5▲

Test yourself

13 This data refers to the reaction of halogenoalkane 2-bromo-2-methylbutane (here represented as R'Br) with hydroxide ions. The results are shown in Table 1.6.

Experiment	[R'Br]/mol dm^{-3}	[OH^-]/mol dm^{-3}	Rate of reaction/mol dm^{-3} s^{-1}
1	0.020	0.020	40.40
2	0.010	0.020	20.19
3	0.010	0.005	20.20

Table 1.6▲

14 Hydrogen gas reacts with nitrogen monoxide gas to form steam and nitrogen. Doubling the concentration of hydrogen doubles the rate of reaction. Tripling the concentration of NO gas increases the rate by a factor of nine.

a) Write the balanced equation for the reaction.

b) Write the rate equation for the reaction.

1.3 The effect of temperature on reaction rates

Raising the temperature often has a dramatic effect on the rate of a reaction, especially reactions which involve the breaking of strong covalent bonds. This explains why the practical procedure for most organic reactions involves heating the reaction mixtures. With the help of collision theory it is possible to make predictions about the effect of temperature changes on rates.

The constant *k* in a rate equation is only a constant at a specified temperature. Generally, the value of the rate constant increases as the temperature rises and this means that the rate of reaction increases.

Collision theory accounts for the effect of temperature on reaction rates by supposing that chemical changes pass through a transition state. The transition state is at a higher energy than the reactants so there is an energy barrier or activation energy (Figure 1.15). Reactant molecules must collide with enough energy to overcome the activation energy barrier. This means that the only collisions which lead to reaction are those with enough energy to break existing bonds and allow the atoms to rearrange to form new bonds in the product molecules.

Definitions

A **transition state** is the state of the reacting atoms, molecules or ions when they are at the top of the activation energy barrier for a reaction step.

A **reaction profile** is a plot that shows how the total energy of the atoms, molecules or ions changes during the progress of a change from reactants to products.

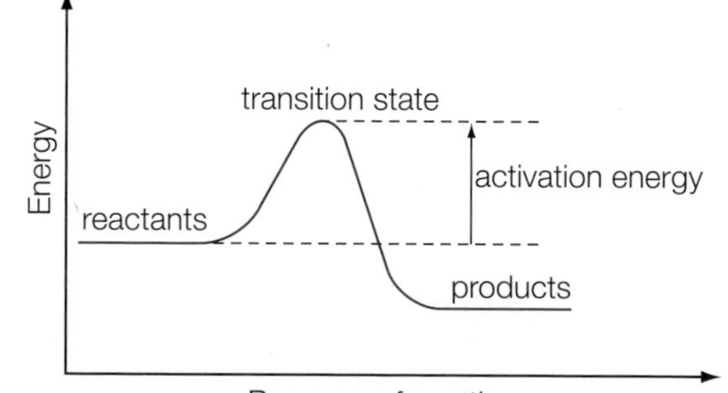

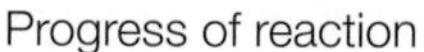

Figure 1.15▶
Reaction profile showing the activation energy for a reaction.

Activation energies account for the fact that reactions go much more slowly than would be expected if every collision in a mixture of chemicals led to reaction. Only a very small proportion of collisions bring about chemical change because molecules can only react if they collide with enough energy to overcome the energy barrier. At around room temperature, for many reactions, only a small proportion of molecules have enough energy to react.

The Maxwell–Boltzmann curve describes the distribution of the kinetic energies of molecules. As Figure 1.16 shows, the proportion of molecules with energies greater than the activation energy is small at around 300 K.

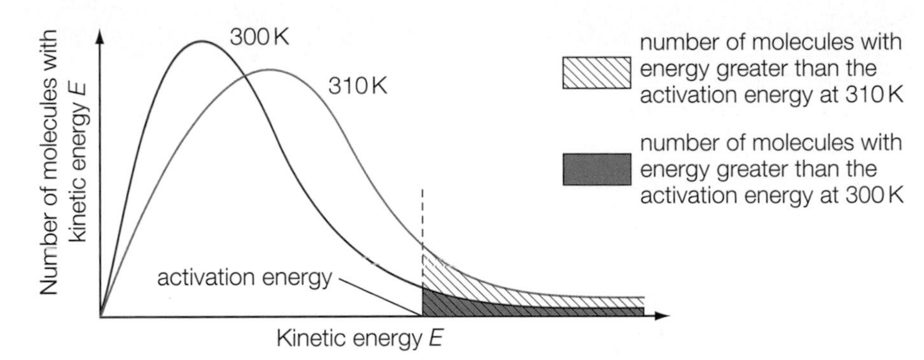

Figure 1.16▲
The Maxwell–Boltzmann distribution of molecular kinetic energies in a gas at two temperatures. The modal energy gets higher as the temperature rises. The area under the curve gives the total number of molecules. This does not change as the temperature rises so the peak height falls as the curve widens.

The shaded areas in Figure 1.16 show the proportions of molecules having at least the activation energy for a reaction at two temperatures. This area is bigger at a higher temperature. So at a higher temperature there are more molecules with enough energy to react when they collide and the reaction goes faster.

Test yourself

15 Why is it that many reactions have activation energies that range between about 50 kJ mol^{-1} and 250 kJ mol^{-1}?

The effect of temperature changes on rate constants

The Swedish physical chemist Svante Arrhenius (1859–1927) found that he obtained a straight line if he plotted the natural logarithm of the rate constant for a reaction against $1/T$ (the inverse of the absolute temperature).
The Arrhenius equation can take this form:

$$\ln k = \frac{-E_a}{R} \times \frac{1}{T} + \text{constant}$$

where k is the rate constant, R is the gas constant, T the absolute temperature and E_a is the activation energy for the reaction.

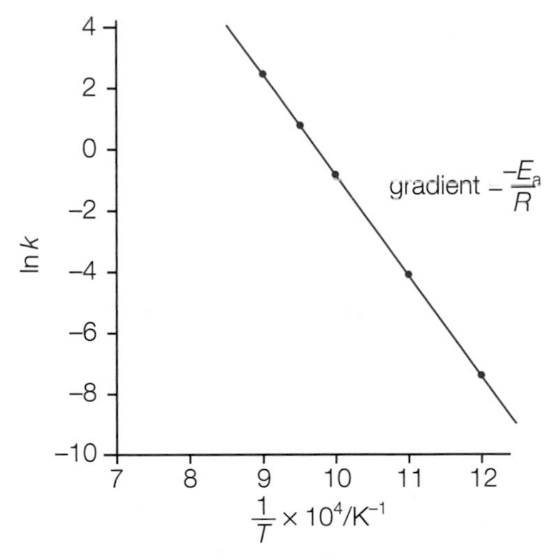

Figure 1.17▲
A plot of $\ln k$ against $1/T$ for a reaction. The activation energy can be calculated from the gradient. The general equation for a straight line is $y = mx + c$ where m is the gradient and c is the intercept on the y-axis. Here c = the constant in the Arrhenius equation and the gradient $m = \frac{-E_a}{R}$.

A useful rough guide, based on the Arrhenius equation, is that the value of the rate constant will double for each 10 degree rise in temperature for a reaction with an activation energy of about 50 kJ mol^{-1}.

Definitions

The **absolute temperature** is the temperature on the Kelvin scale. The absolute zero of temperature is at 0 K which is approximately −273 °C.

The **gas constant** is the constant R in the ideal gas equation $pV = nRT$. The value of the constant depends on the units used for pressure and volume. If all quantities are in SI units, then $R = 8.314\ \text{J K}^{-1}\ \text{mol}^{-1}$.

Logarithms in chemistry are of two kinds – logarithms to base 10 (log) and natural logarithms to base e (ln). Chemists use logarithms to base 10 to handle values which range over several orders of magnitude (see Section 5.3). Natural logarithms appear in relationships in chemical kinetics and thermochemistry. They follow similar mathematical rules as logarithms to base 10.

Test yourself

16 Table 1.7 shows the value of the rate constant for the reaction of benzenediazonium chloride in aqueous solution with water at four temperatures.

Temperature/K	Rate constant/10^{-5} s^{-1}
278	0.15
298	4.1
308	20
323	140

Table 1.7▲

a) What do the units of the rate constant tell you about the form of the rate equation?

b) What do the values tell you about the effect of temperature on the rate of the reaction?

c) What is the effect of a 10 degree rise in temperature on the rate of the reaction?

17 Show that the Arrhenius equation signifies that:

a) the higher the temperature the greater the value of k and hence the faster the reaction

b) a reaction with a relatively high activation energy has a relatively small rate constant.

18 The rate constant for the decomposition of hydrogen peroxide is 4.93×10^{-4} s^{-1} at 295 K. It increases to 1.40×10^{-3} s^{-1} at 305 K. Estimate the activation energy for the reaction.

The effect of catalysts on rate equations

A catalyst works by providing an alternative pathway for a reaction, with a lower activation energy. Lowering the activation energy increases the proportion of molecules with enough energy to react.

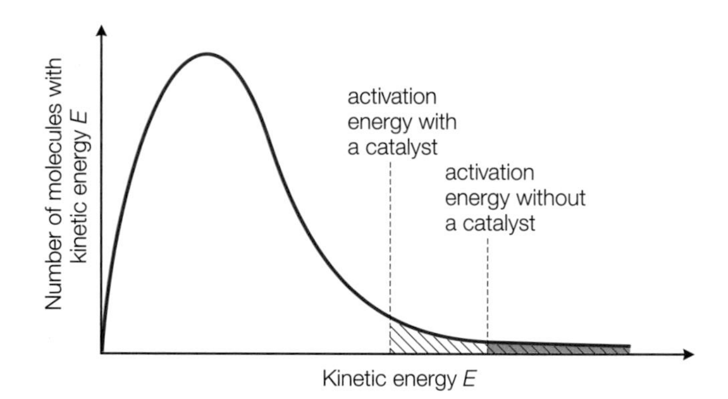

Figure 1.18▲
The distribution of molecular energies in a gas showing how the proportion of molecules able to react increases when a catalyst lowers the activation energy.

Often a catalyst changes the mechanism of a reaction and makes a reaction more productive by increasing the yield of the desired product and reducing waste. This means that the form of the rate equation, the order of reaction and the value of the rate constant are all likely to be different when a catalyst is added to speed up a reaction.

One of the ways in which a catalyst can change the mechanism of a reaction is to combine with the reactants to form an intermediate. The intermediate is a stage in the transition from reactants to products. The intermediate breaks down to give the products and the catalyst is released. This frees up the catalyst to interact with further reactant molecules and the reaction continues.

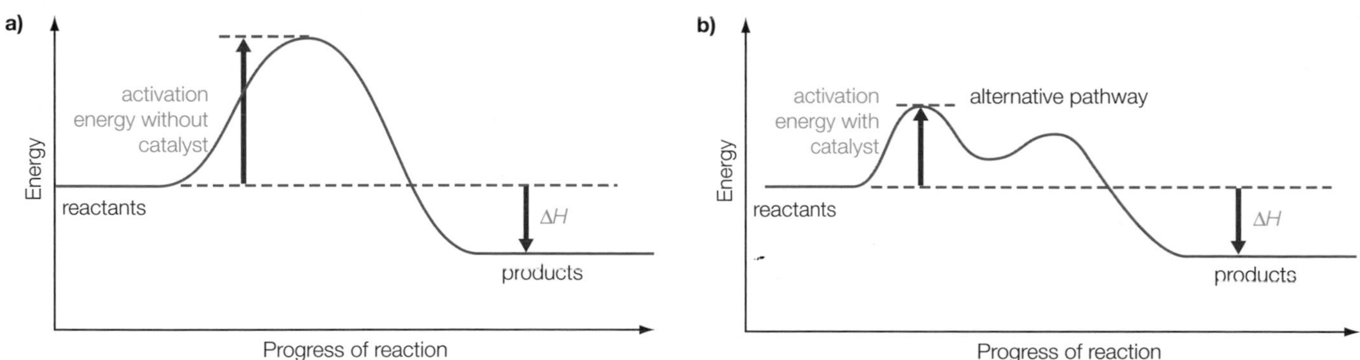

Figure 1.19▲
Reaction profiles for a reaction **a)** without a catalyst and **b)** with a catalyst. The dip in the curve of the pathway with a catalyst shows where an unstable intermediate forms.

In industrial processes it is common to use solids as catalysts and to allow the reacting chemicals to flow through the catalyst and over its solid surfaces. In the reactor for making ammonia, for example, the mixture of nitrogen, hydrogen and ammonia gases flows through layers of an iron catalyst. The iron is an example of a heterogeneous catalyst.

Many laboratory reactions use homogeneous catalysts with the reactants and catalyst all dissolved in the same solvent.

Definitions

A **phase** is one of the three states of matter – solid, liquid or gas. Chemical systems often have more than one phase. Each phase is distinct but need not be pure.

A **homogeneous catalyst** is a catalyst which is in the same phase as the reactants. Typically the reactants and the catalyst are dissolved in the same solution.

A **heterogeneous catalyst** is a catalyst which is in a different phase from the reactants. Generally a heterogeneous catalyst is a solid while the reactants are gases or in solution in a solvent.

Test yourself

19 Classify these catalysts as homogeneous or heterogeneous:
- a) the platinum alloy in a catalytic converter
- b) the sodium hydroxide used to hydrolyse a halogenoalkane
- c) the zeolite use to crack oil fractions
- d) the nickel used to hydrogenate unsaturated fats.

20 The activation energy for the decomposition of ammonia into nitrogen and hydrogen is 335 kJ mol^{-1} in the absence of a catalyst but 162 kJ mol^{-1} in the presence of a tungsten catalyst. Explain the significance of these values in terms of transition state theory.

1.4 Rate equations and reaction mechanisms

Rate equations were one of the first pieces of evidence to set chemists thinking about the mechanism of reactions. They wanted to understand why a rate equation cannot be predicted from the balanced equation for the reaction. They were puzzled that similar reactions turned out to have different rate equations.

Multi-step reactions

The key to understanding reaction mechanisms was the realisation that most reactions do not take place in one step, as suggested by the balanced equation, but in a series of steps.

It is unexpected that the decomposition of ammonia gas in the presence of a hot platinum catalyst is a zero order reaction. How can it be that the concentration of the only reactant does not affect the rate? A possible explanation is illustrated in Figure 1.20.

Ammonia rapidly diffuses to the surface of the metal and is adsorbed onto the surface. This happens fast. Bonds break and atoms rearrange to make new molecules on the surface of the metal. This is the slowest process. Once formed, the nitrogen and hydrogen rapidly break away from the metal into the gas phase.

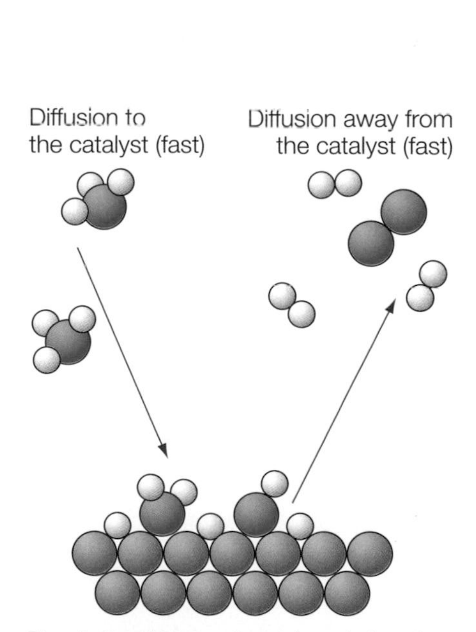

Figure 1.20▲
Three steps in the decomposition of ammonia gas in the presence of a platinum catalyst.

Definitions

The **mechanism of a reaction** describes how the reaction takes, place showing step by step the bonds that break and the new bonds that form.

The **rate-determining step** in a multi-step reaction is the slowest step: the one with the highest activation energy.

Test yourself

21 Give an analogy from the everyday world to explain the idea of a rate-determining step. You could base your example on people getting their meals in a busy self-service canteen, or cars on a motorway with lots of traffic but affected by lane closures.

So there is a rate-determining step which can only happen on the surface of the platinum. The rate of reaction is determined by the surface area of the platinum, which is a constant. This means that the rate of reaction is a constant as long as there is enough ammonia to be adsorbed all over the metal surface. The rate is independent of the ammonia concentration.

Hydrolysis of halogenoalkanes

Another puzzle for chemists was the discovery that there are different rate equations for the reactions between hydroxide ions and two isomers with the formula C_4H_9Br (see questions 13 and 14 in Section 1.2).

Hydrolysis of a primary halogenoalkane, such as 1-bromobutane, is overall second order. The rate equation has the form: rate = $k[C_4H_9Br][OH^-]$.

To account for this chemists have suggested a mechanism showing the C–Br bond breaking at the same time as the nucleophile, OH^-, forms a C–OH bond. In this mechanism both reactants are involved in the single, rate-determining step.

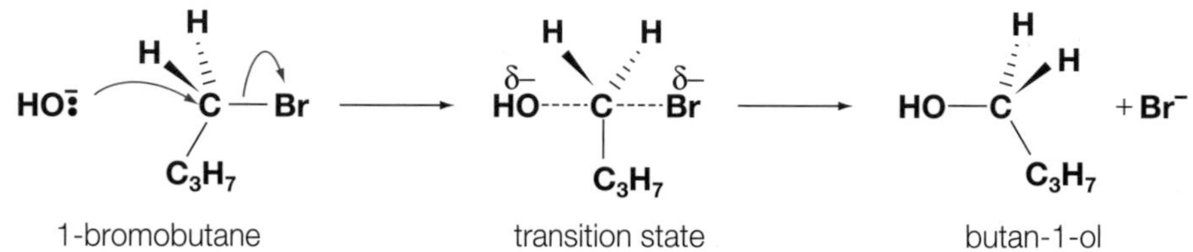

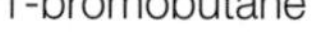

Tutorial 1

Figure 1.21 ▲
A one-step mechanism for the hydrolysis of 1-bromobutane.

In this example of a substitution reaction the nucleophile is the hydroxide ion. Chemists label this mechanism S_N2 where the '2' shows that there are two molecules or ions involved in the rate-determining step.

Hydrolysis of tertiary halogenoalkanes such as 2-bromo-2-methylpropane, however, is overall first order. The rate equation has the form: rate = $k[C_4H_9Br]$.

Tutorial 2

The suggested mechanism shows the C–Br bond breaking first to form an ionic intermediate. Then the nucleophile, OH^-, rapidly forms a new bond with carbon.

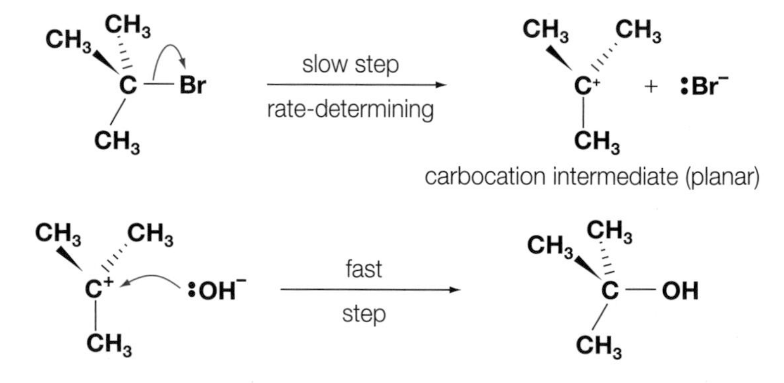

Figure 1.22 ▲
A two-step mechanism for the hydrolysis of 2-bromo-2-methylpropane.

In this example of a substitution reaction the nucleophile is also the hydroxide ion. But in this case chemists label the mechanism S_N1, where the '1' shows that there is just one molecule or ion involved in the rate-determining step. The concentration of the hydroxide ions does not affect the rate of reaction because hydroxide ions are not involved in the rate-determining step.

What these examples show is that it is generally the molecules or ions involved (directly or indirectly) in the rate-determining step that appear in the rate equation for the reaction.

Definitions

An S_N1 **reaction** is a nucleophilic substitution reaction with a mechanism that involves only one molecule or ion in the rate-determining step.

An S_N2 **reaction** is a nucleophilic substitution reaction with a mechanism that involves two molecules or ions in the rate-determining step.

Test yourself

22 Explain, in terms of bonding, why the first step in the S_N1 mechanism is slow while the second step is fast.

23 In the proposed two-step mechanism for the reaction of nitrogen dioxide gas with carbon monoxide gas, the first step is slow while the second step is fast:

$2NO_2(g) \rightarrow NO_3(g) + NO(g)$ *slow*

$NO_3(g) + CO(g) \rightarrow NO_2(g) + CO_2(g)$ *fast*

a) What is the overall equation for the reaction?

b) Suggest a rate equation which is consistent with this mechanism.

c) What, according to your suggested rate equation, is the order of reaction with respect to carbon monoxide?

The reaction of iodine with propanone

The reaction of iodine with propanone is another example which shows that it is not possible to deduce the rate equation from the balanced equation for the reaction.

$$I_2(aq) + CH_3COCH_3(aq) \rightarrow CH_2ICOCH_3(aq) + H^+(aq) + I^-(aq)$$

Experiments show that the reaction is first order with respect to propanone and first order with respect to hydrogen ions but zero order with respect to iodine. The iodine concentration does not affect the rate of reaction. This shows that iodine is not involved in the rate-determining step of the reaction mechanism.

In the presence of hydrogen ions, propanone molecules can react to form an intermediate molecule with a double bond and an –OH group. This involves a slow step followed by a fast step as shown in Figure 1.23. The concentrations of chemical species involved in the rate-determining step appear in the rate-determining step.

In two further steps the intermediate rapidly reacts with iodine. These two steps do not control the overall rate.

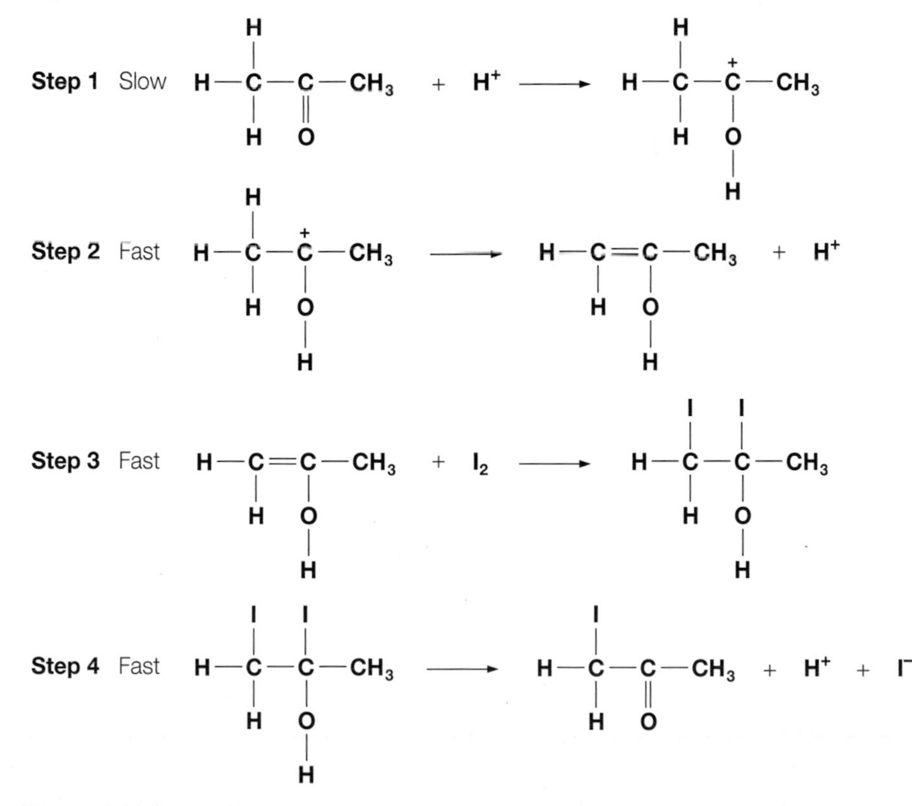

Figure 1.23▲
The suggested mechanism to account for the fact that the iodination of propanone is zero order with respect to iodine.

Test yourself

24 Why do chemists use the term 'enol' to describe the intermediate formed by steps 1 and 2 of the reaction of iodine with propanone?

25 a) Give a rate equation for the reaction of iodine with propanone.

b) Why does the formula for hydrogen ions appear in the rate equation for the iodination of propanone but not as a reactant in the balanced equation for the reaction?

26 Show that the catalyst for the iodination of propanone is not used up in the reaction according to the proposed mechanism in Figure 1.23.

27 Bromine reacts with propanone in a similar way to iodine. The mechanism for the reaction is the same. Explain why bromine reacts with propanone at the same rate as iodine under similar conditions.

Activity

A model for explaining how enzymes work

The lock-and-key model

Enzymes are natural catalysts. They are much more effective than inorganic catalysts and also highly specific. Most enzymes catalyse only one reaction or one type of reaction. An example is catalase. This enzyme occurs in most living tissues, where it dramatically speeds up the decomposition of hydrogen peroxide. It has no effect on any other reactions in cells.

A model proposed to explain how enzymes work is based on an analogy with a lock and a key. Enzymes are proteins. Like all proteins they have a precise molecular shape (see Section 14.3). The lock-and-key model suggests that each enzyme molecule has an active site. Molecules of the main reactant (the substrate) fit into the active site. Typically the active site is a cleft formed by the folding of the protein chain where substrate molecules can be held by intermolecular forces such as hydrogen bonds.

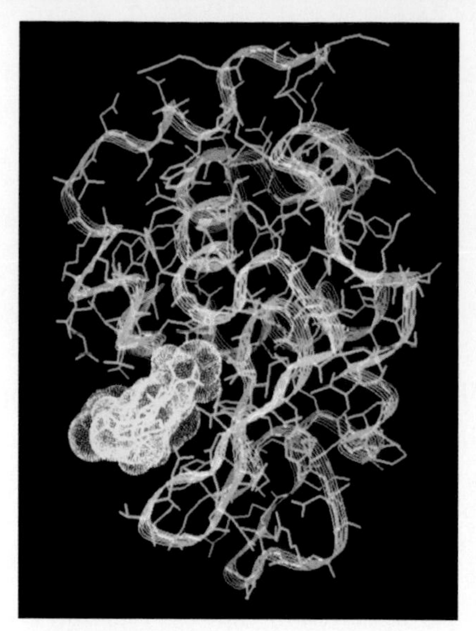

Figure 1.25▲
This computer graphic shows a molecule of the enzyme lysozyme. Lysozyme catalyses the hydrolysis reactions that damage the cell walls of bacteria. In this image the protein is shown in blue. The backbone is traced out by a magenta ribbon. The substrate is shown in yellow, bound to the active site.

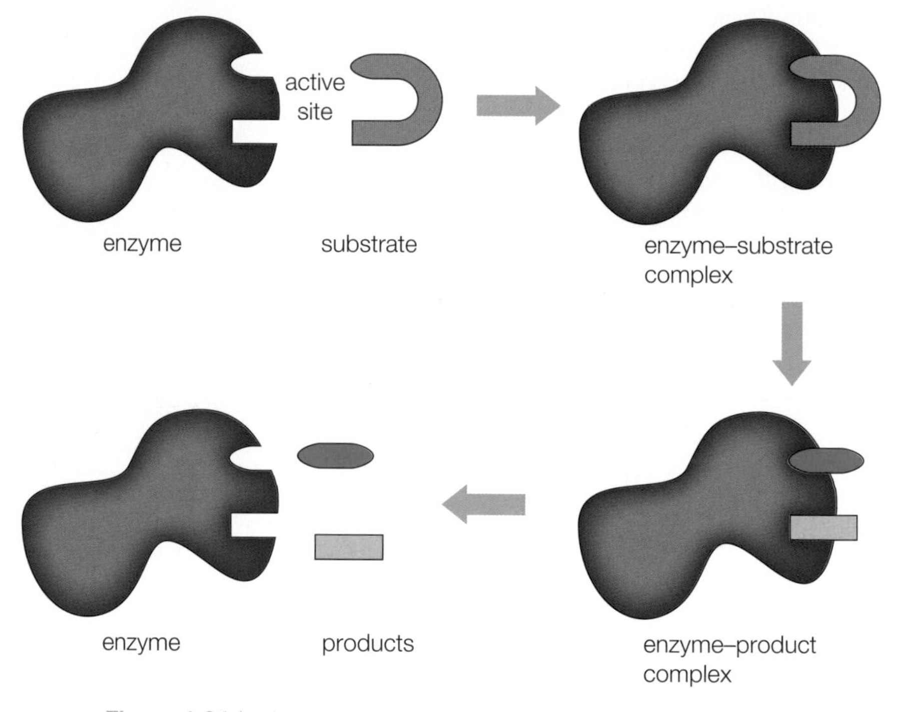

Figure 1.24▲
A diagram to illustrate the lock-and-key model for enzyme action.

The substrate is converted to products at the active site. These products then separate from the enzyme leaving the active site free to accept another molecule of the substrate.

enzyme (E) + substrate (S) → enzyme–substrate complex (ES)
→ enzyme–product complex (EP) → enzyme (E) + products (P)

Investigating urease

The enzyme urease speeds up the reaction of urea ($H_2N{-}CO{-}NH_2$) with water to form carbon dioxide and ammonia. Two series of experiments were carried out to investigate the rate equation for the reaction. The experiments were designed to measure the initial rate of the reaction under different conditions.

Part 1: This part of the investigation involved a series of runs in which the concentration of the enzyme was kept constant but the concentration of the substrate, urea, was varied from one run to the next as shown in Table 1.8.

Part 2: In a second series of runs the concentration of the substrate, urea, was kept constant while the concentration of the enzyme, urease, was varied from one run to the next. This was done by adding different volumes of urease solution to the reaction mixture while keeping the total volume constant by adding water. The results are given in Table 1.9.

Concentration of urea/mol dm^{-3}	Initial rate of reaction /10^{-6} mol dm^{-3} min^{-1}
0.000	0.0
0.005	1.7
0.010	2.3
0.020	3.2
0.050	4.4
0.100	5.9
0.200	7.2
0.300	7.7
0.400	8.0

Table 1.8▲

Volume of urease solution/cm^3	Initial rate of reaction /10^{-6} mol dm^{-3} min^{-1}
0.00	0.0
0.005	0.6
0.10	0.8
0.20	1.8
0.30	3.2
0.50	4.8
1.00	10.4
1.50	14.9
2.00	19.5

Table 1.9▲

1 How can the lock-and-key model explain why enzymes are highly specific?

2 Write an equation for the reaction of urea with water.

3 Plot the rate of reaction against substrate concentration for the results from part 1.

4 What does the graph show about the order of reaction with respect to the substrate urea when:

a) the substrate concentration is high

b) the substrate concentration is low?

5 Plot the rate of reaction against volume of enzyme solution for the results from part 2. (Note the volume of the urease solution is a measure of the enzyme concentration.)

6 What is the order of reaction with respect to the enzyme?

7 Write the full rate equation for the reaction when:

a) the substrate concentration is low

b) the substrate concentration is high.

8 Show how the mechanism for the reaction of enzyme and substrate can account for your answers to **4**, **6** and **7**.

$$E + S \rightarrow ES \rightarrow EP \rightarrow E + P$$

9 Suggest reasons why an understanding of the mechanisms of enzyme-catalysed reactions is important in the development of new drugs.

REVIEW QUESTIONS

Extension questions

1 The data in Table 1.10 refers to the decomposition of hydrogen peroxide, H_2O_2.

Time/10^3 s	$[H_2O_2]/10^{-3}$ mol dm^{-3}
0	20.0
12	16.0
24	13.1
36	10.6
48	8.6
60	6.9
72	5.6
96	3.7
120	2.4

Table 1.10▲

a) Plot a concentration–time graph for the decomposition reaction. (3)

b) Read off three half-lives from the graph and show that this is a first order reaction (3)

c) Draw tangents to the curve in your graph at four different concentrations and hence find the gradient of the curve at each point. (4)

d) Plot a graph of rate against concentration using your results from c) and hence find the value for the rate constant at the temperature of the experiment. (4)

2 The results in Table 1.11 come from a study of the rate of reaction of iodine with a large excess of hex-1-ene dissolved in ethanoic acid.

Time/10^3 s	$[I_2]/10^{-3}$ mol dm^{-3}
0	20.0
1	15.6
2	12.8
3	11.0
4	9.4
5	8.3
6	7.5
7	6.8
8	6.2

Table 1.11▲

a) Plot a concentration–time graph and show that the half-life is not constant. (5)

b) From your graph find the rate of reaction at a series of concentrations. (3)

c) i) Use your results from b) to plot a graph to confirm that the reaction is second order with respect to iodine.

ii) Explain how to interpret the graph you have drawn. (5)

3 Hydrogen peroxide oxidises iodide ions to iodine in the presence of hydrogen ions. The other product is water. The reaction is first order with respect to hydrogen peroxide, first order with respect to iodide ions but zero order with respect to hydrogen ions.

a) Write a balanced equation for the reaction. (1)

b) Write a rate equation for the reaction. (2)

c) What is the overall order of the reaction? (1)

d) A proposed mechanism for the reaction involves three steps:

$H_2O_2 + I^- \rightarrow H_2O + IO^-$

$H^+ + IO^- \rightarrow HIO$

$HIO + H^+ + I^- \rightarrow I_2 + H_2O$

Which step is likely to be the rate-determining step and why? (2)

4 Two gases X and Y react according to this equation:

$X(g) + 2Y(g) \rightarrow XY_2(g)$

This reaction was studied at 400 K giving the results shown in Table 1.12.

Experiment number	Initial concentration of X/mol dm^{-3}	Initial concentration of Y/mol dm^{-3}	Initial rate of formation of XY_2/mol dm^{-3} s^{-1}
1	0.10	0.10	0.0001
2	0.10	0.20	0.0004
3	0.10	0.30	0.0009
4	0.20	0.10	0.0001
5	0.30	0.10	0.0001

Table 1.12▲

a) What is the order of the reaction with respect to:

i) X ii) Y?

Explain your answers. (4)

b) Write a rate equation for the reaction of X with Y. (2)

c) Use the results of the first experiment to calculate a value of the rate constant and give its units. (2)

d) Suggest a possible mechanism for the reaction. (3)

e) Explain why chemists are interested in determining rate equations and measuring rate constants. (5)

5 Table 1.13 shows the results of a series of experiments to determine the activation energy for the oxidation of iodide ions by peroxodisulfate(VI) ions in the presence of iron(III) ions.

$$S_2O_8^{2-}(aq) + 2I^-(aq) \rightarrow 2SO_4^{2-}(aq) + I_2(aq)$$

The results were obtained by the 'clock' method. Each reaction mixture included a small, measured amount of aqueous sodium thiosulfate and a few drops of starch solution.

Temperature, *T*/K	288	292.5	299	308	315
Time, *t*, for the blue colour to appear/s	10.0	7.0	5.0	3.5	2.5

Table 1.13▲

a) Explain why each reaction mixture included a small amount sodium thiosulfate solution and starch solution. **(3)**

b) Analyse the results and plot a graph to find a value for the activation energy in the presence of iron(III) ions. **(8)**

c) In the absence of iron(III) ions the activation energy for the reaction is 52.9 kJ mol^{-1}. Suggest an explanation for the effect of adding iron(III) ions. **(4)**

2 How far? – entropy

Figure 2.1▲
Energy can drive change in the opposite direction. Photosynthesis effectively reverses the changes of respiration. Leaves harness energy from the Sun to convert carbon dioxide and water into carbohydrates.

Chemists want to know how to predict the direction and extent of change. They have found that they can do so by using the laws of thermodynamics. Central to the theory is the concept of entropy which gives us the answer to the question: Why do things happen the way that they do?

Figure 2.2▲
A cheetah hunting its prey in Kenya. The cheetah gets its energy from respiration taking advantage of the natural direction of change. Carbohydrates react with oxygen in muscle cells, forming carbon dioxide and water and giving out energy.

Figure 2.3▲
Theory shows that the change of diamond to graphite is spontaneous. Graphite is more stable than diamond. Fortunately for owners of valuable jewellery the change is very, very slow.

2.1 Entropy changes

Spontaneous changes

A spontaneous reaction is a reaction which tends to go without being driven by any external agency. Spontaneous reactions are the chemical equivalent of water flowing downhill. Any reaction which naturally tends to happen is spontaneous in this sense even if it is very slow, just as water has a tendency to flow down a valley even when held up behind a dam. The chemical equivalent of a dam is a high activation energy for a reaction.

Figure 2.4►
Metals such as magnesium, iron and aluminium react spontaneously with oxygen. They are ingredients of fireworks. They burn when heated by the spontaneous reactions between sulfur, carbon and potassium nitrate in gunpowder.

In practice chemists also use the word spontaneous in its everyday sense as well, to describe reactions that not only tend to go but also go fast on mixing the reactants at room temperature.

Here is a typical example:

'The hydrides of silicon catch fire spontaneously in air, unlike methane which has an ignition temperature of about 500 °C.'

The reactions of the hydrides of carbon, such as methane, with oxygen are also spontaneous in the thermodynamic sense, even at room temperature. However, the activation energies for the reactions are so high that nothing happens until the hydrocarbons are heated with a flame.

The possible ambiguity in the use of the term 'spontaneous' means that chemists often prefer an alternative term. They describe a reaction as 'feasible' if it tends to go naturally.

Definition

A **spontaneous**, or **feasible**, **reaction** is one that naturally tends to happen, even if it is very slow because the activation energy is high.

Test yourself

1 Classify these changes as spontaneous/fast, spontaneous/slow or not spontaneous:

a) ice melting at 5 °C
b) ammonia gas condensing to a liquid at room temperature
c) diamond reacting with oxygen at room temperature
d) water splitting up into hydrogen and oxygen at room temperature
e) sodium reacting with water at room temperature.

Enthalpy changes and spontaneous reactions

Most reactions that are spontaneous are also exothermic. They have a negative standard enthalpy change of reaction. This is such a common pattern that chemists often use the sign of ΔH to decide whether or not a reaction is likely to go.

However, some endothermic processes are spontaneous too. This shows the limitation of using the enthalpy change to decide the likely direction of change. One example is the reaction of citric acid with sodium hydrogencarbonate. The mixture fizzes vigorously while getting colder and colder. ΔH for the reaction is positive. A more colourful example is the reaction of sulfur dichloride oxide with the dark red crystals of cobalt(II) chloride. The reaction looks violent and quantities of fuming hydrogen chloride gas bubble from the mixture while a thermometer dipping into the chemicals shows that the temperature falls sharply.

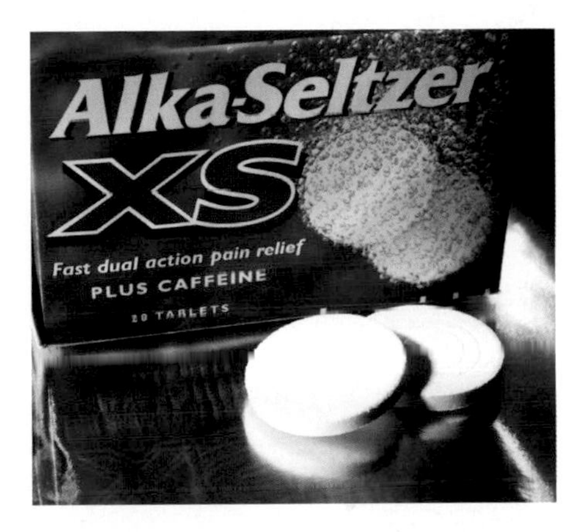

Figure 2.5▲
Alka-Seltzer tablets contain citric acid and sodium hydrogencarbonate. They fizz when added to water. The reaction of the acid and the hydrogencarbonate is a spontaneous endothermic reaction.

Diffusion – a spontaneous change

Open a bottle of perfume in the corner of a room, and it is not long before everyone else in the room can smell it. The perfume molecules spread out naturally and mix with the air molecules. This is diffusion.

The opposite never happens. If there is a bad smell in a room it is impossible to get rid of it by persuading all the smelly molecules to collect together in a small bottle before putting in a stopper.

Figure 2.6◄
The right-hand bottle contains bromine vapour. The left-hand bottle contains air. The two are separated by a barrier.

Figure 2.7◄
After removing the barrier, the bromine diffuses from the right-hand bottle until the bromine molecules are evenly spread between the two bottles. Once mixed the air and bromine never unmix. When the concentrations are the same in both bottles, molecules continue to diffuse between the bottles, but overall there seems to be no change. This is an example of dynamic equilibrium.

Diffusion happens by chance alone. This is shown by the very simple example in Figure 2.8. This shows an imaginary situation with just six molecules of bromine in the right-hand jar (jar R). The left-hand jar is empty and there is a barrier between the jars. This situation can be represented as RRRRRR.

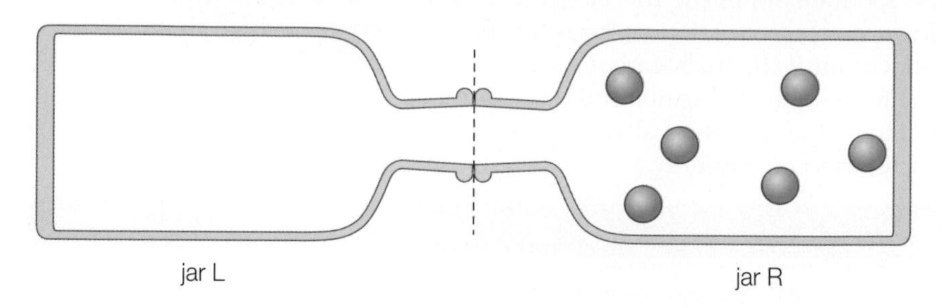

Figure 2.8▲
Two gas jars separated by a barrier with six molecules of bromine in the right-hand jar. The molecules are in rapid, random motion (RRRRRR).

The molecules in jar R are moving around randomly, bumping into each other and the sides of the jar. Figure 2.9 shows what happens immediately after removing the barrier. One molecule has moved into the left-hand jar. This can be represented as RRRRRL.

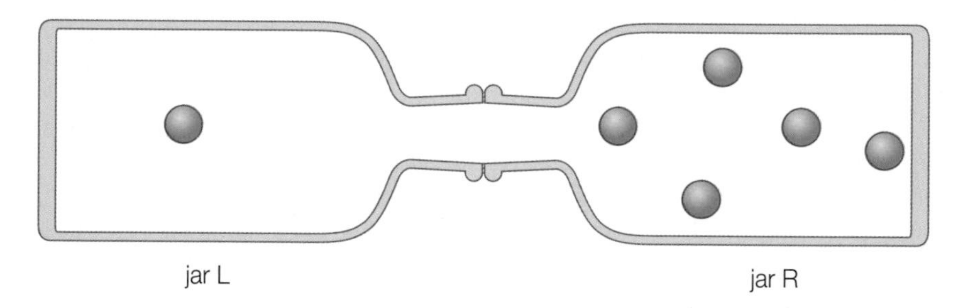

Figure 2.9▲
After removing the barrier one molecule has moved into the left-hand jar. (RRRRRL).

As the molecules move around randomly in the two jars, there are many possible arrangements. Each molecule can either be R or L. So for six molecules moving between the jars there are $2 \times 2 \times 2 \times 2 \times 2 \times 2 = 2^6 = 64$ possible arrangements.

Only one of these arrangements is RRRRRR so there is a 1 in 64 chance that all the molecules will end up back in jar R. There is a much bigger chance that the molecules will be arranged in some other way.

In practice the number of molecules is always many more than six. In Figure 2.6 there are about 10^{22} molecules of bromine in the right-hand jar. The number of ways of distributing the molecules between the two jars is $2^{10^{22}}$. This is a huge number – too big to be imagined. Only one of all this number of arrangements has all the molecules back in the right-hand jar. The chance of this happening is impossibly low.

It is overwhelmingly likely, by chance alone, that bromine molecules will spread out and mix with air molecules unless there is a barrier to stop them. This is a spontaneous change. In general, gas molecules naturally tend to mix up and disperse themselves randomly.

Activity

The number of ways of dissolving

Iodine is soluble in dichloromethane. The reversible process of dissolving is represented schematically in Figure 2.10.

$$I_2(s) \rightleftharpoons I_{2(CH_2Cl_2)}$$

When an iodine molecule dissolves it can be regarded as taking the space previously occupied by a dichloromethane molecule.

1 In the simple situation of Figure 2.10, how many different solvent molecules can the one iodine molecule take the place of as it dissolves?

2 How many different way are there for the iodine molecule in Figure 2.10(b) to return to its vacant space in the crystal lattice?

3 Which is the more probable arrangement for the iodine molecule – in the crystal or in solution?

4 When this example is scaled up to 1 mol I_2(s) surrounded by 1 mol of solvent molecules, how many different arrangements are possible for one iodine molecule?

5 Suppose the one molecule returns to its original place in the lattice. In how many ways can it do this? Is this likely to happen?

6 Why does chance favour dissolving?

7 Some important factors have been ignored in this simplified example. Suggest two.

a) solvent

I_2(s) crystal

b)

Figure 2.10▲
a) A crystal made up of eight iodine molecules in contact with ten molecules of the solvent dichloromethane. **b)** The same crystal and solvent when one iodine molecule has dissolved.

Molecules and energy

It is not only the arrangement of molecules that matters. Even more important is the way that energy is spread out between molecules. The molecules in bromine gas are moving around, spinning and vibrating. The energies of the tiny molecules are quantised, in a similar way to the energies of electrons in atoms. All the time the molecules are bumping into each other and as they do so they lose and gain energy quanta to and from other molecules.

Taking a very simple situation, Table 2.1 shows the number of ways that two molecules can share four energy quanta. In all there are five different ways.

Number of quanta	
Molecule 1	Molecule 2
2	2
3	1
1	3
0	4
4	0

Table 2.1▲
The number of ways that two molecules can share four energy quanta.

Test yourself

2 How many ways can two molecules share five energy quanta?

The more molecules, and the more energy quanta, the more ways there are of sharing the energy between the molecules as shown in Table 2.2.

Number of molecules	Number of quanta	Number of ways of sharing the molecules between the quanta
10	100	$\approx 10^{12}$
100	10	$\approx 10^{13}$
100	100	$\approx 10^{60}$
200	110	$\approx 10^{86}$

Table 2.2▲
The number of ways of sharing energy between molecules.

Increasing the temperature of a substance increases the number of energy quanta in the system and so increases the number of possible ways that the energy quanta can be shared out between the molecules.

Activity

Sharing energy quanta

Figure 2.11 shows two tiny pieces of metal each made up of 100 atoms. The hot piece of metal has 100 energy quanta. The cold piece of metal has only 10 energy quanta.

Answer these questions with the help of Table 2.2.

1. How many ways can the quanta be distributed between the atoms in the hot piece of metal?
2. How many ways can the quanta be distributed between the atoms in the cold piece of metal?
3. Metals are thermal conductors because their atoms can freely share energy quanta. What is the number of ways that the energy quanta can be distributed when the two piece of metal in Figure 2.11 are put in contact with each other?
4. Which is the more probable arrangement – to have 100 quanta in one piece and 10 in the other, or to have the quanta evenly shared between the two?
5. What happens to the temperatures of the two pieces of metal when they are put in contact with each other?

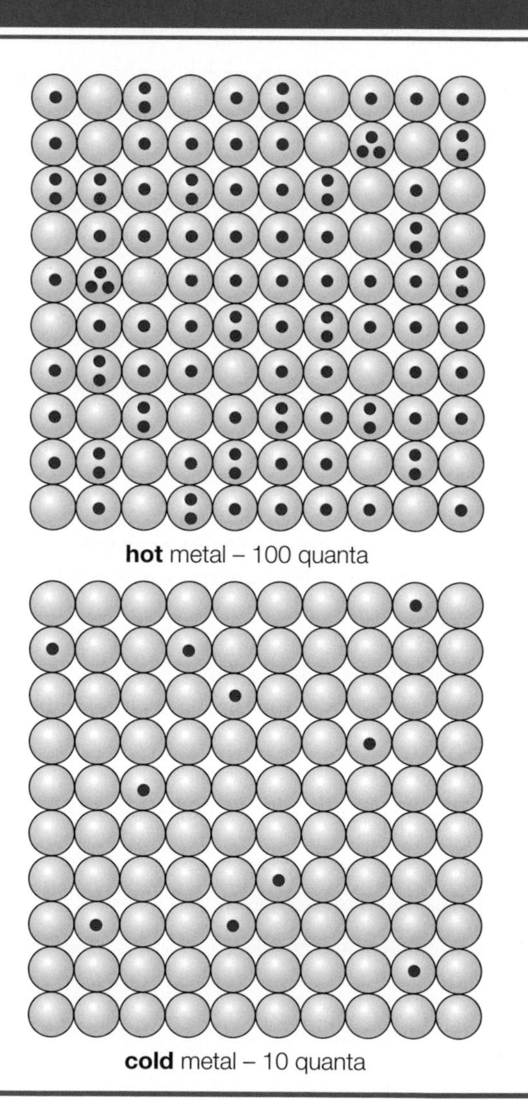

Figure 2.11▶
The blue circles represent the atoms and the red dots the energy quanta. The 100 atoms of the hot piece of metal share 100 energy quanta. The 100 atoms of the cold piece of metal share 10 energy quanta.

Molecules, energy quanta and entropy

Random changes, that happen by chance, always tend to go in the direction that increases the number of ways of distributing the molecules and energy quanta. This is a fundamental principle.

However the number of ways, *W*, of distributing the energy quanta between the molecules in a mole of gas at room temperature is huge. The numbers involved are very hard to deal with. Fortunately, the Austrian physicist Ludwig Boltzmann (1844–1906) showed how the established laws of thermodynamics could be explained in terms of molecules and their energies.

Physicists had already developed the concept of entropy to account for the way that steam engines work. What Boltzmann was able to show was that there is a relationship between this quantity entropy and the number of ways that any chemical system could distribute its molecules and energy. He demonstrated the truth of this formula:

Entropy, $S = k \ln W$

where S is the entropy of the system, *k* is a constant named after Boltzmann and $\ln W$ is the natural logarithm of the number of ways of arranging the particles and energy in the system.

Now chemists can use this quantity called entropy, S, to decide whether or not a reaction is feasible. The formula shows that as *W* increases, S increases. So change happens in the direction which leads to a total increase in entropy.

Chemists sometimes describe entropy as a measure of disorder or randomness. These descriptions have to be interpreted with care because the disorder refers not only to the arrangement of the particles in space but much more significantly to the numbers of ways of distributing the energy of the system across all the available energy levels.

When considering chemical reactions it is essential to calculate the total entropy change in two parts: the entropy change of the system and the entropy change of the surroundings.

$$\Delta S_{total} = \Delta S_{system} + \Delta S_{surroundings}$$

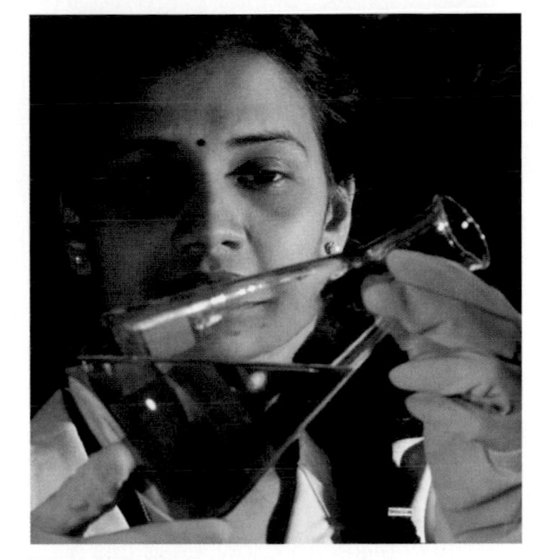

Figure 2.12▲
The reaction mixture in the flask is 'the system'. The air and everything else around the flask makes up the 'surroundings'. It is the total entropy in the system and the surroundings which determines whether or not the reaction is feasible.

Standard molar entropies

In a perfectly ordered crystal at 0 K the entropy is zero. The entropy of a chemical rises as the temperature rises. Increasing the temperature raises the number of energy quanta to share between the atoms and molecules. There are also jumps in the entropy of the chemical wherever there is a change of state. This is because energy is added to change a solid to a liquid or a liquid to a gas (see Figure 2.13). Also there is an increase in the number of ways of arranging the atoms or molecules as a solid changes to a liquid and then to a gas.

Standard molar entropies are quoted for pure chemicals under standard conditions (298 K and 1 atmosphere pressure). Gases generally have higher entropies than comparable liquids, which have higher entropies than similar solids.

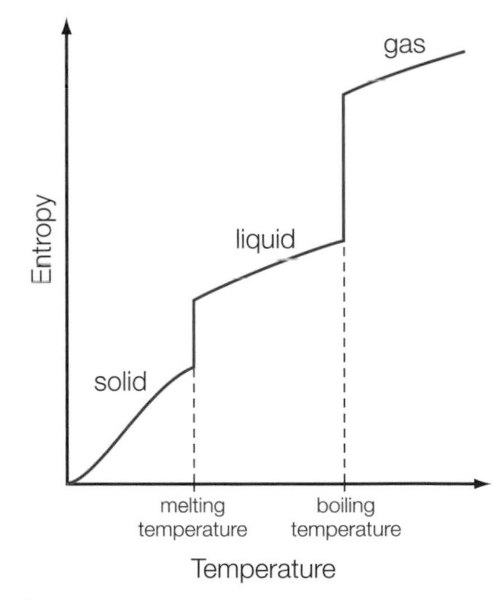

Figure 2.13▲
The entropy of a chemical increases as its temperature rises. There is a jump in the entropy values when the chemical melts or boils.

Definition

Standard molar entropy, *S*, is the entropy per mole for a substance under standard conditions. Chemists use values for standard molar entropies to calculate entropy changes and so to predict the direction and extent of chemical change.

The units for standard molar entropy are joules per mole per kelvin ($J\,mol^{-1}\,K^{-1}$). Note that the units are joules and not kilojoules.

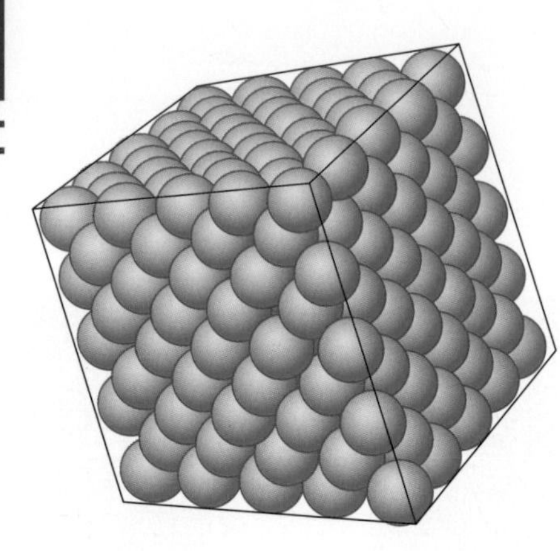

Figure 2.14▲
Structure of a solid. Solids have relatively low values for standard molar entropies. In diamond the carbon atoms are held firmly in place by strong, highly directional covalent bonds. The standard molar entropy of diamond is low. Lead has a higher value for its standard molar entropy because metallic bonds are not directional. The heavier, larger atoms can vibrate more freely and share out their energy in more ways than can carbon atoms in diamond.

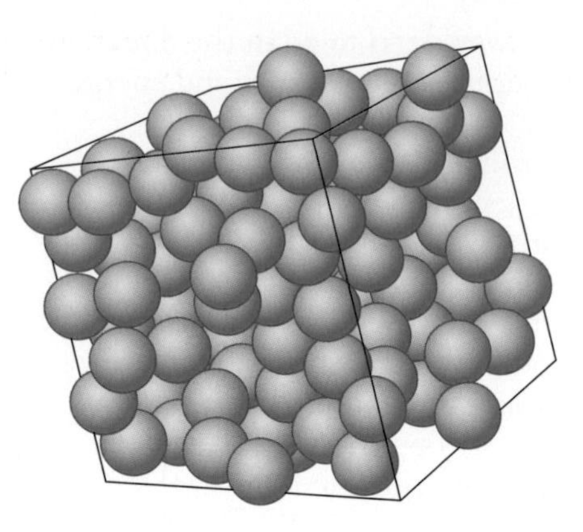

Figure 2.15▲
Structure of a liquid. In general, liquids have higher standard molar entropies than comparable solids because the atoms or molecules are free to move. There are many more ways of distributing the particles and energy – there is more disorder. The standard molar entropy of mercury is higher than that of lead. Molecules with more atoms have higher standard molar entropies because they can vibrate, rotate and arrange themselves in yet more ways.

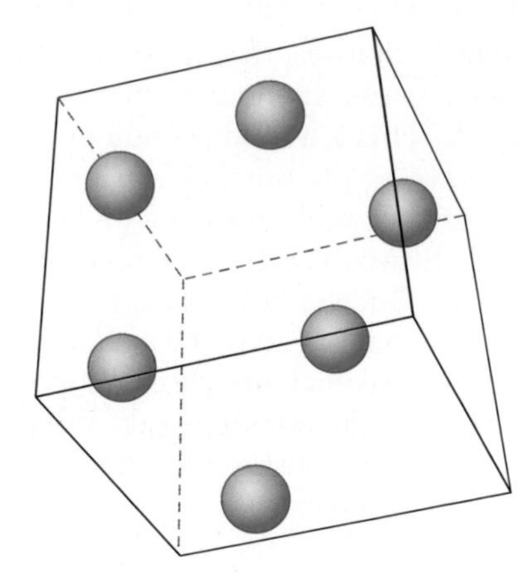

Figure 2.16▲
Atoms of a noble gas. Gases have even higher standard molar entropies than comparable liquids because the atoms or molecules are not only free to move but also widely spaced. There are even more ways of distributing the particles and energy – the disorder is even greater. The standard molar entropy of argon is higher than that of mercury. As with liquids, molecules with more atoms have even higher standard molar entropies because they can vibrate, rotate and arrange themselves in more ways.

Solids	$S^\ominus$/J mol^{-1} K^{-1}	Liquids	$S^\ominus$/J mol^{-1} K^{-1}	Gases	$S^\ominus$/J mol^{-1} K^{-1}
Carbon (diamond)	2.4	Mercury	76.0	Argon	155
Magnesium oxide	26.9	Water	69.9	Ammonia	192
Copper	33.2	Ethanol	160	Carbon dioxide	214
Lead	64.8	Benzene	173	Propane	270

Table 2.3▲
Standard molar entropies for selected solids, liquids and gases.

Test yourself

3 Refer to Table 2.3. Why is the value of the standard molar entropy of:
  a) mercury higher than the value for copper
  b) ammonia higher than the value for water
  c) propane higher than the value for argon?

4 Which substance in each of the following pairs is expected to have the higher standard molar entropy at 298 K?
  a) $Br_2(l)$, $Br_2(g)$
  b) $H_2O(s)$, $H_2O(l)$
  c) $HF(g)$, $NH_3(g)$
  d) $CH_4(g)$, $C_2H_6(g)$
  e) NaCl(s), NaCl(aq)
  f) ethene, poly(ethene)
  g) pentene gas, cyclopentane gas.

The entropy change of the system

Tables of standard molar entropies make it possible to calculate the entropy change of the system of chemicals during a reaction.

$$\Delta S^{\ominus}_{system} = \text{the sum of the standard molar entropies of the products} - \text{the sum of the standard molar entropies of the reactants}$$

Worked example

Calculate the entropy change for the system, $\Delta S^{\ominus}_{system}$, for the synthesis of ammonia from nitrogen and hydrogen. Comment on the value.

Notes on the method

Write the balanced equation for the reaction.

Look up the standard molar entropies on the Data sheet from the Dynamic Learning Student website, taking careful note of the units.

Answer

$$N_2(g) + 3H_2(g) \rightarrow 2NH_3(g)$$

Sum of the standard molar entropies of the products $= 2S^{\ominus}[NH_3(g)]$

$= 2 \times 192.4\,J\,mol^{-1}\,K^{-1} = 384.8\,J\,mol^{-1}\,K^{-1}$

Sum of the standard molar entropies of the reactants

$= S^{\ominus}[N_2(g)] + 3S^{\ominus}[H_2(g)]$

$= 191.6\,J\,mol^{-1}\,K^{-1} + (3 \times 130.6\,J\,mol^{-1}\,K^{-1}) = 583.4\,J\,mol^{-1}\,K^{-1}$

$\Delta S^{\ominus}_{system} = 384.8\,J\,mol^{-1}\,K^{-1} - 583.4\,J\,mol^{-1}\,K^{-1}$

$= -198.6\,J\,mol^{-1}\,K^{-1}$

This shows that the entropy of the system decreases when nitrogen and hydrogen combine to form ammonia. This is not surprising since the change halves the number of molecules so the amount, in moles, of gas decreases.

Test yourself

5 Without doing any calculations predict whether the entropy of the system increases or decreases as a result of these changes:

a) $KCl(s) + aq \rightarrow KCl(aq)$

b) $H_2O(l) \rightarrow H_2O(g)$

c) $Mg(s) + Cl_2(g) \rightarrow MgCl_2(s)$

d) $N_2O_4(g) \rightarrow 2NO_2(g)$

e) $NaHCO_3(s) + HCl(aq) \rightarrow NaCl(aq) + H_2O(l) + CO_2(g)$

6 Use values from the Data sheet to calculate the entropy change for the system, $\Delta S^{\ominus}_{system}$, for the catalytic reaction of ammonia with oxygen to form NO and steam. Comment on the value.

The entropy change of the surroundings

It is not enough to consider only the entropy of the system. What matters is the total entropy change, which is the sum of the entropy changes of the system and the entropy change in the surroundings.

It turns out that the entropy change of the surroundings during a chemical reaction is determined by the size of the enthalpy change, $\Delta H_{reaction}$, and the temperature, T. The relationship is:

$$\Delta S_{surroundings} = -\frac{\Delta H_{reaction}}{T}$$

The minus sign is included because the entropy change is bigger the more the energy transferred to the surroundings. For an exothermic reaction, which transfers energy to the surroundings, ΔH is negative, so $-\Delta H$ is positive.

Figure 2.17▶
The thermite reaction is highly exothermic. It gives out a great deal of energy to its surroundings. The entropy change in the surroundings is large and positive.

What this relationship shows is that the more energy transferred to the surroundings by an exothermic process, the larger the increase in the entropy of the surroundings. It also shows that, for a given quantity of energy, the increase in entropy is greater when the surroundings are cool than when they are hot. Adding energy to molecules in a cool system has a proportionately greater effect on the number of ways of distributing matter and energy than adding the same quantity of energy to a system that is already very hot.

The total entropy change

A reaction is only feasible if the total entropy change, ΔS_{total} is positive:

$$\Delta S_{total} = \Delta S_{system} + \Delta S_{surroundings}$$

It turns out that:

- most exothermic reactions tend to go because at about room temperature the value of $-\frac{\Delta H}{T}$ is much larger and more positive than ΔS_{system} which means that ΔS_{total} is positive.
- an endothermic reaction can be feasible so long as the increase in the entropy of the system is greater than the decrease in the entropy of the surroundings.
- a reaction that does not tend to go at room temperature may become feasible as the temperature rises because $\Delta S_{surroundings}$ decreases in magnitude as T increases.

Table 2.4 summarises the possibilities.

Entropy change of the system	Enthalpy change	Entropy change of the surroundings	Is the reaction feasible?
Increase ΔS_{system} positive	Exothermic $\Delta H_{reaction}$ negative	Increase $\Delta S_{surroundings}$ positive	Yes: ΔS_{total} is positive
Decrease ΔS_{system} negative	Exothermic $\Delta H_{reaction}$ negative	Increase $\Delta S_{surroundings}$ positive	Maybe: ΔS_{total} is positive if $\Delta S_{surroundings}$ is more positive than ΔS_{system} is negative
Increase ΔS_{system} positive	Endothermic $\Delta H_{reaction}$ positive	Decrease $\Delta S_{surroundings}$ negative	Maybe: ΔS_{total} is positive if ΔS_{system} is more positive than $\Delta S_{surroundings}$ is negative
Decrease ΔS_{system} negative	Endothermic $\Delta H_{reaction}$ positive	Decrease $\Delta S_{surroundings}$ negative	No: ΔS_{total} is negative

Table 2.4▲

Worked example

Calculate the total entropy change for the reaction of calcium oxide with ammonium chloride under standard conditions to determine whether or not the reaction is feasible under standard conditions.

$$CaO(s) + 2NH_4Cl(s) \rightarrow CaCl_2(s) + 2NH_3(g) + H_2O(l)$$

Comment on the result.

Notes on the method

Get the data you need from the Data sheet.

www Data

Calculate the standard entropy change for the reaction using the formula:

$$\Delta S^{\ominus}_{system} = \Sigma S^{\ominus}[\text{products}] - \Sigma S^{\ominus}[\text{reactants}]$$

Calculate the enthalpy change using a Hess' law cycle or the formula:

$$\Delta H^{\ominus}_{reaction} = \Sigma \Delta H^{\ominus}_f[\text{products}] - \Sigma \Delta H^{\ominus}_f[\text{reactants}]$$

Hence calculate $\Delta S^{\ominus}_{surroundings}$ remembering to convert the value for the enthalpy change from $kJ\,mol^{-1}$ to $J\,mol^{-1}$. Under standard conditions the temperature is 298 K.

Use your two calculated entropy values to calculate $\Delta S^{\ominus}_{total}$.

Answer

$$\Sigma S^{\ominus}[\text{products}] = 105\,J\,mol^{-1}\,K^{-1} + (2 \times 192\,J\,mol^{-1}\,K^{-1}) + 69.9\,J\,mol^{-1}\,K^{-1} = 559\,J\,mol^{-1}\,K^{-1}$$

$$\Sigma S^{\ominus}[\text{reactants}] = 39.7\,J\,mol^{-1}\,K^{-1} + (2 \times 94.6\,J\,mol^{-1}\,K^{-1}) = 229\,J\,mol^{-1}\,K^{-1}$$

$$\Delta S^{\ominus}_{system} = 559\,J\,mol^{-1}\,K^{-1} - 229\,J\,mol^{-1}\,K^{-1} = 330\,J\,mol^{-1}\,K^{-1}$$

$$\Sigma \Delta H_f^{\ominus}[\text{products}] = -796\,kJ\,mol^{-1} + 2 \times (-46.1\,kJ\,mol^{-1}) - 286.1\,kJ\,mol^{-1} = -1174\,kJ\,mol^{-1}$$

$$\Sigma \Delta H_f^{\ominus}[\text{reactants}] = -635\,kJ\,mol^{-1} + 2 \times (-314\,kJ\,mol^{-1}) = -1263\,kJ\,mol^{-1}$$

$$\Delta H^{\ominus}_{reaction} = (-1174\,kJ\,mol^{-1}) - (-1263\,kJ\,mol^{-1}) = +89\,kJ\,mol^{-1} = +89\,000\,J\,mol^{-1}$$

$$\Delta S^{\ominus}_{surroundings} = -\frac{89\,000\,J\,mol^{-1}}{298\,K} = -299\,J\,mol^{-1}\,K^{-1}$$

$$\Delta S^{\ominus}_{total} = \Delta S^{\ominus}_{system} + \Delta S^{\ominus}_{surroundings} = 330\,J\,mol^{-1}\,K^{-1} - 299\,J\,mol^{-1}\,K^{-1} = +31\,J\,mol^{-1}\,K^{-1}$$

The reaction is endothermic but the total entropy change is positive so the reaction is feasible at 298 K. The formation of a solid with a gas and a liquid from two solids means that there is a substantial increase in the entropy of the system. More than enough to compensate for the decrease in the entropy of the surroundings.

A change that is not feasible at a lower temperature may become feasible if the temperature rises. This is because $\Delta S_{surroundings}$ gets smaller as T rises because:

$$\Delta S_{surroundings} = -\frac{\Delta H_{reaction}}{T}$$

Generally the values of ΔH and ΔS_{system} do not change markedly with temperature, and so it is possible to estimate the point at which a reaction which is not feasible at room temperature becomes feasible as the temperature rises.

Test yourself

7 Can an exothermic reaction which is not feasible at room temperature become feasible at a higher temperature if the entropy change for the reaction is negative?

8 Consider the reaction of magnesium with oxygen:

$2Mg(s) + O_2(g) \rightarrow 2MgO(s)$ $\quad \Delta H^{\ominus} = -602\,kJ\,mol^{-1}$, $\quad \Delta S^{\ominus}_{system} = -217\,J\,mol^{-1}\,K^{-1}$

a) Why does the entropy of the system decrease?

b) Show why the reaction of magnesium with oxygen is feasible at 298 K despite the decrease in the entropy.

Figure 2.18▲
Roofer fitting copper to a roof in Boston, USA. Copper in air and water: stable or inert?

Figure 2.19▲
Coils of aluminium at a processing plant. Aluminium in air: stable or inert?

Test yourself

9 Consider the reaction of ammonium chloride with barium hydroxide

$2NH_4Cl(s) + Ba(OH)_2(s) \rightarrow BaCl_2.2H_2O(s) + 2NH_3(g)$

a) Use the Data sheet to calculate the entropy change for the system and comment on the value you get.
b) Use the Data sheet to calculate the enthalpy change for the reaction.
c) Calculate the entropy change of the surroundings at 298 K.
d) Work out the total entropy change for the reaction and decide whether or not it is feasible under standard conditions. Comment on your answer.

10 Consider the reduction of iron(III) oxide by carbon:

$2Fe_2O_3(s) + 3C(s) \rightarrow 4Fe(s) + 3CO_2(g)$

$\Delta H^{\circ} = +468\ \text{kJ mol}^{-1}$ $\quad \Delta S^{\circ}_{system} = +558\ \text{J mol}^{-1}\ \text{K}^{-1}$

a) Why does the entropy of the system increase for this reaction?
b) Show that the reaction is not feasible at room temperature (298 K).
c) Assuming that ΔH° and $\Delta S^{\circ}_{system}$ do not vary with temperature, estimate the temperature at which the decomposition becomes feasible.

Definitions

A chemical or mixture of chemicals is **thermodynamically stable** if there is no tendency for a reaction. If the total entropy change, ΔS_{total}, is negative, this indicates that the reaction does not tend to occur.

Kinetic inertness is a term used when a reaction does not go even though it appears to be feasible. The reaction tends to go as indicated by a positive total entropy change, yet nothing happens. There is no change because the rate of reaction is too slow to be noticeable. There is a barrier preventing change – usually a high activation energy. The compound or mixture is inert.

2.2 Stable or inert?

The study of energetics (thermochemistry) and rates of reaction (kinetics) helps to explain why some chemicals are stable while others react rapidly.

Compounds are stable if they have no tendency to decompose into their elements or into other compounds. Magnesium oxide, for example, has no tendency to split up into magnesium and oxygen. However, a compound which is stable at room temperature and pressure may become more, or less, stable as conditions change

A chemical is inert if it has no tendency to react even when the reaction is feasible. The lighter noble gases, helium and neon, live up to the original name for group 0 (8). They are inert towards all other reagents.

Nitrogen is a relatively unreactive gas which can be used to create an 'inert atmosphere' free of oxygen (which is much more reactive). However, nitrogen is not inert in all circumstances. It reacts with hydrogen in the Haber process to form ammonia and with oxygen at high temperatures to form nitrogen oxides.

A compound such as the gas N_2O, for example, is thermodynamically unstable but it continues to exist at room temperature because it is kinetically inert. The total entropy change for the decomposition reaction is positive so the compound tends to decompose into its elements, but the rate is very slow under normal conditions.

Examples of kinetic inertness are:

- a mixture of hydrogen and oxygen at room temperature
- a solution of hydrogen peroxide in the absence of a catalyst
- aluminium metal in dilute hydrochloric acid.

ΔS_{total}	Activation energy	Change observed
Negative	High	No reaction – reactants stable relative to products
Negative	High	No reaction – reactants unstable relative to products but kinetically inert
Negative	Low	No reaction – reactants stable relative to products
Positive	Low	Fast reaction – reactants unstable relative to products

Table 2.5▲

Sometimes there is no tendency for a reaction to go because the reactants are stable. This is so if the total entropy change for the reaction is negative. Sometimes there is no reaction even though thermochemistry suggests that it should go. The total entropy change is positive so the change is feasible, but a high activation energy means that the rate of reaction is very, very slow.

Test yourself

11 Suggest examples of reactions to illustrate each of the four possibilities in Table 2.5.

12 Draw a reaction profile to show the energy changes from reactants to products for reactants which are thermodynamically unstable relative to the products, but kinetically inert.

Activity

The thermal stability of group 2 carbonates

The decomposition of group 2 metal carbonates is used on a large scale to make oxides such as magnesium and calcium oxides.

$$MgCO_3(s) \rightarrow MgO(s) + CO_2(g)$$
$$\Delta H^{\ominus} = +117\ \text{kJ mol}^{-1}, \quad \Delta S^{\ominus}_{system} = +175\ \text{J mol}^{-1}\ \text{K}^{-1}$$

$$BaCO_3(s) \rightarrow BaO(s) + CO_2(g)$$
$$\Delta H^{\ominus} = +268\ \text{kJ mol}^{-1}, \quad \Delta S^{\ominus}_{system} = +172\ \text{J mol}^{-1}\ \text{K}^{-1}$$

The carbonates of group 2 metals do not decompose at room temperature. They do decompose on heating.

1 Why does the entropy of the system increase when a group 2 carbonate decomposes?

2 **a)** Calculate the total entropy change for the decomposition of:

   i) magnesium carbonate

   ii) barium carbonate.

   b) Are these two compounds stable or unstable relative to decomposition into their oxide and carbon dioxide at room temperature (298 K)?

3 Assuming that ΔH and ΔS_{system} for the reactions do not vary with temperature, estimate the temperatures at which the two decomposition reactions become feasible.

Figure 2.20▲
Crystals of dolomite (foreground) and magnesite (background) from Brazil. Dolomite is a calcium magnesium carbonate rock. Magnesite is pure magnesium carbonate.

4 Down group 2, do the metal carbonates become more or less stable relative to decomposition into the oxide and carbon dioxide?

Chemists seek to explain the trend in thermal stability of the group 2 carbonates by analysing the energy changes.

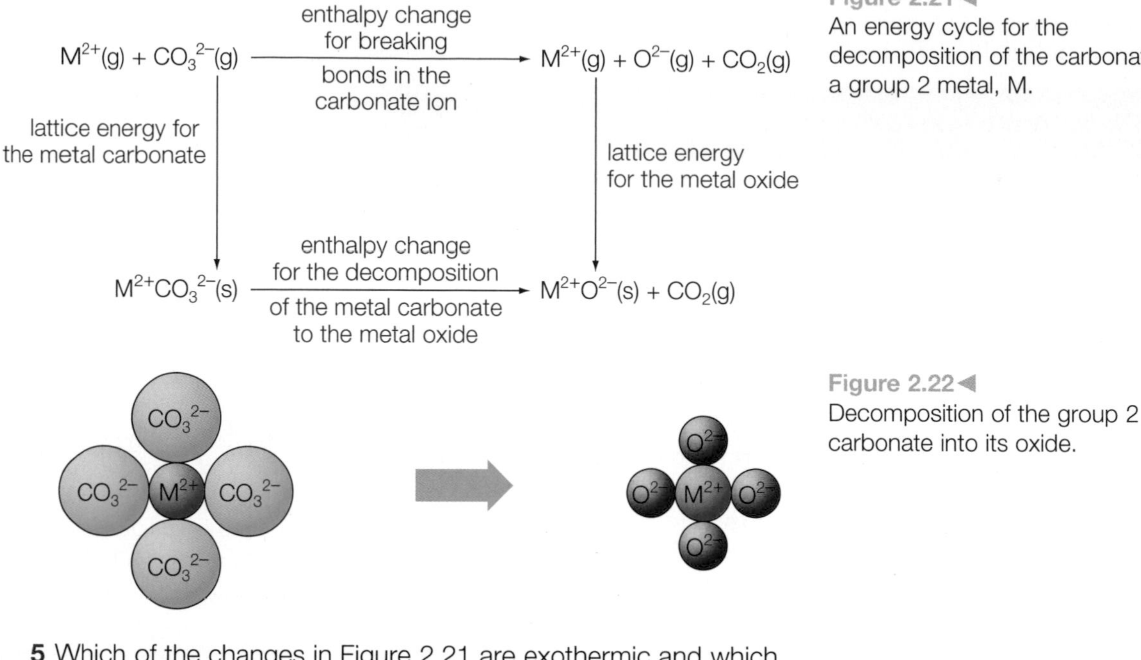

Figure 2.21 ◀ An energy cycle for the decomposition of the carbonate of a group 2 metal, M.

Figure 2.22 ◀ Decomposition of the group 2 carbonate into its oxide.

5 Which of the changes in Figure 2.21 are exothermic and which are endothermic?

6 Why is the lattice energy of magnesium oxide more negative than the lattice energy of barium oxide?

7 Why is the lattice energy of magnesium oxide more negative than the lattice energy of magnesium carbonate?

8 Why is the difference between the lattice energies of the metal carbonates and oxides significant in explaining the trend in thermal stability of the group 2 carbonates?

9 How does the trend in thermal stability of the metal carbonates down group 2 relate to the polarising power of the metal ions?

2.3 The solubility of ionic crystals

Enthalpy changes during dissolving

Why do ionic crystals dissolve in water even though ions in the lattice are strongly attracted to each other? What, in general, are the factors which determine the extent to which an ionic salt dissolves in water? Chemists look for answers to questions of this kind by analysing the enthalpy and entropy changes which take place as crystals dissolve.

An ionic compound such as sodium chloride does not dissolve in non-polar solvents like hexane, but it does dissolve in a polar solvent like water. When one mole of sodium chloride dissolves in water to produce a solution of concentration $1\ mol\,dm^{-3}$ under standard conditions, there is an enthalpy change of $+3.8\ kJ\,mol^{-1}$. This enthalpy change is described as the standard enthalpy change of solution of sodium chloride.

Figure 2.23◄
The concentration of salts in the Dead Sea in Israel is so high that they crystallise out in some places.

The process can be summarised by the equation

$$NaCl(s) + aq \rightarrow Na^+(aq) + Cl^-(aq) \qquad \Delta H^\ominus_{solution} = +3.8\,kJ\,mol^{-1}$$

or simply as $\Delta H^\ominus_{solution}[NaCl(s)] = +3.8\,kJ\,mol^{-1}$
In the equation, '+ aq' is short for the addition of water.

Enthalpy changes of solution are often measured and tabulated for one mole of a compound dissolving in different amounts of water under standard conditions.

For example, if 1 mole of sodium chloride is dissolved in 100 moles of water under standard conditions, the enthalpy change is represented by the equation

$$NaCl(s) + 100H_2O(l) \rightarrow NaCl(aq, 100H_2O) \qquad \Delta H^\ominus_{solution} = +4.1\,kJ\,mol^{-1}$$

It is not immediately obvious why the charged ions in a crystal, such as sodium chloride, separate and go into solution in water. Where does the energy come from to overcome the attractive forces between oppositely charged ions? The explanation depends on the fact that that the ions are hydrated by polar water molecules. Energy is given out as the ions are hydrated.

Figure 2.24 shows the structures of hydrated sodium and chloride ions. In water molecules, there is a δ+ charge in the region between the hydrogen atoms and a δ– charge on the oxygen atoms. This means that the polar water molecules are attracted to both positive cations and negative anions. The bond between the ions and the water molecules is an electrostatic attraction.

With cations, the electrostatic attraction involves the positive charge on the cations and the δ– charges on the oxygen atoms of the water molecules. In contrast, with anions the attraction involves the negative charge on the anions and the δ+ charge between the hydrogen atoms in the water molecules.

When sodium chloride dissolves in water, the overall process can be analysed into two stages and these are shown in Figure 2.25.

In the first stage, Na^+ and Cl^- ions must be separated from the solid NaCl crystals to form well spaced ions in the gaseous state, $Na^+(g)$ and $Cl^-(g)$. This is the reverse of the lattice energy and is labelled is $-\Delta H^\ominus_{lattice} = +787\,kJ\,mol^{-1}$ in Figure 2.25.

In the second stage, gaseous $Na^+(g)$ and $Cl^-(g)$ ions are hydrated by polar water molecules forming a solution of sodium chloride, $Na^+(aq) + Cl^-(aq)$. Under standard conditions, this process is the sum of the standard enthalpy changes of hydration of $Na^+(g)$ and $Cl^-(g)$. This is written as $\Delta H^\ominus_{hyd}[Na^+] + \Delta H^\ominus_{hyd}[Cl^-] = -784\,kJ\,mol^{-1}$ in Figure 2.25.

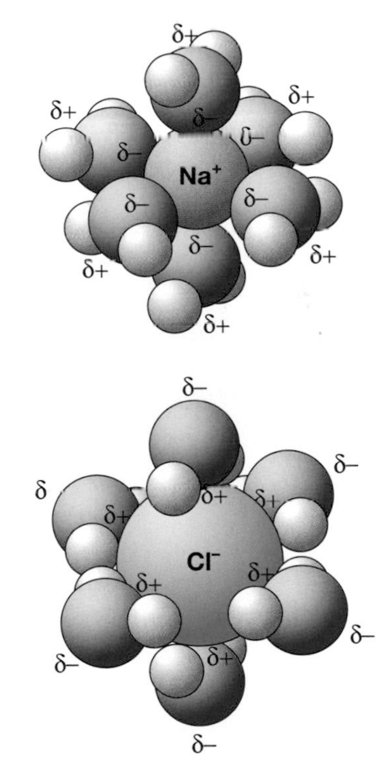

Figure 2.24▲
Sodium and chloride ions are hydrated when they dissolve in water. Polar water molecules are attracted to both cations and anions.

Definitions

The **standard enthalpy change of solution**, $\Delta H^\ominus_{solution}$, is the enthalpy change when one mole of a compound dissolves to form a solution containing 1 mol dm^{-3} under standard conditions. Enthalpy changes of solution are often calculated and quoted for 1 mole of a compound dissolving in specified amounts of water.

The standard enthalpy change of hydration is the enthalpy change when one mole of gaseous ions is hydrated under standard conditions to form a solution in which the concentration of ions is 1 mol dm^{-3}.

For example, for sodium ions:

$$Na^+(g) + aq \rightarrow Na^+(aq) \quad \Delta H^\ominus_{hyd} = -444\ \text{kJ mol}^{-1}$$

Enthalpy changes of hydration are sometimes just called hydration enthalpies.

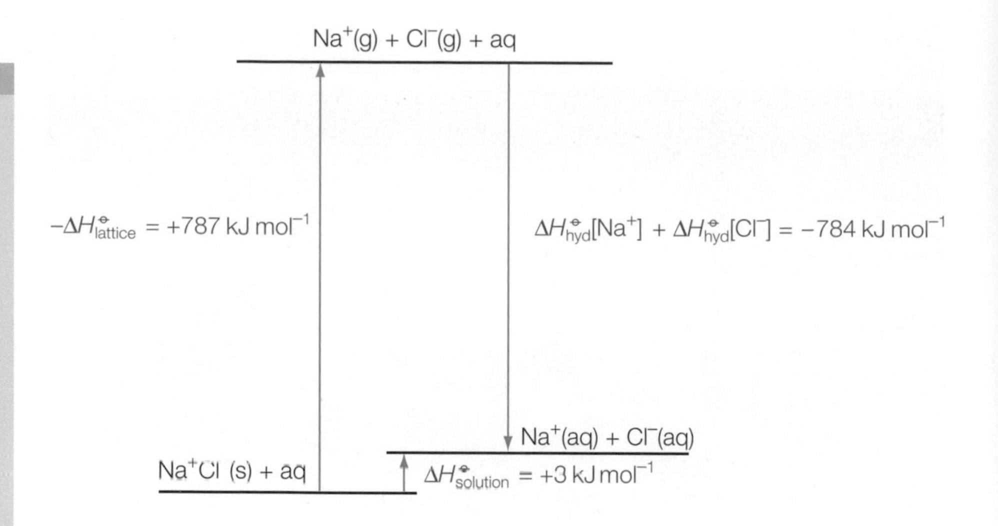

Figure 2.25▲
An energy cycle for sodium chloride dissolving in water.

It is now possible to see from Figure 2.25 why it is possible for sodium chloride to dissolve in water. The explanation is that the ions, Na^+ and Cl^-, are so strongly hydrated by the polar water molecules that the exothermic enthalpy changes of hydration nearly balance the energy required to separate the ions (the reverse lattice enthalpy).

The enthalpy change of solution is the difference between the energy needed to separate the ions from the crystal lattice (–lattice energy) and the energy given out as the ions are hydrated (the sum of the hydration enthalpies).

Test yourself

13 a) Use the equation

energy change = $m \times c \times \Delta T$

to calculate the temperature change when one mole of sodium chloride is used to make 1 dm^3 of a solution containing 1.0 mol dm^{-3} ($\Delta H^\ominus_{sol}[NaCl(s)] = +3.8\ \text{kJ mol}^{-1}$).
(Assume that the density and specific heat capacity, c, of the solution are the same as for water.)

b) Explain how the temperature of the mixture changes as sodium chloride dissolves in water.

Factors affecting the enthalpy change of hydration and lattice energy

Ionic charge

Enthalpy changes of hydration and lattice energies both involve electrostatic attractions between opposite charges. Because of this, both processes are exothermic.

In addition, both processes become more exothermic as the charge on an ion increases because the charge density of the ion increases and therefore its attraction for any opposite charge increases.

This is illustrated very well by the enthalpy changes of hydration for Na^+ and Mg^{2+} and the lattice energies of NaF and MgO in Table 2.6.

Enthalpy change of hydration/kJ mol^{-1}		Lattice energy/kJ mol^{-1}	
Na^+	−444	NaF	−918
Mg^{2+}	−2003	MgO	−3791
Li^+	−559	LiF	−1031
K^+	−361	KF	−817

Table 2.6▲
Comparing some enthalpy changes of hydration and lattice energies.

Test yourself

14 Refer to the Data sheet: 'Enthalpy changes of hydration' on the Dynamic Learning Student website.
a) How is the enthalpy change of hydration affected by increasing ionic charge?
b) List an appropriate series of ions and their enthalpy changes of hydration to illustrate your conclusion in part a).

Ionic radius

As the radius of an ion increases, its charge density decreases. This results in a weaker attraction for oppositely charged ions and for the δ+ or δ− charges on polar molecules such as water. So, an increase in ionic radius leads to less exothermic values for enthalpy changes of hydration and lattice energies. This point is neatly illustrated by the enthalpy changes of hydration for Li^+ and K^+ and the lattice energies of LiF and KF in Table 2.6.

Ion	Li^+	Na^+	K^+	Mg^{2+}	Al^{3+}
Ionic radius/nm	0.074	0.102	0.138	0.072	0.053
Ion	N^{3-}	O^{2-}	F^-		
Ionic radius/nm	0.171	0.140	0.133		

Table 2.7▲
The ionic radii of some ions.

Test yourself

15 Look at Table 2.7, which shows the radii of some ions, and the Data sheet 'Enthalpy changes of hydration' on the Dynamic Learning Student website.
a) How is the enthalpy change of hydration affected by increasing ionic radius?
b) List an appropriate series of ions, their radii and their enthalpy changes of hydration to illustrate your conclusion in part a).

16 a) Use the Data sheets: 'Lattice energies of some ionic compounds' and 'Enthalpy changes of hydration' on the Dynamic Learning Student website to calculate the enthalpy changes of solution of lithium fluoride and lithium iodide.
b) Account for the relative values of the lattice energies and hydration enthalpies of the two compounds in terms of ionic radii.
c) To what extent, if at all, can your answers to part a) explain the differences in the solubilities of the two compounds?
(Solubilities: LiF = 5 × 10^{-5} mol in 100 g water; LiI = 1.21 mol in 100 g water.)

Data 1

Data 2

Entropy changes during dissolving

Sodium chloride dissolves readily in water despite the fact that the process is slightly endothermic. This is another example which shows that the sign of ΔH is not a reliable guide to whether or not a process will happen. This is particularly the case when the magnitude of ΔH is small.

When sodium chloride dissolves, the disorder increases as the ions leave the regular lattice and mix with the molecules of liquid water. This means that the entropy of the system increases as a salt dissolves. The entropy change in the surroundings is negative for this endothermic change, but the increase in the entropy of the system is more than enough to compensate for this and so the total entropy change is positive.

Worked example

Use the values in Table 2.8 to calculate the total entropy change when sodium chloride dissolves in water under standard conditions.

$$NaCl(s) \rightarrow Na^+(aq) + Cl^-(aq) \qquad \Delta H^{\ominus}_{solution} = +3.8\,kJ\,mol^{-1}$$

	$S^{\ominus}$/J mol^{-1} K^{-1}
NaCl(s)	+72.1
Na^+(aq)	+321
Cl^-(aq)	+56.5

Table 2.8◀

Notes on the method

Calculate the entropy change of the surroundings using the formula:

$$\Delta S^{\ominus}_{surroundings} = -\frac{\Delta H^{\ominus}}{T}$$

Remember to convert the value of the enthalpy change to J mol ^{1}.

Answer

$$\Delta S^{\ominus}_{surroundings} = -\frac{3800\,J\,mol^{-1}}{298\,K} = -12.8\,J\,mol^{-1}\,K^{-1}$$

$$\Delta S^{\ominus}_{system} = \Sigma S^{\ominus}[products] - \Sigma S^{\ominus}[reactants]$$

$$= 321\,J\,mol^{-1}\,K^{-1} + 56.5\,J\,mol^{-1}\,K^{-1} - 72.1\,J\,mol^{-1}\,K^{-1}$$

$$= +305\,J\,mol^{-1}\,K^{-1}$$

$$\Delta S^{\ominus}_{total} = \Delta S^{\ominus}_{system} + \Delta S^{\ominus}_{surroundings}$$

$$= 305\,J\,mol^{-1}\,K^{-1} - 12.8\,J\,mol^{-1}\,K^{-1} = +292\,J\,mol^{-1}\,K^{-1}$$

The total entropy change is positive. Sodium chloride dissolving in water is a spontaneous process.

Dissolving depends on the balance between the change in entropy of the solution and the change in entropy of the surroundings. This balance starts to alter as soon as a salt starts to dissolve and the concentration of the solution rises. In a more concentrated solution the ions are closer together and there are fewer free water molecules to hydrate the ions. These changes modify the values of the hydration energies and the entropies of the ions in solution. Overall this means that once a certain amount of a salt has dissolved, the processes ceases to be spontaneous and no more solid dissolves. At this point the solution is saturated. The salt crystals and the saturated solution are in equilibrium.

	$S^{\ominus}$/J mol^{-1} K^{-1}
NH_4NO_3(s)	151
NH_4^+(aq)	113
NO_3^-(aq)	146

Table 2.9▲

Test yourself

17 a) Use the values in Table 2.9 to calculate the total entropy change when ammonium nitrate dissolves in water under standard conditions.
$NH_4NO_3(s) \rightarrow NH_4^+(aq) + NO_3^-(aq) \qquad \Delta H^{\ominus} = +28.1\,kJ\,mol^{-1}$

b) Comment on the fact that ammonium nitrate dissolves in water even though the process is endothermic.

REVIEW QUESTIONS

Extension questions

1 The graph in Figure 2.26 shows how the entropy of water changes from 0 K to 450 K

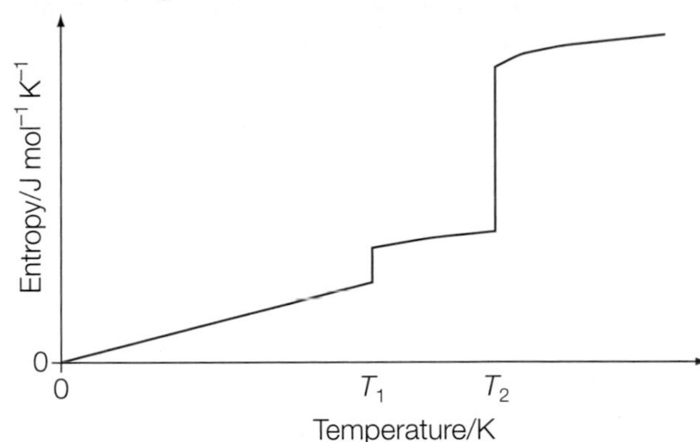

Figure 2.26▲

a) Explain why the molar entropy of water is zero at 0 K. (2)

b) Account for the entropy changes at temperatures T_1 and T_2. Explain why one is bigger. (3)

c) Why does $\Delta S_{total} = 0\,J\,mol^{-1}$ for the change of water to steam at 373 K and 1 atmosphere pressure? (2)

d) The enthalpy change of vaporisation of water is 41.1 kJ mol^{-1} at 373 K. Calculate the value of ΔS_{system} for the conversion of water into steam at 373 K and 1 atmosphere pressure. (5)

2 Electrolysis is usually used to extract aluminium metal. In principle it should be possible to extract the metal by heating its oxide with carbon. Use the data in the table to answer the questions.

	$Al_2O_3(s)$	C(s)	Al(s)	CO(g)
$\Delta H_f^{\ominus}$/kJ mol^{-1}	−1669	0	0	−111
$S^{\ominus}$/J mol^{-1} K^{-1}	50.9	5.7	28.3	198

a) Write the equation for the reduction of aluminium oxide by carbon assuming that the carbon is converted to carbon monoxide. (1)

b) Calculate the standard enthalpy change for the reduction reaction. (3)

c) Calculate the standard entropy change for the reduction reaction. (3)

d) i) Calculate the total entropy change for the reduction at 298 K. (2)

ii) What does you answer to **d) i)** tell you about the feasibility of the reaction at 298 K? (1)

e) Calculate the minimum temperature at which the reduction of aluminium oxide by carbon becomes feasible. (2)

f) Suggest a reason why the reaction between aluminium oxide and carbon does not happen to a significant extent until the temperature is about 1000 degrees higher than your answer to e). (1)

3 Carbon monoxide is one of the products when methane burns in a limited supply of air.

$$CH_4(g) + \tfrac{3}{2}O_2(g) \rightarrow CO(g) + 2H_2O(g)$$

$\Delta H^{\ominus} = -519\,kJ\,mol^{-1}$ $\quad \Delta S^{\ominus}_{system} = +82\,J\,mol^{-1}\,K^{-1}$

a) i) Calculate the total entropy change for the reaction under standard conditions. (2)

ii) Is the reaction of methane with oxygen to form carbon monoxide and steam feasible at 298 K? (1)

iii) Why does methane not burn in air at 298 K? (2)

b) Sooty carbon is another of the products when methane burns in a limited supply of air.

$$CH_4(g) + O_2(g) \rightarrow C(s) + 2H_2O(g)$$

$\Delta S^{\ominus} = -8\,J\,mol^{-1}\,K^{-1}$

How do you account in the difference in the values for $\Delta S^{\ominus}_{reaction}$ for the reaction producing carbon monoxide and the reaction that forms soot? (4)

4 When calcium chloride dissolves in water, the process can be represented by the equation:

$$Ca^{2+}(Cl^-)_2(s) + aq \rightarrow Ca^{2+}(aq) + 2Cl^-(aq)$$

The enthalpy change for this process is called the enthalpy change of solution. Its value can be calculated from an energy cycle using the following data.

Lattice energy of calcium chloride = −2258 kJ mol^{-1}

Enthalpy change of hydration of $Ca^{2+}(g)$ = −1657 kJ mol^{-1}

Enthalpy change of hydration of $Cl^-(g)$ = −340 kJ mol^{-1}

a) Draw and label the cycle linking the enthalpy change of solution of calcium chloride with the enthalpy changes in the data above. (4)

b) Use your cycle to calculate the enthalpy change of solution of calcium chloride. (3)

c) What factors affect the size of the enthalpy change of hydration of $Ca^{2+}(g)$ compared with that of $Li^+(g)$? (2)

d) Why are the hydration enthalpies of both anions and cations negative? (3)

3 Equilibria

Figure 3.1 ▲
Red blood cells flowing through a blood vessel magnified ×2500. The protein haemoglobin has just the right properties to take up oxygen in the lungs and release it to cells throughout the body. The position of equilibrium of this reversible process varies with the concentration of oxygen.

All chemical reactions tend towards a state of dynamic equilibrium. Chemists have discovered a law which allows them to predict the concentrations of chemicals expected in equilibrium mixtures. This law is one of several approaches which chemists use to answer the questions 'How far?' and 'In which direction?'. An understanding of equilibrium ideas helps to explain changes in the natural environment, the biochemistry of living things and the conditions used in the chemical industry to manufacture new products.

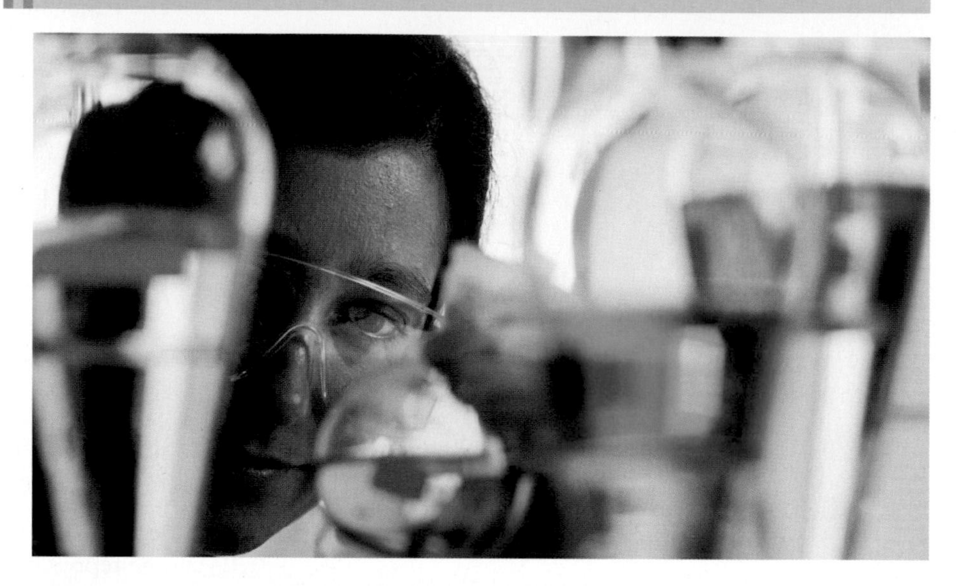

Figure 3.2 ▶
Solvent extraction is used to separate and purify chemicals. A compound dissolved in one liquid can be extracted into another liquid in which it is more soluble. After shaking the mixture in a tap funnel the two solutions are in equilibrium, but with more of the solute in one of the solvents.

3.1 The equilibrium law

Reversibility and equilibrium

Reversible reactions tend towards a state of balance. They reach equilibrium when neither the forward change nor the backward change is complete but both changes are still going on at equal rates. They cancel each other out and there is no overall change. This is dynamic equilibrium.

Under given conditions the same equilibrium state can be reached either by starting with the chemicals on one side of the equation for a reaction or by starting with the chemicals on the other side. Figures 3.3 and 3.4 illustrate this for the reversible reaction between hydrogen and iodine.

$$H_2(g) + I_2(g) \rightleftharpoons 2HI(g)$$

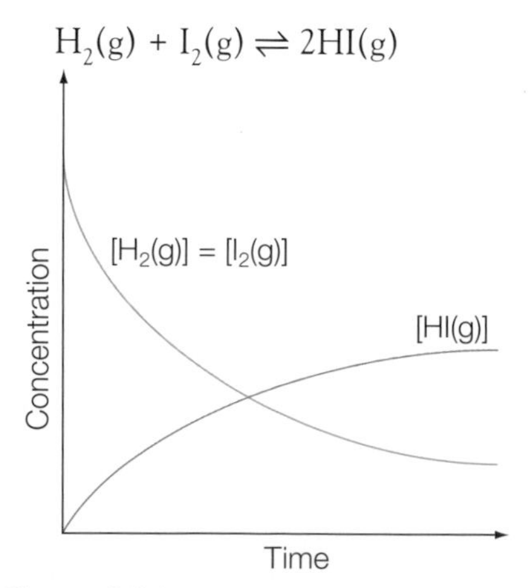

Figure 3.3 ▲
Reaching an equilibrium state by the reaction of equal amounts of hydrogen gas and iodine gas.

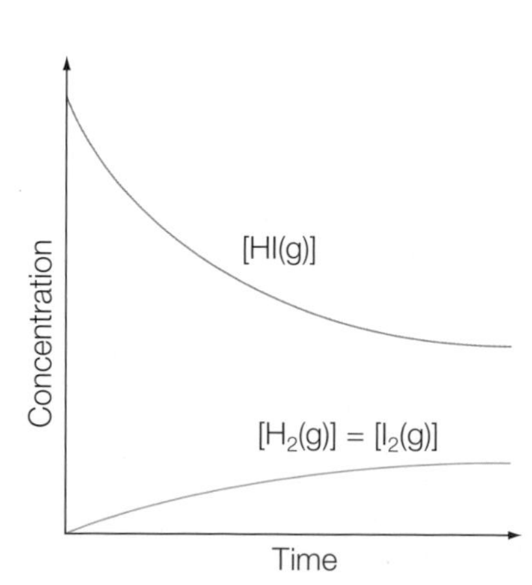

Figure 3.4 ▲
Reaching the same equilibrium state by the decomposition of hydrogen iodide, HI(g), under the same conditions as for Figure 3.3.

Definition

In **dynamic equilibrium** the forward and backward changes continue at equal rates so that overall there is no change. At the molecular level there is continuous activity. At the macroscopic level nothing appears to be happening.

An empirical law

The equilibrium law has been established by experiment. It is a quantitative law for predicting the amounts of reactants and products when a reversible reaction reaches a state of dynamic equilibrium.

In general, for a reversible reaction at equilibrium:

$$aA + bB \rightleftharpoons cC + dD \qquad K_c = \frac{[C]^c[D]^d}{[A]^a[B]^b}$$

This is the form for the equilibrium constant, K_c, when the concentrations of the reactants and products are measured in moles per cubic decimetre. [A], [B] and so on are the equilibrium concentrations, sometimes written as $[A]_{eqm}$ and $[B]_{eqm}$ to make this clear.

The concentrations of the chemicals on the right-hand side of the equation appear on the top line of the expression. The concentrations of reactants on the left appear on the bottom line. Each concentration term is raised to the power of the number in front of its formula in the equation.

Worked example

Calculate the value of K_c for the reaction which forms an ester from ethanoic acid and ethanol (see Section 8.3) from this data.

$$CH_3COOH(l) + C_2H_5OH(l) \rightleftharpoons CH_3COOC_2H_5(l) + H_2O(l)$$

When 0.50 mol of $CH_3COOH(l)$ is dissolved in 0.5 dm^3 of an organic solvent with 0.09 mol $C_2H_5OH(l)$ and allowed to come to equilibrium at 293 K the amount of the ester $CH_3COOC_2H_5(l)$ formed at equilibrium is 0.086 mol.

Note on the method

Write down the equation. Underneath write first the initial amounts, then write the amounts at equilibrium. Use the equation to calculate the amounts not given.

Calculate the equilibrium concentrations given the volume of the solution. Substitute the values and units in the expression for K_c.

Answer

	$CH_3COOH(l)$ +	$C_2H_5OH(l) \rightleftharpoons$	$CH_3COOC_2H_5(l)$ +	$H_2O(l)$
Equation:	$CH_3COOH(l)$	$C_2H_5OH(l)$	$CH_3COOC_2H_5(l)$	$H_2O(l)$
Initial amounts/mol	0.50	0.09	0	0
At equilibrium:				
amount given/mol:			0.086	
amounts calculated/mol:	(0.50 − 0.086) = 0.414	(0.09 − 0.086) = 0.004		0.086
Equilibrium concentrations/mol dm^{-3}	0.414 ÷ 0.5 = 0.828	0.004 ÷ 0.5 = 0.008	0.086 ÷ 0.5 = 0.172	0.086 ÷ 0.5 = 0.172

$$K_c = \frac{[CH_3COOC_2H_5(l)][H_2O(l)]}{[CH_3CO_2H(l)][C_2H_5OH(l)]} = \frac{0.172\ \text{mol dm}^{-3} \times 0.172\ \text{mol dm}^{-3}}{0.828\ \text{mol dm}^{-3} \times 0.008\ \text{mol dm}^{-3}}$$

Hence K_c = 4.47.

In this example the units cancel, so K_c has no units.

Test yourself

3 Calculate K_c for the reaction: $PCl_5(g) \rightleftharpoons PCl_3(g) + Cl_2(g)$ given that on mixing 1.68 mol $PCl_5(g)$ with 0.36 mol $PCl_3(g)$ in a 2.0 dm^3 container and allowing the mixture to reach equilibrium, the amount of PCl_5 in the equilibrium mixture was 1.44 mol.

4 K_c = 170 $dm^3\ mol^{-1}$ at 298 K for the equilibrium system: $2NO_2(g) \rightleftharpoons N_2O_4(g)$. If a 5 dm^3 flask contains 1.0×10^{-3} mol of NO_2 and 7.5×10^{-4} mol N_2O_4, is the system at equilibrium? Is there any tendency for the concentration of NO_2 to change and, if so, does it tend to increase or decrease?

Test yourself

1 Give examples of dynamic equilibrium involving:
- a) a solid and a liquid
- b) a solid and a solution
- c) two solutions
- d) chemical change.

2 Describe what is happening to the molecules in the gas mixtures from time zero to the time at which each mixture reaches equilibrium as described by
a) Figure 3.3 and b) Figure 3.4.

Note

Equilibrium constants are constant for a particular temperature.

Unlike rate equations, the form of the expression for an equilibrium constant can be deduced from the balanced chemical equation.

Practical guidance

Equilibrium constants and balanced equations

An equilibrium constant always applies to a particular chemical equation and can be deduced directly from the equation.

There are two common ways of writing the reaction of sulfur dioxide with oxygen. As a result there are two forms for the equilibrium constant, which have different values. So long as the matching equation and equilibrium constant are used in any calculation, the predictions based on the equilibrium law are the same. For:

$$2SO_2(g) + O_2(g) \rightleftharpoons 2SO_3(g) \qquad \text{Equation 1}$$

$$K_c = \frac{[SO_3(g)]^2}{[SO_2(g)]^2[O_2(g)]}$$

But for:

$$SO_2(g) + \tfrac{1}{2}O_2(g) \rightleftharpoons SO_3(g) \qquad \text{Equation 2}$$

$$K_c = \frac{[SO_3(g)]}{[SO_2(g)][O_2(g)]^{1/2}}$$

So it is important to write the balanced equation and the equilibrium constant together.

Reversing the equation also changes the form of the equilibrium constant because the concentration terms for the chemicals on the right-hand side of the equation always appear on the top of the expression for K_c.

So for:

$$2SO_3(g) \rightleftharpoons 2SO_2(g) + O_2(g) \qquad \text{Equation 3}$$

$$K_c = \frac{[SO_2(g)]^2[O_2(g)]}{[SO_3(g)]^2}$$

Test yourself

5 Consider the equilibrium between sulfur dioxide, oxygen and sulfur trioxide.
  a) Show that the units for the equilibrium constant, K_c, for equation 1 are $dm^3\,mol^{-1}$.
  b) $K_c = 1.6 \times 10^6\,dm^3\,mol^{-1}$ for equation 1 at a particular temperature. What is the value of K_c for equation 2?
  c) What is the value of K_c for equation 3?

6 Write the expression for K_c for these equations and state the units of the equilibrium constant for each example.
  a) $CO_2(g) + H_2(g) \rightleftharpoons CO(g) + H_2O(g)$
  b) $N_2(g) + 3H_2(g) \rightleftharpoons 2NH_3(g)$
  c) $4PF_5(g) \rightleftharpoons P_4(g) + 10F_2(g)$

Activity

Testing the equilibrium law

The reversible reaction involving hydrogen, iodine and hydrogen iodide has been used to test the equilibrium law experimentally. In a series of six experiments, samples of the chemicals were sealed in reaction tubes and then heated at 731 K until the mixtures reached equilibrium. Four of the tubes started with different mixtures of hydrogen and iodine. Two of the tubes started with just hydrogen iodide.

Once the tubes had reached equilibrium they were rapidly cooled to stop the reactions. Then the contents of the tubes were analysed to find the compositions of the equilibrium mixture. The results for six of the tubes are shown in Table 3.1

Table 3.1

Tube	Initial concentrations/10^{-2} mol dm^{-3}			Equilibrium concentrations/10^{-2} mol dm^{-3}		
	$[H_2(g)]$	$[I_2(g)]$	$[HI(g)]$	$[H_2(g)]$	$[I_2(g)]$	$[HI(g)]$
1	2.40	1.38	0	1.14	0.12	2.52
2	2.40	1.68	0	0.92	0.20	2.96
3	2.44	1.98	0	0.77	0.31	3.34
4	2.46	1.76	0	0.92	0.22	3.08
5	0	0	3.04	0.345	0.345	2.35
6	0	0	7.58	0.86	0.86	5.86

1 Write the equation for the reversible reaction to form hydrogen iodide from hydrogen and iodine.

2 Show that the equilibrium concentration of:

a) hydrogen in tube 1 is as expected given the value of $[I_2(g)]_{eqm}$

b) hydrogen iodide in tube 2 is as expected given the value of $[I_2(g)]_{eqm}$.

3 Explain why $[H_2(g)]_{eqm} = [I_2(g)]_{eqm}$ for tubes 5 and 6.

4 For each of the tubes work out the value of:

a) $\dfrac{[HI(g)]_{eqm}}{[H_2(g)]_{eqm}[I_2(g)]_{eqm}}$

b) $\dfrac{[HI(g)]^2_{eqm}}{[H_2(g)]_{eqm}[I_2(g)]_{eqm}}$

c) Enter your values in a table and comment on the results.

5 What is the value of K_c for the reaction of hydrogen with iodine at 731 K?

Heterogeneous equilibria

In some equilibrium systems the substances involved are not all in the same phase. An example is the equilibrium state involving two solids and a gas formed on heating calcium carbonate in a closed container.

$$CaCO_3(s) \rightleftharpoons CaO(s) + CO_2(g)$$

The concentrations of solids do not appear in the expression for the equilibrium constant. Pure solids have, in effect, a constant 'concentration'.

So $K_c = [CO_2(g)]$

The same applies to heterogeneous systems which have a separate pure liquid phase as one of the reactants or products.

Definition

A **phase** is one of the three states of matter – solid, liquid or gas. Chemical systems often have more than one phase. Each phase is distinct but need not be pure:

- a solid in equilibrium with its saturated solution is a two-phase system
- in the reactor for ammonia manufacture the mixture of nitrogen, hydrogen and ammonia gases is one phase with the iron catalyst being a separate solid phase.

Figure 3.5▲
Stalactites growing in the Florida Caverns, Marianna, Florida. The stalactites grow very slowly because this heterogeneous system is almost at equilibrium.

Test yourself

7 Write the expression for K_c for these equilibria:
a) $3Fe(s) + 4H_2O(g) \rightleftharpoons Fe_3O_4(s) + 4H_2(g)$
b) $H_2(g) + S(l) \rightleftharpoons H_2S(g)$
c) $Ag^+(aq) + Fe^{2+}(aq) \rightleftharpoons Fe^{3+}(aq) + Ag(s)$

8 Calculate the concentration of water in water (in mol dm^{-3}) to show that it is reasonable to regard the concentration of water as a constant when writing the expression for K_c for equilibria in dilute aqueous solution.

Another example is the equilibrium state between solid calcium carbonate and a dilute solution containing dissolved carbon dioxide and calcium hydrogencarbonate.

$$CaCO_3(s) + CO_2(aq) + H_2O(l) \rightleftharpoons Ca^{2+}(aq) + 2HCO_3^-(aq)$$

This example illustrates another general rule. The K_c expression for dilute solutions does not include a concentration term for water. There is so much water present that its concentration is effectively constant.

So the expression for the equilibrium constant becomes:

$$K_c = \frac{[Ca^{2+}(aq)]\,[HCO_3^-(aq)]^2}{[CO_2(aq)]}$$

3.2 Gaseous equilibria

Many important industrial processes involve reversible reactions between gases. Applying the equilibrium law to these reactions helps to determine the optimal conditions for manufacturing chemicals. When it comes to gas reactions it is often easier to measure pressures rather than concentrations and to use a modified form of the equilibrium law.

Figure 3.6▲
Many industrial processes involve reactions between gases under pressure.

Definitions

Pressure is defined as force per unit area. The SI unit of pressure is the pascal (Pa) which is a pressure of one newton per square metre (1 N m^{-2}). The pascal is a very small unit so pressures are often quoted in kilopascals, kPa.

Standard pressure for definitions in thermodynamics is now 1 bar which is 100 000 N m^{-2} = 100 kPa

In accounts of chemical processes, multiples of **atmospheric pressure** give an indication of the extent to which gases are compressed. The average atmospheric pressure at sea level is 101.3 kPa (which is very close to 1 bar).

The **partial pressure** of a gas is a measure of its concentration in a mixture of gases.

Gas mixtures and partial pressures

In any mixture of gases the total pressure of the mixture can be 'shared out' between the gases. The contribution each gas makes to the total pressure is its partial pressure. It is possible to calculate a partial pressure for each gas in the mixture. In a mixture of gases A, B and C, the sum of the three partial pressures equals the total pressure.

$$p_A + p_B + p_B = p_{total}$$

Partial pressures are a useful alternative to concentrations when studying mixtures of gases and gas reactions.

In gas mixtures it is the amounts in moles that matter and not the chemical nature of the molecules. This means that the total pressure is shared between the gases simply according to their mole fractions in the mixture.

In a mixture of n_A moles of A with n_B moles of B and n_C moles of C, the total amount in moles is $(n_A + n_B + n_C)$. The mole fractions (symbol X) are given by the following:

$$X_A = \frac{n_A}{n_A + n_B + n_C} \qquad X_B = \frac{n_B}{n_A + n_B + n_C} \qquad X_C = \frac{n_C}{n_A + n_B + n_C}$$

So the mole fraction of A is the fraction of the total number of molecules which are molecules of A.

The sum of all the mole fractions is 1, so $X_A + X_B + X_C = 1$.

On this basis the partial pressures of three gases A, B and C in a gas mixture with total pressure p are:

$$p_A = X_A p, \qquad p_B = X_B p \qquad \text{and} \qquad p_C = X_C p$$

The partial pressure for each gas is the pressure it would exert if it was the only gas in the container under the same conditions. The partial pressure of a gas is proportional to the concentration of the gas in the mixture. This makes it possible to work in partial pressures when applying the equilibrium law to gas reactions.

3.3 K_p

K_p is the symbol for the equilibrium constant for an equilibrium involving gases when the concentrations are measured by partial pressures. The rules for writing equilibrium expressions are the same for K_p as for K_c with partial pressures replacing concentrations as shown in Table 3.2.

Note

Do not use square brackets when writing K_p expressions. In the context of the equilibrium law, square brackets signify concentrations in moles per dm^3.

Equilibrium	K_p	Units of K_p
$H_2(g) + I_2(g) \rightleftharpoons 2HI(g)$	$K_p = \frac{(p_{HI})^2}{p_{H_2} \times p_{I_2}}$	No units
$N_2(g) + 3H_2(g) \rightleftharpoons 2NH_3(g)$	$K_p = \frac{(p_{NH_3})^2}{p_{N_2} \times (p_{H_2})^3}$	Pa^{-2} or atm^{-2}
$N_2O_4(g) \rightleftharpoons 2NO_2(g)$	$K_p = \frac{p_{NO_2}{}^2}{p_{N_2O_4}}$	Pa or atm
$HCl(g) + LiH(s) \rightleftharpoons H_2(g) + LiCl(s)$	$K_p = \frac{p_{H_2}}{p_{HCl}}$	No units

Table 3.2▲
Examples of equilibrium expressions for K_p. Note that when writing an expression for K_p for a heterogeneous reaction the same rules apply as for K_c. The expression does not include terms for any separate pure solid phases.

Note

Changing the total pressure or the composition of the gas mixture has no effect on the value of K_p so long as the temperature stays constant.

Worked example

An experimental study of the equilibrium between $N_2(g)$, $H_2(g)$ and $NH_3(g)$ found that one equilibrium mixture contained 2.15 mol of $N_2(g)$, 6.75 mol of $H_2(g)$ and 1.41 mol of $NH_3(g)$ at a total pressure 1000 kPa. Calculate the value for K_p under the conditions that the measurements were taken.

Notes on the method

First work out the mole fractions of the gases.

Multiply the total pressure by the mole fractions to get the partial pressures.

Check that the sum of the partial pressures equals the total pressure.

Finally substitute in the expression for K_p and give the units.

Answer

$$\text{Total number of moles} = 2.15\,\text{mol} + 6.75\,\text{mol} + 1.41\,\text{mol} = 10.31\,\text{mol}$$

$$\text{Mole fraction of } N_2(g) = \frac{2.15\,\text{mol}}{10.31\,\text{mol}} = 0.208$$

$$\text{Mole fraction of } H_2(g) = \frac{6.75\,\text{mol}}{10.31\,\text{mol}} = 0.655$$

$$\text{Mole fraction of } NH_3(g) = \frac{1.41\,\text{mol}}{10.31\,\text{mol}} = 0.137$$

$$\text{Partial pressure of } N_2(g) = 0.208 \times 1000\,\text{kPa} = 208\,\text{kPa}$$

$$\text{Partial pressure of } H_2(g) = 0.655 \times 1000\,\text{kPa} = 655\,\text{kPa}$$

$$\text{Partial pressure of } NH_3(g) = 0.137 \times 1000\,\text{kPa} = 137\,\text{kPa}$$

For the equilibrium:

$$N_2(g) + 3H_2(g) \rightleftharpoons 2NH_3(g)$$

$$K_p = \frac{(p_{NH_3})^2}{p_{N_2} \times (p_{H_2})^3} = \frac{(137\,\text{kPa})^2}{(208\,\text{kPa})(655\,\text{kPa})^3} = 3.2 \times 10^{-7}\,\text{kPa}^{-2}$$

Test yourself

9 Write the expression for K_p for these equilibria and give the units with pressures measured in pascals, Pa:
a) $2SO_2(g) + O_2(g) \rightleftharpoons 2SO_3(g)$
b) $4NH_3(g) + 3O_2(g) \rightleftharpoons 2N_2(g) + 6H_2O(g)$
c) $CaCO_3(s) \rightleftharpoons CaO(s) + CO_2(g)$

10 Calculate K_p at 330 K and 120 kPa pressure for this equilibrium mixture:
$N_2O_4(g) \rightleftharpoons 2NO_2(g)$
Under these conditions a sample of the gas mixture consists of 8.1 mol $N_2O_4(g)$ and 3.8 mol $NO_2(g)$.

11 Calculate K_p for this reversible reaction at 1000 K and 180 kPa pressure:
$C_2H_6(g) \rightleftharpoons C_2H_4(g) + H_2(g)$
given that under these conditions starting with just 5 mol ethane yields an equilibrium mixture containing 1.8 mol ethene.

3.4 Predicting the direction of change

Equilibrium constants and entropy

This topic has shown that the direction and extent of a change can be described by an equilibrium constant. Topic 2 introduced the idea that the spontaneous direction of change is determined by the total entropy change for a reaction. There has to be some connection between these two quantities, and the theory of thermodynamics has established that they are related by this formula, where R is a constant:

$$\Delta S_{total} = R \ln K$$

Table 3.3 shows the equivalent values of the total entropy change and K_c based on this formula, and relates them to the direction and extent of change.

Reaction does not go	Reaction reaches an equilibrium in which the reactants predominate	Roughly equal amounts of reactants and products at equilibrium	Reaction reaches an equilibrium in which the products predominate	Reaction goes to completion
$K_c < 1 \times 10^{-10}$	$K_c \approx 0.01$	$K_c = 1$	$K_c \approx 100$	$K_c > 1 \times 10^{10}$
$\Delta S_{total} < -200$	$\Delta S_{total} \approx -40$	$\Delta S_{total} = 0$	$\Delta S_{total} \approx +40$	$\Delta S_{total} > +200$

Table 3.3▲

Table 3.3 shows that if the value of an equilibrium constant is large then the position of equilibrium is over to the right-hand side of the equation. Conversely if the value of an equilibrium constant is small then the position of equilibrium is over to the left-hand side of the equation. If the value of K_c is close to 1 then there are significant quantities of both reactants and products present at equilibrium.

It is very important to keep in mind that neither equilibrium constants nor the total entropy change say anything about the time it takes for a reaction mixture to reach equilibrium. The system may reach equilibrium rapidly or slowly. The value of K_c for the reaction of hydrogen with chlorine to make hydrogen chloride, for example, is about 1×10^{31} at room temperature, but in the absence of a catalyst or ultraviolet light there is no reaction.

Test yourself

12 What can you conclude about the direction and extent of change in each of these examples?

a) $Zn(s) + Cu^{2+}(aq) \rightleftharpoons Zn^{2+}(aq) + Cu(s)$ $K_c = 1 \times 10^{37}$ at 298 K

b) $2HBr(g) \rightleftharpoons H_2(g) + Br_2(g)$ $K_c = 1 \times 10^{-10}$ at 298 K

c) $N_2(g) + 3H_2(g) \rightleftharpoons 2NH_3(g)$ $K_c = 2.2$ at 623 K

13 In general if the equilibrium constant for a forward reaction is large, what is the size of the equilibrium constant for the reverse of this reaction?

The effect of temperature

The total entropy change for a reaction is made up of the entropy change of the system and the entropy change in the surroundings.

$$\Delta S_{total} = \Delta S_{system} + \Delta S_{surroundings}$$

As shown in Section 2.1, the entropy change in the surroundings is determined by the enthalpy change of the reaction and the temperature, so that the relationship can be rewritten:

$$\Delta S_{total} = \Delta S_{system} - \frac{\Delta H_{reaction}}{T}$$

The entropy change of the system and the enthalpy change of reaction do not change to any great extent with temperature, but this relationship shows that the total entropy change does vary with temperature.

If the reaction is exothermic, $\Delta H_{reaction}$ is negative. This means that $\Delta S_{surroundings}$ is positive and the value of ΔS_{total} gets smaller as the temperature rises. Since $\Delta S_{total} = R \ln K$, this also means that the value of K gets smaller as the temperature rises. This is as predicted qualitatively by le Chatelier's principle which shows that a system at equilibrium shifts to the left for an exothermic reaction as the temperature rises.

The converse is the case if a reaction is endothermic. This means that $\Delta H_{reaction}$ is positive and that $\Delta S_{surroundings}$ is negative. The consequence is that ΔS_{total} gets larger as the temperature rises and so the value of K gets larger. Again, this is as predicted by le Chatelier's principle.

Figure 3.7▲
Sealed tubes containing equilibrium mixtures of $NO_2(g)$ which is orange-brown and $N_2O_4(g)$ which is colourless. The tube on the left is in hot water and the tube on the right in ice.

Test yourself

14 a) Write an equation for the dissociation of $N_2O_4(g)$ to $NO_2(g)$.

b) Use the Data sheet to calculate the enthalpy change for the dissociation of $N_2O_4(g)$.

c) What is the effect of raising, or lowering, the temperature on the total entropy change of the reaction?

d) Show how your answer to c) can account for the observations illustrated by Figure 3.7.

e) Show that your answer to d) is consistent with the predictions of le Chatelier's principle.

Data

REVIEW QUESTIONS

www
Extension questions

1 a) A flask contains an equilibrium mixture of hydrogen gas (0.01 mol dm^{-3}), iodine gas (0.01 mol dm^{-3}) and hydrogen iodide gas (0.07 mol dm^{-3}) at a constant temperature. Calculate K_c for the reaction of hydrogen with iodine to form hydrogen iodide. (3)

b) Enough hydrogen is added to the mixture in **a)** to suddenly double the hydrogen concentration in the flask to 0.02 mol dm^{-3}. After a while the mixture settles down with a new iodine concentration of 0.007 mol dm^{-3} at the same temperature as before.

i) What are the new concentrations of hydrogen and hydrogen iodide? (2)

ii) Show that the new mixture is at equilibrium. (2)

c) How does a sudden doubling of the hydrogen concentration affect the position of equilibrium? (2)

2 A solution of ammonia in water was shaken with an equal volume of an organic solvent until the system reached equilibrium with the ammonia distributed between the two solvents. In a series of titrations, 10 cm^3 of the aqueous layer was neutralised by an average of 17.0 cm^3 of 0.50 mol dm^{-3} hydrochloric acid. In a second series of titrations, 10 cm^3 of the organic layer was neutralised by 6.0 cm^3 of 0.010 mol dm^{-3} hydrochloric acid.

a) What was the concentration of the ammonia in the aqueous layer at equilibrium? (2)

b) What was the concentration of the ammonia in the organic solvent at equilibrium? (2)

c) What is the value of K_c for the equilibrium: $NH_3(org) \rightleftharpoons NH_3(aq)$? (2)

3 At 298 K the value of K_c for the following equilibrium is 1×10^{10}:

$Sn^{2+}(aq) + 2Fe^{3+}(aq) \rightleftharpoons Sn^{4+}(aq) + 2Fe^{2+}(aq)$

a) i) Write the expression for K_c. (2)

ii) What are the units of K_c for this reaction? Explain your answer. (2)

b) What is the value of K_c for:

i) $Sn^{4+}(aq) + 2Fe^{2+}(aq) \rightleftharpoons Sn^{2+}(aq) + 2Fe^{3+}(aq)$ (2)

ii) $\frac{1}{2}Sn^{2+}(aq) + Fe^{3+}(aq) \rightleftharpoons \frac{1}{2}Sn^{4+}(aq) + Fe^{2+}(aq)$ (2)

4 Nitrogen and hydrogen react to form ammonia when heated under pressure in the presence of a catalyst.

$N_2(g) + 3H_2(g) \rightleftharpoons 2NH_3(g) \qquad \Delta H = -92\ kJ\ mol^{-1}$

Analysis of an equilibrium mixture of the gases at 10 atmospheres and 650 K found that it contained 1.41 mol NH_3, 6.75 mol H_2 and 2.15 mol N_2.

a) Explain, in this context, the term dynamic equilibrium. (2)

b) i) Calculate the mole fraction of each of the three gases in the mixture at 10 atm and 650 K. (3)

ii) Calculate the partial pressures of the three gases. (2)

iii) Calculate a value for K_p and give the units. (3)

c) Explain, in terms of entropy, the effect of an increase in temperature on the value of K_p, and hence on the proportions of the three gases in the equilibrium mixture. (5)

4 Application of rates and equilibrium ideas

The chemical industry is being reinvented to make it more sustainable. It is no longer acceptable to operate processes that make inefficient use of valuable resources and create large quantities of hazardous wastes. Chemists and chemical engineers are devising new methods by applying the theoretical ideas and models that explain how fast reactions go, in which direction and how far.

Figure 4.1▲
This vast catalytic cracker in Germany is used to make ethene from natural gas or oil. Controlling chemical reactions carried out on such a large scale requires precise application of chemical principles.

Figure 4.2▲
Careful checking of the conditions of temperature and pressure is essential for any chemical process. This control panel, at the Fawley petrochemical plant near Southampton, is part of the monitoring system for a catalytic method for purifying chemicals from oil.

Figure 4.3▲
On the left, a yellow solution of chromate(VI) ions in water. On the right, a solution of chromate(VI) ions in water after adding a few drops of strong acid – the solution has turned orange as more dichromate(VI) ions form.

4.1 Factors affecting systems at equilibrium

The effect of changing concentrations on systems at equilibrium

The equilibrium law makes it possible to explain the effect of changing the concentration of one of the chemicals in an equilibrium mixture.

An example is the equilibrium in solution involving chromate(VI) and dichromate(VI) ions in water:

$$\underset{\text{yellow}}{2CrO_4^{2-}(aq)} + 2H^+(aq) \rightleftharpoons \underset{\text{orange}}{Cr_2O_7^{2-}(aq)} + H_2O(l)$$

$$\text{At equilibrium: } K_c = \frac{[Cr_2O_7^{2-}(aq)]}{[CrO_4^{2-}(aq)]^2[H^+(aq)]^2}$$

where these are equilibrium concentrations.

Adding a few drops of concentrated acid increases the concentration of $H^+(aq)$ on the left-hand side of the equation.

This briefly upsets the equilibrium. For an instant after adding acid:

$$\frac{[Cr_2O_7^{2-}(aq)]}{[CrO_4^{2-}(aq)]^2[H^+(aq)]^2} < K_c$$

Note

Changing the concentrations does not alter the value of the equilibrium constant so long as the temperature stays constant.

Note

In dilute solution $[H_2O(l)]$ is constant so it does not appear in the equilibrium law expression.

The system restores equilibrium as chromate(VI) ions react with hydrogen ions to produce more of the products. There is very soon a new equilibrium. Once again:

$$\frac{[Cr_2O_7^{2-}(aq)]}{[CrO_4^{2-}(aq)]^2[H^+(aq)]^2} = K_c$$

but now with new values for the various equilibrium concentrations.

Chemists sometimes say that adding acid makes the 'position of equilibrium shift to the right'. The effect is visible because the yellow colour of the chromate(VI) ions turns to the orange colour of dichromate(VI) ions. This is as le Chatelier's principle predicts. The advantage of using K_c is that it makes quantitative predictions possible.

Test yourself

1 Describe and explain the effect of adding alkali to a solution of dichromate(VI) ions.

2 a) Use the equilibrium law to predict and explain the effect of adding pure ethanol to an equilibrium mixture of ethanoic acid, ethanol, ethyl ethanoate and water:

$CH_3COOH(l) + C_2H_5OH(l) \rightleftharpoons CH_3COOC_2H_5(l) + H_2O(l)$

b) Show that your prediction is consistent with le Chatelier's principle.

The effects of pressure changes on systems at equilibrium

Changes of pressure are not generally significant for equilibria that involve only solids or liquids, but they have a marked effect on gaseous equilibria. Fundamentally these effects can be interpreted in terms of entropy. Lowering the pressure of a gas leads to an increase in volume which means that there are more ways of distributing the molecules. As a result the entropy increases (Section 2.1). In general this means that lowering the pressure on a system at equilibrium favours the direction of change that produces more molecules, while increasing the pressure favours the change that produces fewer molecules. This is what le Chatelier's principle predicts.

$$\begin{array}{c} \text{low pressure} \longrightarrow \\ \text{fewer molecules} \rightleftharpoons \text{more molecules} \\ \longleftarrow \text{high pressure} \end{array}$$

The effects of increasing or decreasing the total pressure of a gas mixture at equilibrium can be predicted quantitatively with the help of the equilibrium law. Take the example of the reaction used to make ammonia in the Haber process:

$$N_2(g) + 3H_2(g) \rightleftharpoons 2NH_3(g)$$

The equilibrium law expression in terms of partial pressures is:

$$K_p = \frac{(p_{NH_3})^2}{p_{N_2} \times (p_{H_2})^3}$$

Note

Changing pressure does not alter the value of the equilibrium constant K_p so long as the temperature stays constant.

Suppose that the equilibrium partial pressures of nitrogen, hydrogen and ammonia are a atm, b atm and c atm, respectively. Substituting in the expression for K_p gives:

$$K_p = \frac{c^2}{ab^3}\,\text{atm}^{-2}$$

Now suppose that the total pressure is suddenly doubled. At that instant all the partial pressures double so that $p_{N_2} = 2a$ atm, $p_{H_2} = 2b$ atm and $p_{NH_3} = 2c$ atm. Substituting these values in the 'equilibrium constant ratio' gives:

$$\frac{(p_{NH_3})^2}{p_{N_2} \times (p_{H_2})^3} = \frac{(2c)^2}{2a \times (2b)^3} = \frac{1}{4} \times \frac{c^2}{ab^3}\,\text{atm}^{-2}$$

So at that moment the 'equilibrium constant ratio' is one-quarter of the value of K_p. The system is not at equilibrium. In order to restore equilibrium, some of the nitrogen and hydrogen must react to decrease their partial pressures and form more ammonia to increase its partial pressure. This happens until the values are such that the 'equilibrium constant ratio' again equals K_p. In other words, increasing the pressure causes the equilibrium to shift to the right. In this way the equilibrium law makes it possible to predict not only the direction but also the extent of the shift (Table 4.1).

Total pressure/atm	10	50	100	200
Percentage by volume of ammonia at equilibrium at 773 K	1.2	5.6	10.6	18.3

Table 4.1 ▲

Test yourself

3 Predict the effect of increasing the pressure on these systems at equilibrium:
a) $2SO_2(g) + O_2(g) \rightleftharpoons 2SO_3(g)$
b) $CH_4(g) + H_2O(g) \rightleftharpoons CO(g) + 3H_2(g)$
c) $N_2(g) + O_2(g) \rightleftharpoons 2NO(g)$

4 For the reaction $N_2O_4(g) \rightleftharpoons 2NO_2(g)$, the value of K_p is 0.11 atm at 298 K. Is a mixture containing $N_2O_4(g)$ with a partial pressure of 2.4 atm and $NO_2(g)$ with a partial pressure of 1.2 atm at equilibrium at 298 K? If not, which gas tends to increase its partial pressure?

5 Use the expression for K_p to predict the effect of the following changes on an equilibrium mixture of hydrogen, carbon monoxide and methanol:
$2H_2(g) + CO(g) \rightleftharpoons CH_3OH(g)$
a) adding more hydrogen to the gas mixture at constant total pressure
b) compressing the mixture to increase the total pressure
c) adding an inert gas such as argon while keeping the total pressure constant.

Activity

Changing pressure and the equilibrium between the gases H_2, I_2 and HI

Consider the reaction:

$$H_2(g) + I_2(g) \rightleftharpoons 2HI(g)$$

Suppose that the equilibrium partial pressures of hydrogen, iodine and hydrogen iodide are respectively x atm, y atm and z atm, at a particular temperature.

1 Write an expression for the equilibrium constant in terms of x, y and z.

2 Suppose the overall pressure of the equilibrium mixture is doubled. What are the new partial pressures of the three gases?

3 At the instant that the pressure is doubled, what is the value of the 'equilibrium constant ratio'?

4 Use the equilibrium law to explain the effect of doubling the pressure on the position of equilibrium.

5 Show that your answer to **4** is consistent with predictions based on le Chatelier's principle.

Note

If ΔH for the forward reaction is negative, then ΔH for the reverse reaction has the same magnitude but the opposite sign. So if the forward reaction is exothermic, then the reverse reaction is endothermic.

Note

The value of an equilibrium constant for an exothermic reaction becomes smaller as the temperature rises.

The value of the equilibrium constant for an endothermic reaction rises as the temperature rises.

The effects of temperature changes on systems at equilibrium

Le Chatelier's principle predicts that raising the temperature makes the equilibrium shift in the direction which is endothermic. For example, for the reaction which produces sulfur trioxide during the manufacture of sulfuric acid, raising the temperature lowers the percentage of sulfur trioxide at equilibrium.

$$2SO_2(g) + O_2(g) \rightleftharpoons 2SO_3(g) \qquad \Delta H = -98\,\text{kJ mol}^{-1}$$

The equilibrium shifts to the left as the temperature rises because this is the direction in which the reaction is endothermic.

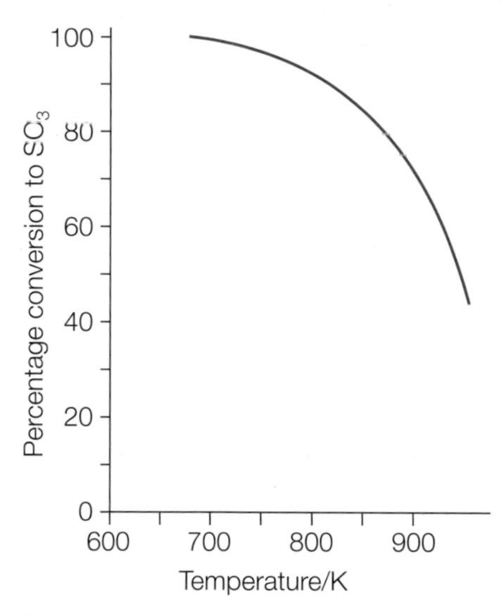

Figure 4.4▲
The effect of raising the temperature on the equilibrium between SO_2, O_2 and SO_3.

Temperature/K	K_p/atm^{-1}
298	4.0×10^{24}
500	2.5×10^{10}
700	3.0×10^{4}
1100	1.3×10^{-1}

Table 4.2▲
Values of K_p at four temperatures for the equilibrium $2SO_2(g) + O_2(g) \rightleftharpoons 2SO_3(g)$

The reason that temperature changes cause a shift in the position of equilibrium is that the value of the equilibrium constant changes. This is illustrated by the values in Table 4.2.

Tutorial

Test yourself

6 Show that the values in Table 4.2 are consistent with predictions for the equilibrium based on le Chatelier's principles.

7 The value of K_p for the equilibrium $N_2O_4(g) \rightleftharpoons 2NO_2(g)$ is 4.79 atm at 400 K and 347 atm at 500 K.
   a) What is the effect of raising the temperature on the position of equilibrium?
   b) How can your answer to a) account for the appearance of the gas mixtures in Figure 3.7 in Section 3.4?
   c) What is the sign of ΔH for the reaction?

8 For the reaction between hydrogen and iodine to form hydrogen iodide, the value of K_p is 794 at 298 K but 54 at 700 K. What can you deduce from this information?

The effects of catalysts on systems at equilibrium

Catalysts speed up reactions but are not used up as they do so. It is important to note that while a catalyst speeds up the rate at which a reaction gets to an equilibrium state, it has no effect on the final position of equilibrium. In other words, a catalyst provides a faster route to the same equilibrium state. The alternative route with a catalyst has a lower activation energy but speeds up the forward and back reactions to the same extent, so that the dynamic equilibrium is unchanged.

Figure 4.5◄
Crystals of palladium seen under an electron microscope. Palladium is a rare and precious metal which is used to catalyse hydrogenation reactions.

4.2 Efficiency and sustainability in the chemical industry

Synthesis is at the heart of any process to make new chemicals. In the past, important synthetic reactions have often been based on toxic reagents and produced noxious waste. Now the principles of green chemistry are being applied to reinvent old processes and to create new ones that are based on renewable resources, avoid hazardous reagents and minimise waste. These developments are often based on insights into reaction mechanisms that lead to the discovery of new catalysts. Greener processes make more efficient use of energy and can add to the profitability of industry.

Refining an established process

Even well established processes, such as the Haber process for making ammonia, are benefiting from new research. Ammonia is an important bulk chemical because it is used to make fertilisers and other chemicals, including those needed to make polymers, dyes and explosives. Worldwide the production of ammonia exceeds 100 million tonnes per year, with 80 per cent of production being used to make fertilisers.

Ammonia is produced on such a large scale that small improvements to the efficiency of the process can be valuable. Most ammonia plants are large and often capable of making more than 1000 tonnes of ammonia per day.

Figure 4.6◄
Myanmar farmers spreading fertiliser into the rice fields by hand in Taikkyi, about 90 kilometres north of Yangon. Intensive farming all over the world depends on high inputs of nitrogen fertilisers made from ammonia. Common fertilisers are ammonium nitrate, ammonium sulfate and urea.

Activity

The Haber process

The Haber process has been used to make ammonia for about 100 years. The main raw material is usually natural gas but can be one of the fractions from distilling oil. The hydrocarbons react with steam and air, in a series of steps, to produce a mixture of nitrogen and hydrogen. The only by-product is carbon dioxide, which can be released into the air or used to convert some of the ammonia into urea, $CO(NH_2)_2$.

The mixture of nitrogen and hydrogen is heated to a temperature in the range 650–720 K and compressed to a pressure in the range 100–250 atmospheres. Then the gases pass through a series of catalyst beds where the two gases react to make ammonia.

$$N_2(g) + 3H_2(g) \rightleftharpoons 2NH_3(g) \qquad \Delta H = -92\,\text{kJ mol}^{-1}$$

The conditions chosen are a compromise to balance chemical efficiency with cost and safety. The higher the pressure, the higher the yield of ammonia at equilibrium, but high pressure plants are expensive to build and run. They can also be more hazardous for the plant operators. The lower the temperature, the higher the yield at equilibrium, but the reaction is too slow to be economic if the temperature is too low.

In the traditional process the catalyst is iron mixed with promoters: aluminium oxide to prevent the pores in iron collapsing; potassium oxide to increase the chemical activity of the iron (Table 4.3).

The ammonia produced is cooled and condensed. Unreacted nitrogen and hydrogen are fed back into the reactor. With continuous recycling of unreacted gases, yields of up to 98% ammonia are produced.

Catalyst	Promoter	Percentage ammonia in the gases leaving the reactor
Fe	None	3–5
Fe	K_2O	8–9
Fe	K_2O and Al_2O_3	13–14

Table 4.3▲
The effect of promoters on the efficiency of iron as a catalyst for the Haber process at 200 atm and 673 K.

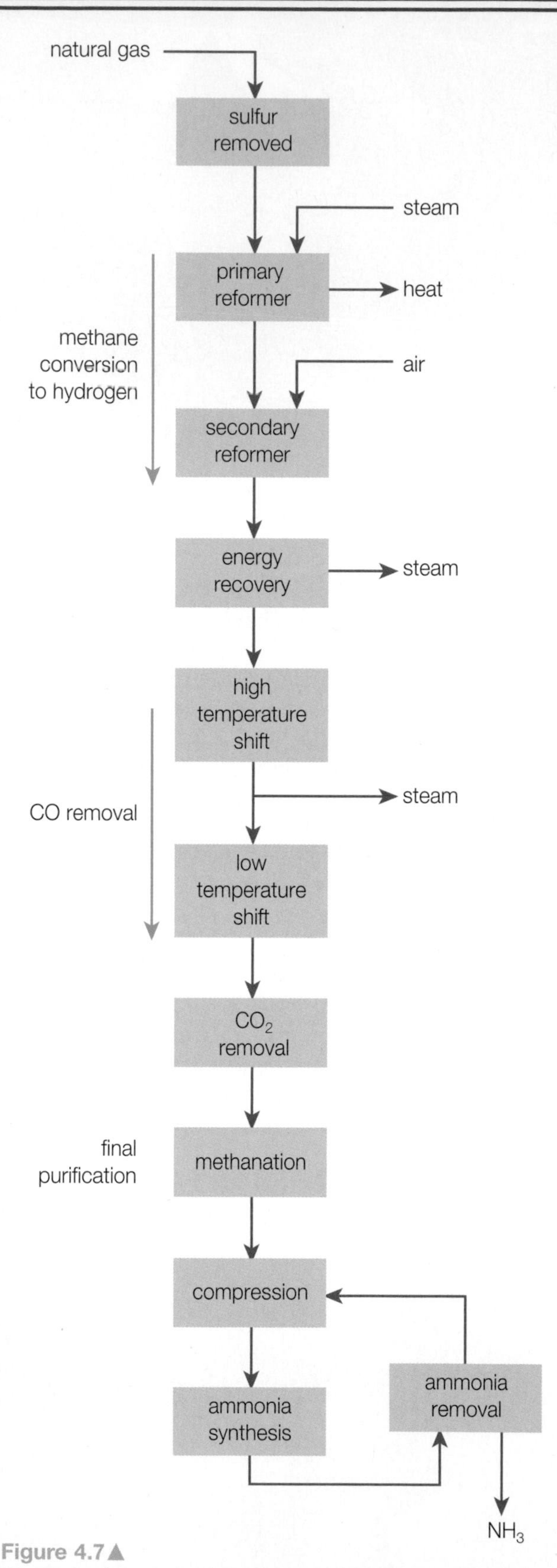

Figure 4.7▲
Stages in the production of ammonia from natural gas.

1 Suggest reasons why ammonia manufacture has, in the past decade, shifted from countries such as the UK to countries such as Qatar, Saudi Arabia, India and China.

2 Electricity from hydroelectric plants in Iceland has been used to make hydrogen for the Haber process by electrolysis. Why is this change to 'greener' hydrogen not something that can happen on a large scale?

3 Suggest reasons why it is important to remove the sulfur from the natural gas used to make ammonia.

4 Methane reacts with steam in the presence of a nickel oxide catalyst to make hydrogen.

$$CH_4(g) + H_2O(g) \rightleftharpoons CO(g) + 3H_2(g) \qquad \Delta H = +210\,\text{kJ mol}^{-1}$$

What conditions of temperature and pressure favour the formation of products in this reaction?

5 The so-called 'shift reaction' uses steam in the presence of a catalyst to get rid of carbon monoxide. The reaction is exothermic. Write an equation for the reaction and state the conditions of temperature and pressure that favour the conversion of carbon monoxide to carbon dioxide.

6 An organic base is used to absorb and remove carbon dioxide. Suggest advantages of using the carbon dioxide to make urea rather than releasing it into the air when the base is regenerated.

7 Explain the advantage of using heterogeneous catalysts in all the stages of making ammonia.

8 Suggest reasons why the finely divided iron catalyst in the reactor for making ammonia is less efficient in the absence of the aluminium oxide promoter.

9 The mixture of nitrogen and hydrogen passes through two to four catalyst beds in the reactor. The gas mixture is cooled between each bed. Explain why this cooling is necessary.

10 At 200 atm and 673 K the equilibrium percentage of ammonia is 38.8%. Account for the fact that the maximum percentage of ammonia in the gas mixture leaving the reactor is about 15%.

Even after 100 years of experience, there is much about the catalytic process in the Haber process that is not fully understood. In recent years it has been possible to use new experimental techniques and computer modelling to begin to understand the nature of the active sites on the surface of the catalyst. This in turn has led to the development of new catalysts that allow the process to operate at lower temperatures and pressures. A new catalyst based on ruthenium can be installed in existing ammonia plants, which are then able to operate at 40 atmospheres.

Another advantage of the ruthenium catalyst is that its effectiveness is little changed as the partial pressure of ammonia in the reaction mixture rises. This contrasts with the traditional iron catalyst, which becomes significantly less effective when there is more ammonia in the gas mixture.

Test yourself

9 Explain the advantages of using a new catalyst for the Haber process that:
   a) can be installed in existing manufacturing plants
   b) allows the plant to run at lower pressures
   c) is effective at lower temperatures
   d) retains its effectiveness as the partial pressure of ammonia rises.

10 Also in the research stage is a system to circulate an absorbent through the reactor where the nitrogen and hydrogen react to form ammonia. The absorbent is designed to remove the ammonia from the gas mixture in the reactor. Explain why this could increase the efficiency of the process.

Devising new processes

Traditionally chemists have used the percentage yield as the measure of the efficiency of a chemical process. However the yield is calculated only with reference to the one main product. It takes no account of by-products and wastes. A reaction can have a high percentage yield while at the same time producing much waste.

In 1991, Professor Barry Trost of Stanford University published a paper putting forward the idea of atom economy as an alternative measure of the efficiency of a synthesis. Atom economy shows the extent to which the atoms of the reactants get built into the final product rather than ending up as waste.

Test yourself

11 a) What is the molecular formula of ibuprofen?
   b) What was the atom economy of the original synthesis of ibuprofen?
   c) What is the atom economy of the newer process?
   d) The yields of the three steps in the newer process, in order, are: 90%, 98% and 98%. What is the overall yield?
   e) The catalyst in the second step of the newer process is a heterogeneous nickel catalyst. What is the advantage of using a heterogeneous catalyst in a synthesis?

Definition

Chemists define atom economy as:

$$\text{percentage atom economy} = \frac{\text{mass of atoms in wanted product}}{\text{total mass of atoms in all reactants}} \times 100$$

The companies that manufacture chemicals increasingly aim to develop more efficient processes with high atom economies. A classic example is the invention of a new method of making the painkiller ibuprofen. The original synthesis was a six-step process in which less than half of the mass of all the atoms in the reagents ended up in the product with the rest going to waste. One of the main catalysts used was aluminium chloride. This is a homogeneous catalyst which could not be recovered and reused, but which added to the waste.

Once the original patent on ibuprofen had expired, there was competition to develop alternative syntheses for the drug. The most successful of these was a three-step process with a much higher atom economy (Figure 4.8).

Original synthesis
- six steps
- uses aluminium chloride as a catalyst which cannot be recovered
- atoms in all the reactants: 20 C, 42 H, N, 10 O, 4 Cl, Na, Al

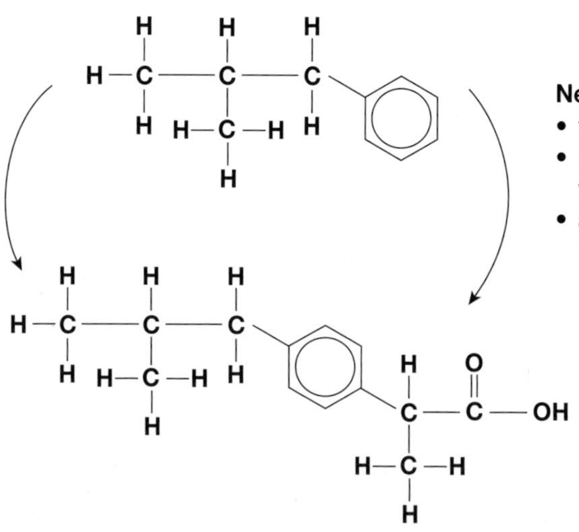

New synthesis
- three steps
- uses three catalysts, each of which can be recovered
- atoms in all the reactants: 15 C, 22 H, 4 O

Figure 4.8▲
Two routes to ibuprofen.

Activity

Two routes to epoxyethane

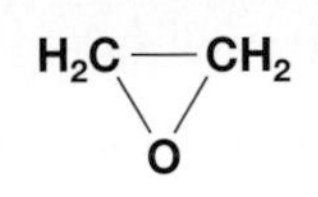

Figure 4.9▲
The structure of epoxyethane.

Epoxyethane is a very reactive compound that is used to make many other chemicals. The chemical industry worldwide produces about 15 million tones of the chemical each year. Products made with epoxyethane include detergents, polymers, lubricants and solvents.

The first synthesis of epoxyethane was a two-step process. In the first step ethene reacted with aqueous chlorine to form 2-chloroethanol. In the second step, 2-chloroethanol was converted to epoxyethane using calcium hydroxide, with the formation of the waste product aqueous calcium chloride. This process was phased out in the mid 1970s when a new process was adopted everywhere.

The new process uses finely divided silver on an inert support to catalyse the reaction of ethene with oxygen under pressure at 520–600 K.

$$CH_2{=}CH_2(g) + \tfrac{1}{2}O_2(g) \rightarrow \underset{\text{O}}{H_2C\text{—}CH_2} \qquad \Delta H = -107\ \text{kJ mol}^{-1}$$

The reactants spend just 1–4 seconds in the reactor. The process has to be carefully controlled because if the temperature rises the ethene burns in oxygen.

1 a) Write two equations for the original process used to make epoxyethane.

b) Calculate the atom economy of the process.

2 Calculate the atom economy of the new process.

3 Suggest three reasons why the older process was phased out and replaced with the new process.

4 Temperature control is critically important in the newer process for making epoxyethane.

a) Why does the temperature in the reactor tend to rise?

b) Why is this a problem?

5 About a third of all epoxyethane is converted to ethane-1,2-diol to make antifreeze and polyester. This is done by reaction with excess water in the presence of an acid catalyst.

a) Write an equation for the reaction.

b) Suggest a reason why excess water is needed to prevent the formation of other products.

6 The chemicals that react with epoxyethane include water, ammonia, alcohols and amines. What do these molecules have in common? What type of reagent are they when they react with epoxyethane and open up the ring?

Figure 4.10▲
Doh Ho Suh's exhibit 'Staircase -V, 2008' made of polyester and stainless steel at the Hayward Gallery in 2008. Epoxyethane is converted to the ethane-1,2-diol needed to make polyester.

REVIEW QUESTIONS

Extension questions

1 At 473 K, the value of K_c for the decomposition of PCl_5 is 8×10^{-3} mol dm^{-3}.

$PCl_5(g) \rightleftharpoons PCl_3(g) + Cl_2(g) \qquad \Delta H = +124\,kJ\,mol^{-1}$

a) Write the expression for K_c for the reaction. (1)

b) What is the value of K_c for the reverse reaction at 473 K and what are its units? (2)

c) A sample of pure PCl_5 is heated to 473 K in a vessel containing no other chemicals. At equilibrium the concentration of PCl_5 is 5×10^{-2} mol dm^{-3}. What are the equilibrium concentrations of PCl_3 and Cl_2? (3)

d) Explain how the concentrations of PCl_5, PCl_3 and Cl_2 change in the equilibrium mixture if:

i) more PCl_5 is added (2)

ii) the pressure is increased (2)

iii) the temperature is increased. (2)

e) What is the effect on the value of K_c if:

i) more PCl_5 is added (1)

ii) the pressure is increased (1)

iii) the temperature is increased? (2)

2 Explain what is wrong with each of the following statements. To what extent, if at all, is there any truth in each of the statements?

a) Once a reaction mixture reaches equilibrium there is no further reaction. (3)

b) Adding more of one of the reactants to an equilibrium mixture increases the yield of products because the value of the equilibrium constant increases. (3)

c) Adding a catalyst to make a reaction go faster can increase the amount of product at equilibrium. (3)

d) Raising the temperature to make a reaction go faster can increase the amount of product at equilibrium. (3)

e) Adding a catalyst can mean that a reaction that is only feasible at a high temperature becomes feasible at a much lower temperature. (3)

3 Hydrogen is made from natural gas by partial oxidation with steam. This involves this reaction:

$$CH_4(g) + H_2O(g) \rightleftharpoons CO(g) + 3H_2(g) \qquad \Delta H = +210\,kJ\,mol^{-1}$$

a) Write an expression for K_p for this reaction. (2)

b) How is the value of K_p affected by:

i) increasing the pressure (1)

ii) increasing the temperature (1)

iii) using a catalyst? (1)

c) How does the composition of an equilibrium mixture of the gases change when:

i) the pressure rises (1)

ii) the temperature rises (1)

iii) a catalyst is added? (1)

4 Ammonia is converted to nitric acid on a large scale. In the first step ammonia is mixed with air and compressed before passing through a reactor containing catalyst gauzes. The catalyst is an alloy of platinum and rhodium. The reversible, exothermic reaction produces nitrogen monoxide (NO) and steam.

a) Write an equation for the reaction of ammonia with oxygen to form nitrogen monoxide. (2)

b) Predict, qualitatively, the conditions which favour a high yield of NO in the equilibrium mixture. (3)

c) The industrial process typically runs at 1175 K and a pressure of about 7 atm with a mixture of 10% ammonia and 90% air. How and why are these conditions similar to, or different from, those you predicted in **b)**? (3)

d) Explain the advantage of using a heterogeneous catalyst in this process. (1)

e) Why does the gas mixture leaving the reactor not contain as high a percentage of NO as predicted by the equilibrium law? (2)

f) The hot gas mixture leaving the reactor has to be cooled before the next step. Suggest how this might be done in a way that improves the overall energy efficiency of the process. (2)

5 Explain these statements and illustrate them with examples.

a) Syntheses based on addition reactions have higher atom economies than syntheses based on substitution reactions. (5)

b) A synthesis can give a 100% yield and yet have a low atom economy. (5)

c) Temperature control is important in ensuring a good yield of a product at an economic rate. (5)

5 Acid–base equilibria

Acids and bases are very common not only in laboratories but also in living things, in the home and in the natural environment. Acid–base reactions are reversible and governed by the equilibrium law. This means that chemists are able to predict reliably and quantitatively how acids and bases behave. This is important for the supply of safe drinking water, the care of patients in hospital, the formulation of shampoos and cosmetics, as well as the processing of food and many other aspects of life.

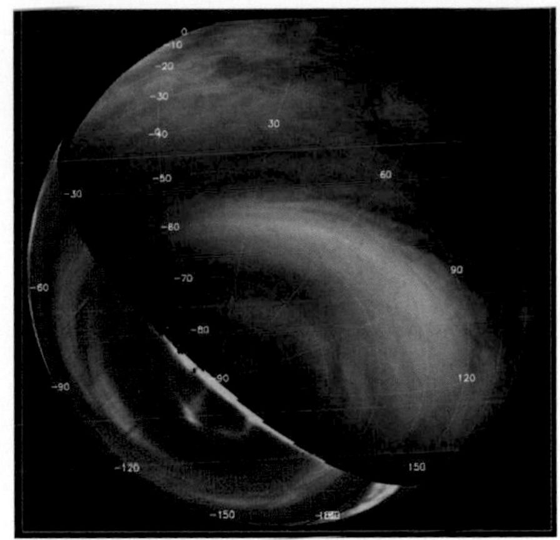

Figure 5.1▲
A composite, false-colour image produced by combining ultraviolet and infrared images from instruments on the European Space Agency's Venus Express spacecraft in July 2007. There is a very thick cloud layer at around 60 kilometres altitude around the planet. This lies between the lower and middle layers of Venus's atmosphere. Scientists now know that the upper part of this layer is mostly composed of tiny droplets of sulfuric acid.

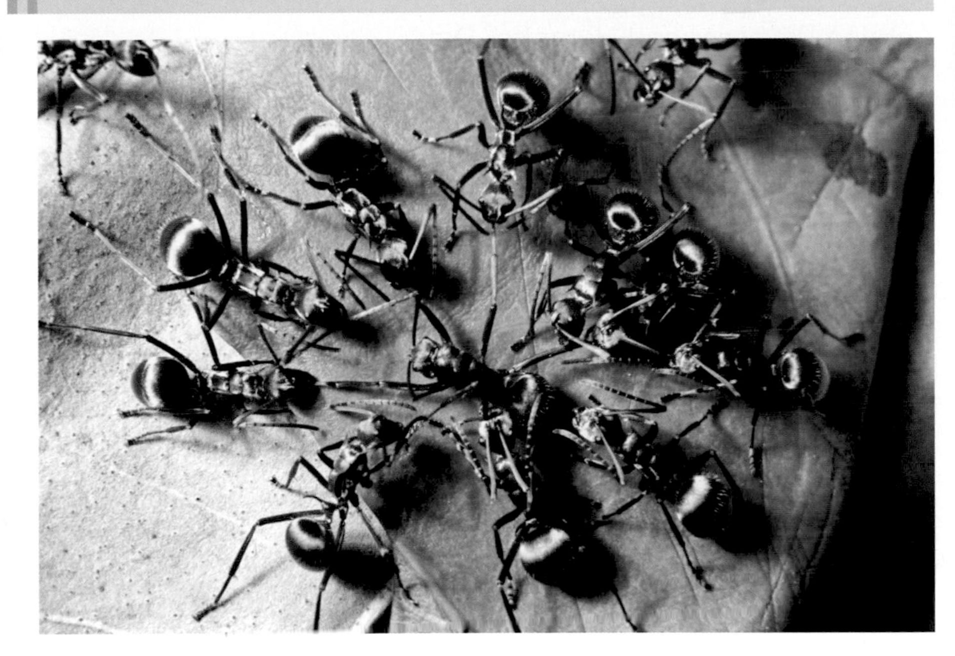

Figure 5.2◀
Polyrhachis laboriosa ants attacking an intruder from a different colony. They are about to kill the intruder by spraying a jet of methanoic acid from their abdomens. This species of ant lives in trees at the edges of tropical forests in Cameroon.

5.1 Theories of acidity

Jabir ibn-Hayyan

The mineral acids that we now take for granted were discovered by Jabir ibn-Hayyan (c.722–c.815) who worked in the alchemical tradition but pioneered experimental chemistry. He developed the techniques of crystallisation and distillation and used them to discover sulfuric, hydrochloric and nitric acids. He also studied ethanoic acid in vinegar and tartaric acid in wine.

Note

All the names for acids in this chapter are the modern chemical names and not the names generally used at the time when these chemicals were first discovered.

Figure 5.3◀
Jabir ibn-Hayyan in a coloured engraving, published in 1883, which shows him teaching at the school at Edessa in Mesopotamia (now Sanliurfa in Turkey). He played a key part in turning chemistry from a mystical practice (alchemy) into a science. He pioneered experimental techniques and invented much of the equipment that is still commonly used in laboratories.

Acids were first recognised by their chemical properties. Acidic solutions have a sour taste; they tend to corrode metals and they change the colour of indicators.

Test yourself

1 Jabir ibn-Hayyan discovered sulfuric acid and then studied its reactions on heating with salts such as sodium chloride and potassium nitrate. How could these studies, and his improved methods of distillation, lead to the discovery of hydrochloric and nitric acids?

Figure 5.4▲
The portrait of Antoine Lavoisier and his wife, Marie-Anne, painted by Jacques-Louis David in 1788. Marie-Anne contributed in many ways to Antoine Lavoisier's work.

Antoine Lavoisier and Humphry Davy

The scientist who laid the foundations of modern chemical theory was the French nobleman Antoine Lavoiser (1743–1794). His experiments and insights led to the oxygen-theory of burning and a systematic approach to quantitative chemistry.

The name 'oxygen' means 'acid former'. Lavoisier gave the gas this name because he thought that all acids were compounds containing this element.

The generalisation that all acids contain oxygen was disproved by a series of experiments carried out by Humphry Davy between 1809 and 1810. He heated hydrogen chloride (then called muriatic acid) to high temperatures with a range of metals and non-metals. He could find no trace of oxygen in the compound. After further work, Davy proposed, in 1816, that what all acids have in common is that they contain hydrogen.

Test yourself

2 Why is it not surprising that Lavoiser thought that all acids contain oxygen?
3 Identify three acids, other than hydrogen chloride, which do not contain oxygen.
4 Give examples which show that:
a) acids contain hydrogen
b) not all compounds that contain hydrogen are acids.

Svante Arrhenius

As a young man in his mid-20s, the Swedish chemist Svante Arrhenius wrote a doctoral thesis which proposed that some compounds are ionised in solution all the time. This was the start of the ionic theory of solutions that we now take for granted. In 1884 it was highly controversial. At the time, Arrhenius was bitterly disappointed to be awarded the bottom grade for his paper. Later he was vindicated and awarded the Nobel prize for chemistry in 1903.

Arrhenius used his ionic theory to come up with an explanation of why it is that all acids have similar properties when dissolved in water. His theory could also account for what happens when an acid is neutralised by an alkali and explain the difference between strong and weak acids.

In 1887 Arrhenius defined an acid as a compound that could produce hydrogen ions when dissolved in water, and an alkali as a compound that could produce hydroxide ions in water. According to Arrhenius's theory, hydrochloric acid is a strong acid which is fully ionised when dissolved in water.

$$HCl(aq) \rightarrow H^+(aq) + Cl^-(aq)$$

Ethanoic acid is a weak acid which is only slightly ionised.

$$CH_3COOH(aq) \rightleftharpoons CH_3CO_2^-(aq) + H^+(aq)$$

Arrhenius's theory was a big advance in its time. It could account for the similarities between acids. In this theory the typical reactions of dilute acids in water are the reactions of aqueous hydrogen ions.

With metals: $Mg(s) + 2H^+(aq) \rightarrow Mg^{2+}(aq) + H_2(g)$
With carbonates: $CO_3^{2-}(s) + 2H^+(aq) \rightarrow CO_2(g) + H_2O(l)$
With bases: $O^{2-}(s) + 2H^+(aq) \rightarrow H_2O(l)$

The theory is still useful today but it has a number of weaknesses, one of which is that it is limited to aqueous solutions.

Test yourself

5 What, according to Arrhenius's theory, happens when an acid neutralises an alkali?
6 Suggest a simple practical demonstration of the difference between equimolar solutions of a strong acid and of a weak acid.
7 Write ionic equations to show how Arrhenius's theory describes the reactions of nitric acid with:
a) zinc
b) potassium carbonate
c) calcium oxide
d) lithium hydroxide.

Johannes Brønsted and Thomas Lowry

The preferred theory for discussing acid–base equilibria today was put forward independently in 1923 by the Danish chemist Johannes Brønsted and the English chemist Thomas Lowry. This theory describes acids as proton donors and bases as proton acceptors.

5.2 Acids, bases and proton transfer

Acids as proton donors

According to the Brønsted–Lowry theory, hydrogen chloride molecules give hydrogen ions (protons) to water molecules when they dissolve in water, producing hydrated hydrogen ions called oxonium ions. The water acts as a base.

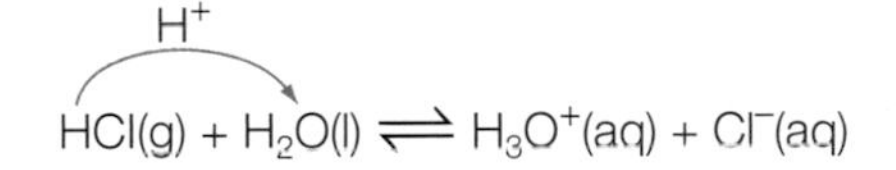

Figure 5.5▲
Proton transfer between hydrogen chloride molecules and water molecules. This is reversible. A proton from the oxonium ion can transfer back to the chloride ion to give hydrogen chloride and water.

Definition

A **proton** is the nucleus of a hydrogen atom, so a hydrogen ion, H^+, is just a proton.

Note

It is more correct to represent hydrogen ions in aqueous solution as $H_3O^+(aq)$; however, chemists commonly use a shorter symbol for hydrated protons, $H^+(aq)$.

Hydrogen chloride is a strong acid. What this means is that it readily gives up its protons to water molecules and the equilibrium in solution lies well over to the right. Hydrogen chloride is effectively completely ionised in solution. Other examples of strong acids are sulfuric acid and nitric acid.

Definitions

A **strong acid** is an acid which is fully ionised when it dissolves in water.

Monobasic acids are acids that can give away (donate) one proton per molecule. Examples are hydrochloric acid, HCl; nitric acid, HNO_3; and ethanoic acid, CH_3COOH. These are also called monoprotic acids.

Dibasic acids are acids that can give away (donate) two protons per molecule. Examples are sulfuric acid, H_2SO_4; and ethanedioic acid, HOOC–COOH. These are also called diprotic acids.

Test yourself

8 a) What type of bond links the water molecule to a proton in an oxonium ion?
b) Draw a dot-and-cross diagram to show the bonding in an oxonium ion.
c) Predict the shape of an oxonium ion.
9 Write a balanced ionic equation for the reaction of 1 mol ethanedioic acid with 2 mol NaOH showing the structural formulae for the acid and for the ethanedioate ion formed.
10 a) Identify the products of the reaction when concentrated sulfuric acid reacts with sodium chloride.
b) Show that this is a proton transfer reaction.
c) Account for the fact that this reaction can give a good yield of hydrogen chloride gas.

Bases as proton acceptors

According to the Brønsted–Lowry theory, a base is a molecule or ion which can accept a hydrogen ion (proton) from an acid. A base has a lone pair of electrons which can form a dative covalent bond with a proton.

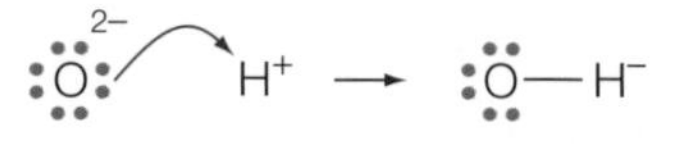

Figure 5.6▲
Oxide ions have lone pairs of electrons which can form dative covalent bonds with hydrogen ions.

An ionic oxide, such as calcium oxide, reacts completely with water to form calcium hydroxide. The calcium ions do not change; but the oxide ions, which are powerful proton acceptors, all take protons from water molecules. An oxide ion is a strong base. Common bases include the oxide and hydroxide ions, ammonia, amines, as well as the carbonate and hydrogencarbonate ions.

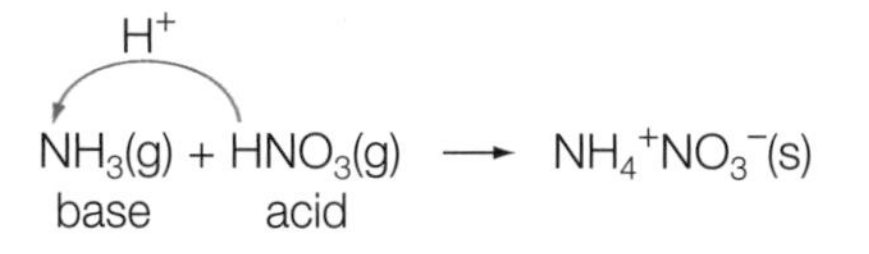

Figure 5.7▲
There is a lone pair on the nitrogen atom of ammonia which allows it to act as a base.

In biochemistry the term base often refers to one of the five nitrogenous bases which make up nucleotides and the nucleic acids DNA and RNA. These compounds (adenine, guanine, cytosine, uracil and thymine) are bases in the chemical sense because they have lone pairs on nitrogen atoms which can accept hydrogen ions.

Note

Many compounds of the group 1 and group 2 metals form alkaline solutions. This is because metals such as sodium, potassium, magnesium and calcium (unlike other metals) form oxides, hydroxides and carbonates which are soluble (to a greater or lesser extent) in water. It is important to realise that it is the oxide, hydroxide or carbonate ions in these compounds that are bases, and not the metal ions.

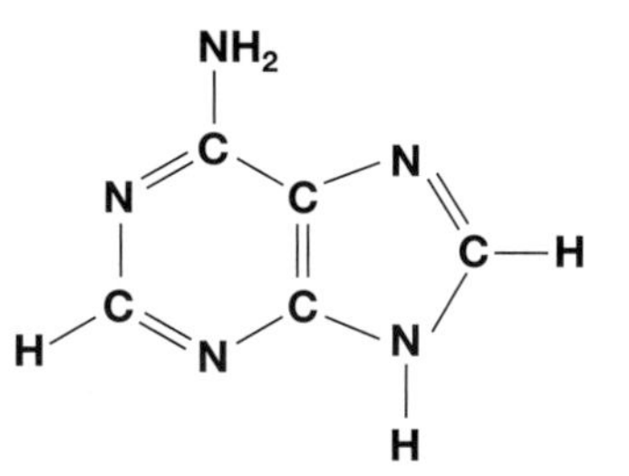

Figure 5.8▲
The displayed formula of adenine. This is one of the bases in DNA.

Test yourself

11 Show that the reactions between these pairs of compounds are acid–base reactions and identify as precisely as possible the molecules or ions which are the acid and the base in each example.
   a) $MgO + HCl$
   b) $H_2SO_4 + NH_3$
   c) $NH_4NO_3 + NaOH$
   d) $HCl + Na_2CO_3$

12 a) What type of bond links the ammonia molecule to a proton in an ammonium ion?
   b) Draw a dot-and-cross diagram to show the bonding in an ammonium ion.
   c) Predict the shape of an ammonium ion.

Conjugate acid–base pairs

Any acid–base reaction involves competition for protons. This is illustrated by a solution of an ammonium salt, such as ammonium chloride, in water.

$$\underset{\text{acid 1}}{NH_4^+(aq)} + \underset{\text{base 2}}{H_2O(l)} \rightleftharpoons \underset{\text{base 1}}{NH_3(aq)} + \underset{\text{acid 2}}{H_3O^+(aq)}$$

In this example there is competition for protons between ammonia molecules and water molecules. On the left-hand side of the equation the protons are held by lone pairs on the ammonia molecules. On the right-hand side they are held by lone pairs on water molecules. The position of equilibrium shows which of the two bases has the stronger hold on the protons.

Chemists use the term conjugate acid–base pair to describe a pair of molecules or ions which can be converted from one to the other by the gain or loss of a proton. The equilibrium in a solution of the ammonium salt above involves two conjugate acid–base pairs:

NH_4^+ and NH_3
H_3O^+ and H_2O.

Definitions

An acid turns into its **conjugate base** when it loses a proton. A base turns into its **conjugate acid** when it gains a proton. Any pair of compounds made up of an acid and a base that can be converted from one to the other by proton transfer is a **conjugate acid–base pair**.

Test yourself

13 Identify and name the conjugate bases of these acids: HNO_3, CH_3COOH, H_2SO_4, HCO_3^-.
14 Identify and name the conjugate acids of these bases: O^{2-}, OH^-, NH_3, CO_3^{2-}, HCO_3^-, SO_4^{2-}.
15 Explain and illustrate these two statements:
a) the stronger the acid, the weaker its conjugate base
b) the stronger the base, the weaker its conjugate acid.

5.3 The pH scale

The concentration of hydrogen ions in aqueous solutions commonly ranges from 2 mol dm^{-3} to 1×10^{-14} mol dm^{-3}. The concentration of aqueous hydrogen ions in dilute hydrochloric acid is about 100 000 000 000 000 times greater than the concentration of hydrogen ions in dilute sodium hydroxide solution.

Given such a wide range of concentrations, chemists find it convenient to use a logarithmic scale to measure the concentration of aqueous hydrogen ions in acidic or alkaline solutions. This is the pH scale.

The definition of pH is:

$$pH = -\log[H^+(aq)]$$

Figure 5.9▲
A scientist measuring the pH of glacier melt water during research into air and water pollution.

pH	0	1	2	3	4	5	6	7	8	9	10	11	12	13	14
$[H^+(aq)]$/mol dm^{-3}	10^0	10^{-1}	10^{-2}	10^{-3}	10^{-4}	10^{-5}	10^{-6}	10^{-7}	10^{-8}	10^{-9}	10^{-10}	10^{-11}	10^{-12}	10^{-13}	10^{-14}

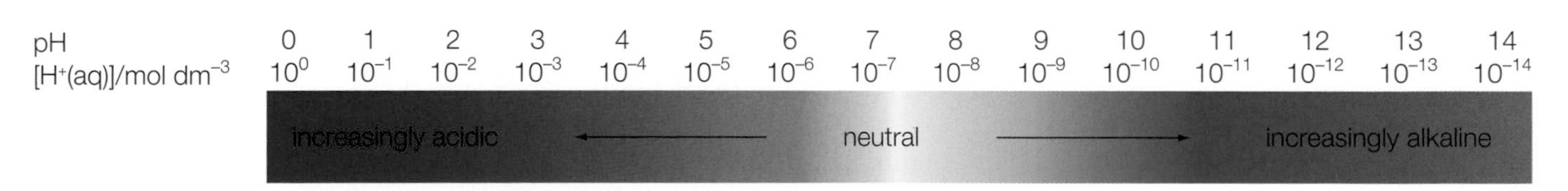

Figure 5.10▲
The pH scale showing the colours of full-range indicator at the different pH values.

Note

Strictly speaking the definition of pH is given by:

$$pH = -\log_{10}\left(\frac{[H^+(aq)]}{1\ \text{mol dm}^{-3}}\right)$$

because mathematically it is only possible to take logarithms of numbers and not of quantities with units. Dividing by a standard concentration of 1 mol dm^{-3} does not change the value but cancels the units.

Definition

Chemists use **logarithms to base 10** to handle values which range over several orders of magnitude. Logarithms to base 10 are defined such that:

$\log 10^3 = 3$
$\log 10^2 = 2$
$\log 10^1 = 1$
$\log 10^0 = \log 1 = 0$
$\log 10^{-1} = \log 0.1 = -1$
$\log 10^{-2} = \log 0.01 = -2$

In general: $\log 10^x = x$
For example: $2 = 10^{0.301}$,
so $\log_{10} 2 = 0.301$

By definition:

$\log xy = \log x + \log y$
and
$\log x^n = n \log x$

so it follows that:

$\log \frac{1}{x} = \log x^{-1} = -\log x$.

Worked examples

1 What is the pH of 0.020 mol dm^{-3} hydrochloric acid?

Notes on the method
Hydrochloric acid is a strong acid so it is fully ionised. Note that 1 mol HCl gives 1 mol $H^+(aq)$. Use the log button on your calculator. Do not forget the minus sign in the definition of pH.

Answer
$[H^+(aq)] = [HCl(aq)] = 0.020\ mol\ dm^{-3}$
$pH = -\log(0.020) = 1.70$

2 The pH of human blood is 7.40. What is the aqueous hydrogen ion concentration in blood?

Notes on the method
$pH = -\log[H^+(aq)]$
From the definition of logarithms this rearranges to $[H^+(aq)] = 10^{-pH}$
Use the inverse log button (10^x) on your calculator. Do not forget the minus sign in the definition of pH.

Answer
$pH = 7.4$
$[H^+(aq)] = 10^{-7.4} = 4.0 \times 10^{-8}\ mol\ dm^{-3}$

Test yourself

16 What is the pH of solutions of hydrochloric acid with these concentrations?
  a) 0.1 mol dm^{-3}
  b) 0.01 mol dm^{-3}
  c) 0.001 mol dm^{-3}
17 Calculate the pH of a 0.08 mol dm^{-3} solution of nitric acid.
18 What is the concentration of hydrogen ions in these solutions?
  a) orange juice with a pH of 3.3
  b) coffee with a pH of 5.4
  c) saliva with a pH of 6.7
  d) a suspension of an antacid in water with a pH of 10.5.

The ionic product of water

There are hydrogen and hydroxide ions even in pure water because of a transfer of hydrogen ions between water molecules. This only happens to a very slight extent.

$$H_2O(l) + H_2O(l) \rightleftharpoons H_3O^+(aq) + OH^-(aq)$$

Which can be written more simply as:

$$H_2O(l) \rightleftharpoons H^+(aq) + OH^-(aq)$$

The equilibrium constant $K_c = \frac{[H^+(aq)][OH^-(aq)]}{[H_2O(l)]}$

There is such a large excess of water that $[H_2O(l)]$ is a constant, so the relationship simplifies to:

$$K_w = [H^+(aq)][OH^-(aq)]$$

where K_w is the ionic product of water.

The pH of pure water at 298 K is 7. So the hydrogen ion concentration at equilibrium is:

$$[H^+(aq)] = 1 \times 10^{-7}\ mol\ dm^{-3}$$

Also, in pure water $[H^+(aq)] = [OH^-(aq)]$

So: $[OH^-(aq)] = 1 \times 10^{-7}\,mol\,dm^{-3}$

Hence $K_w = 1 \times 10^{-14}\,mol^2\,dm^{-6}$

K_w is a constant in all aqueous solutions at 298 K. This makes it possible to calculate the pH of alkalis.

Worked example

What is the pH of a $0.05\,mol\,dm^{-3}$ solution of sodium hydroxide?

Notes on the method
Sodium hydroxide is fully ionised in solution. So in this solution:
$[OH^-(aq)] = 0.05\,mol\,dm^{-3}$

$pH = -\log [H^+(aq)]$

Answer
For this solution:

$K_w = [H^+(aq)] \times 0.05\,mol\,dm^{-3} = 1 \times 10^{-14}\,mol^2\,dm^{-6}$

$$\text{So } [H^+(aq)] = \frac{1 \times 10^{-14}\,mol^2\,dm^{-6}}{0.05\,mol\,dm^{-3}}$$

$$= 2 \times 10^{-13}\,mol\,dm^{-3}$$

Hence $pH = -\log(2 \times 10^{-13}) = 12.7$

Figure 5.11▲
Tufa towers at Lake Mono, California, USA, which were first exposed when the water level in the lake dropped because water was diverted from rivers feeding the lake to Los Angeles. These pillars of calcium carbonate were created underwater when water from underground streams rich in calcium salts mixed with the lake water, which was alkaline because it was rich in hydrogencarbonates.

Working in logarithms

The logarithmic form of equilibrium constants is particularly useful for pH calculations. Taking logarithms produces a convenient small range of values.

$$K_w = [H^+(aq)][OH^-(aq)] = 10^{-14} \text{ at } 298\,K$$

Taking logarithms, and applying the rule that $\log xy = \log x + \log y$, gives:

$$\log K_w = \log [H^+(aq)] + \log [OH^-(aq)] = \log 10^{-14} = -14$$

Test yourself

19 The value of K_w varies with temperature. At 273 K its value is $1.1 \times 10^{-15}\,mol^2\,dm^{-6}$ while at 303 K it is $1.5 \times 10^{-14}\,mol^2\,dm^{-6}$.
- **a)** Is the ionisation of water an exothermic or an endothermic process?
- **b)** What happens to the hydrogen ion concentration in pure water, and hence the pH, as the temperature rises?
- **c)** Does pure water stop being neutral if its temperature is above or below 298 K?

20 Calculate the pH of these solutions:
- **a)** $1.0\,mol\,dm^{-3}$ NaOH
- **b)** $0.02\,mol\,dm^{-3}$ KOH
- **c)** $0.001\,mol\,dm^{-3}$ $Ba(OH)_2$.

Note

Do not try to remember the formula pH = 14 – pOH. Only use it if you can work it out quickly for yourself from the definition of K_w.

Multiplying through by –1 reverses the signs:

$$-\log K_w = -\log[H^+(aq)] - \log[OH^-(aq)] = 14$$

Hence: $pK_w = pH + pOH = 14$, where, by analogy with pH:

pK_w is defined as $-\log K_w$, and

pOH is defined as $-\log[OH^-(aq)]$.

So: pH = 14 – pOH, which makes it easy to calculate the pH of alkaline solutions.

Worked example

What is the pH of a 0.05 mol dm^{-3} solution of sodium hydroxide?

Note on the method
Sodium hydroxide, NaOH, is a strong base so it is fully ionised.
Find the values of logarithms with the log button of a calculator.

Answer
$[OH^-(aq)] = 0.05\ mol\ dm^{-3}$

$pOH = -\log 0.05 = 1.3$

$pH = 14 - pOH = 14 - 1.3 = 12.7$

Figure 5.12▲
Bacteria added to milk ferment the lactose sugar and turn it into lactic acid. The acid turns the milk into yogurt and also restricts the growth of food poisoning bacteria.

Test yourself

21 Explain why measuring the pH of a solution of an acid does not provide enough evidence to show whether or not the acid is strong or weak.

22 Explain why it takes the same amount of sodium hydroxide to neutralise 25 cm^3 of 0.1 mol dm^{-3} ethanoic acid as it does to neutralise 25 cm^3 of 0.1 mol dm^{-3} hydrochloric acid.

5.4 Weak acids and bases

Most organic acids and bases ionise to only a slight extent in aqueous solution. Carboxylic acids (see Section 8.1) such as ethanoic acid in vinegar, citric acid in fruit juices and lactic acid in sour milk are all weak acids. Ammonia and amines (see Section 13.1) are weak bases.

Weak acids

In a 0.1 mol dm^{-3} solution of ethanoic acid only about one in a hundred molecules ionise to produce hydrogen ions.

$$CH_3COOH(aq) \rightleftharpoons CH_3CO_2^-(aq) + H^+(aq)$$

This means that the pH of a 0.1 mol dm^{-3} solution of ethanoic acid is 2.9 and not 1, as it would be if it were a strong acid.

There is a very important distinction between acid strength and concentration. Strength is the extent of ionisation. Concentration is the amount in moles of acid in a cubic decimetre. It takes just as much sodium hydroxide to neutralise 25 cm^3 of a 0.1 mol dm^{-3} solution of a weak acid such as ethanoic acid as it does to neutralise 25 cm^3 of a 0.1 mol dm^{-3} solution of a strong acid such as hydrochloric acid.

Weak bases

Weak bases are only slightly ionised when they dissolve in water. In a 0.1 mol dm^{-3} solution of ammonia, for example, 99 in every 100 molecules do not react but remain as dissolved molecules. Only 1 molecule in a 100 reacts to form ammonium ions.

$$NH_3(aq) + H_2O(l) \rightleftharpoons NH_4^+(aq) + OH^-(aq)$$

As with weak acids, it is important to distinguish between strength and concentration.

Acid dissociation constants

Chemists use the equilibrium constant for the reversible ionisation of a weak acid as a measure of its strength. The equilibrium constant shows the extent to which acids dissociate into ions in solution.

A weak acid can be represented by the general formula HA, where A^- is the ion produced when the acid ionises.

$$HA(aq) \rightleftharpoons H^+(aq) + A^-(aq)$$

According to the equilibrium law, the equilibrium constant takes this form:

$$K_a = \frac{[H^+(aq)][A^-(aq)]}{[HA(aq)]}$$

In this context the equilibrium constant K_a is called the acid dissociation constant. Given the value for K_a it is possible to calculate the pH of a solution of a weak acid.

Test yourself

23 If a weak acid is shown as HA, what is A^- in the particular case of
- **a)** hydrogen fluoride
- **b)** methanoic acid
- **c)** phenol?

Worked example

Calculate the hydrogen ion concentration and the pH of a 0.01 mol dm^{-3} solution of propanoic acid. K_a for the acid is 1.3×10^{-5} mol dm^{-3}.

Practical guidance

Notes on the method

Two approximations simplify the calculation.

1. The first assumption is that at equilibrium $[H^+(aq)] = [A^-(aq)]$. In this example A^- is the propanoate ion $CH_3CH_2CO_2^-$. This assumption seems obvious from the equation for the ionisation of a weak acid but it ignores the hydrogen ions from the ionisation of water. Water produces far fewer hydrogen ions than most weak acids so its ionisation can usually be ignored. This assumption is acceptable so long as the pH of the acid is below 6.
2. The second assumption is that so little of the propanoic acid ionises in water that at equilibrium $[HA(aq)] \approx 0.01$ mol dm^{-3}. Here HA represents propanoic acid. This is a riskier assumption which has to be checked because in very dilute solutions the degree of ionisation may become quite large relative to the amount of acid in the solution. Chemists generally accept that this assumption is acceptable so long as less than 5% of the acid ionises.

Answer

$$CH_3CH_2COOH(aq) \rightleftharpoons H^+(aq) + CH_3CH_2CO_2^-(aq)$$

$$K_a = \frac{[H^+(aq)][CH_3CH_2CO_2^-(aq)]}{[CH_3CH_2COOH(aq)]} = \frac{[H^+(aq)]^2}{0.01\ \text{mol dm}^{-3}}$$

$$= 1.3 \times 10^{-5}\ \text{mol dm}^{-3}$$

Therefore $[H^+(aq)]^2 = 1.3 \times 10^{-7}$ mol^2 dm^{-6}

So $[H^+(aq)] = 3.6 \times 10^{-4}$ mol dm^{-3}

$$pH = -\log [H^+(aq)]$$
$$= -\log (3.6 \times 10^{-4})$$
$$= 3.4$$

Check the second assumption: in this case less than 0.0004 mol dm^{-3} of the 0.0100 mol dm^{-3} of acid (4%) has ionised. In this instance the degree of ionisation is small enough to justify the assumption that $[HA(aq)] \approx$ the concentration of un-ionised acid.

One method which can, in principle, be used to measure K_a for a weak acid is to measure the pH of a solution when the concentration of the acid is accurately known. This is not a good method for determining the size of K_a because the pH values of dilute solutions are very susceptible to contamination – for example by dissolved carbon dioxide from the air.

Test yourself

24 Calculate the pH of a 0.01 mol dm^{-3} solution of hydrogen cyanide given that $K_a = 4.9 \times 10^{-10}$ mol dm^{-3}.
25 Calculate the pH of a 0.05 mol dm^{-3} solution of ethanoic acid given that $K_a = 1.7 \times 10^{-5}$ mol dm^{-3}.
26 Calculate K_a for methanoic acid given that pH = 2.55 for a 0.050 mol dm^{-3} solution of the acid.
27 Calculate K_a for butanoic acid, C_3H_7COOH, given that pH = 3.42 for a 0.01 mol dm^{-3} solution of the acid.

Worked example

Calculate the K_a of lactic acid given that pH = 2.43 for a 0.10 mol dm^{-3} solution of the acid.

Notes on the method

The same two approximations simplify the calculation.

1 Assume that $[H^+(aq)] = [A^-(aq)]$ where $A^-(aq)$ here represents the aqueous lactate ion. Since the pH is well below 6 this is certainly justified.
2 Also assume that so little of the lactic acid ionises in water that at equilibrium $[HA(aq)] \approx 0.1$ mol dm^{-3}. Here HA represents lactic acid. This is a riskier assumption which again can be checked during the calculation.

Answer

pH = 2.43

$[H^+(aq)] = 10^{-2.43} = 3.72 \times 10^{-3}$ mol dm^{-3}

$[H^+(aq)] = [A^-(aq)] = 3.72 \times 10^{-3}$ mol dm^{-3}

In this example less than 5% of the acid is ionised (less than 0.004 out of 0.100 mol in each litre).

So $[HA(aq)] \approx 0.1$ mol dm^{-3}

Substituting in the expression for K_a:

$$K_a = \frac{[H^+(aq)][A^-(aq)]}{[HA(aq)]} = \frac{(3.72 \times 10^{-3}\ \text{mol dm}^{-3})^2}{0.1\ \text{mol dm}^{-3}}$$

$K_a = 1.38 \times 10^{-4}$ mol dm^{-3}

Working in logarithms

Chemists find it convenient to define a quantity $pK_a = -\log K_a$ when working with weak acids. This definition means that hydrocyanic acid, HCN, with a pK_a value of 9.3, is a much weaker acid than nitrous acid, HNO_2, with a pK_a value of 3.3.

Books of data tabulate pK_a values. The relationship between acid strength and pH can be expressed more simply because both are logarithmic quantities.

$$K_a = \frac{[H^+(aq)][A^-(aq)]}{[HA(aq)]}$$

The two common assumptions when using this expression in calculations are that:

$$[H^+(aq)] = [A^-(aq)]$$

$[HA(aq)] = c_A$, where c_A = the concentration of the un-ionised acid.

Substituting in the expression for K_a gives:

$$K_a = [H^+(aq)]^2 / c_A$$

Hence: $K_a \times c_A = [H^+(aq)]^2$

Taking logarithms:

$$\log (K_a \times c_A) = \log [H^+(aq)]^2$$

Applying the rules that $\log xy = \log x + \log y$ and that $\log x^n = n \log x$, gives:

$$\log K_a + \log c_A = 2 \log [H^+(aq)]$$

which on multiplying by –1 becomes:

$$-\log K_a - \log c_A = -2 \log [H^+(aq)]$$

Hence $pK_a - \log c_A = 2 \times pH$

Note

Do not try to remember this logarithmic form of the equilibrium law. The relationship is easy to use, but only apply it if you can derive it quickly from first principles as shown here. Do not forget that this form of the law has two built-in assumptions so it only applies when these assumptions are acceptable.

This shows that for a solution of a weak acid which is less than 5% ionised:

$$pH = \frac{1}{2}(pK_a - \log c_A) \quad \text{which rearranges to} \quad pK_a = 2pH + \log c_A.$$

Test yourself

28 What is the value of pK_a for methanoic acid given that $K_a = 1.6 \times 10^{-4}$ mol dm^{-3}?

29 What is the value of K_a for benzoic acid given that $pK_a = 4.2$?

30 Show that the logarithmic relationship $pK_a = 2pH + \log c_A$, gives the same answers from the data as the methods used in the worked examples on pages 65 and 66.

Activity

The acid strength of halogenated carboxylic acids

Introducing halogen atoms into the structure of carboxylic acids can have a marked effect on their acid strength. This is illustrated by the values in Table 5.1.

Acid	K_a/mol dm^{-3}	pK_a
Ethanoic acid	1.7×10^{-5}	4.8
Fluoroethanoic acid	2.2×10^{-3}	2.7
Chloroethanoic acid	1.3×10^{-3}	2.9
Iodoethanoic acid	7.6×10^{-4}	3.1
Dichloroethanoic acid	5.0×10^{-2}	1.3
Trichloroethanoic acid	2.3×10^{-1}	0.7
Butanoic acid	1.5×10^{-5}	4.8
2-chlorobutanoic acid	1.4×10^{-3}	2.8
3-chlorobutanoic acid	8.7×10^{-5}	4.0
4-bromobutanoic acid	3.0×10^{-5}	4.5

Table 5.1 ▲

1 Calculate the pH of a 0.1 mol dm^{-3} solution of:

a) butanoic acid

b) trichloroethanoic acid.

2 a) What is the pattern in the acid strength of fluoro-, chloro- and iodo-ethanoic acids when compared with the value for ethanoic acid?

b) Suggest an explanation for the pattern.

3 a) What is the pattern in the acid strength of chloro-, dichloro- and trichloro-ethanoic acids when compared with the value for ethanoic acid?

b) Is the pattern consistent with your suggested explanation in **2b)**?

4 a) What is the pattern in the acid strength of the chlorinated butanoic acids when compared with the value for butanoic acid?

b) Suggest an explanation for the pattern.

5.5 Acid–base titrations

The equilibrium law helps to explain what happens during acid–base titrations and it provides a rationale for the selection of the right indicator for a titration.

Figure 5.13▲
Chemist operating a computer controlled automatic titrator.

The pH changes during a titration as a solution of an alkali runs from a burette and mixes with an acid in a flask. Plotting pH against volume of alkali added gives a graph whose shape is determined by the nature of the acid and the base. Usually there is a marked change in pH near the equivalence point and it is this which makes it possible to detect the end-point of the titration with an indicator.

It is common to use an indicator to find the end-point during a titration. This is the point at which the colour change of the indicator shows that enough of the solution in the burette has been added to react with the amount of the chemical in the flask. In a well planned titration the colour change observed at the end-point corresponds exactly with the equivalence point.

The equivalence point is the point during any titration when the amount in moles of one reactant added from a burette is just enough to react exactly with all of the measured amount of chemical in the flask as shown by the balanced equation.

An alternative method of following the course of a titration and determining the equivalence point is to use a pH meter.

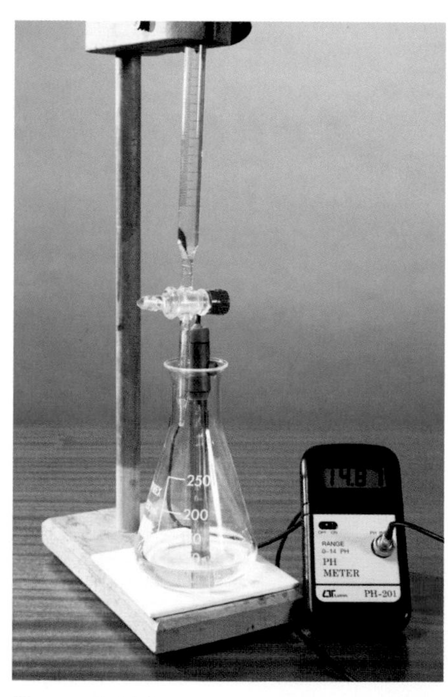

Figure 5.14▲
Apparatus for measuring the changes in pH during an acid–base titration.

Activity

Titration of a strong acid with a strong base

Strong acids and bases are fully ionised in solution. Figure 5.15 shows the shape of the pH curve for the titration of a strong acid, such as hydrochloric acid, with a strong base, such as sodium hydroxide.

1 Show that pH = 1 for a solution of 0.1 mol dm^{-3} HCl(aq).

2 Why does pH = 7 at the equivalence point of a titration of a strong acid with a strong base?

3 Calculate the pH of 25 cm^3 of a solution of sodium chloride after adding:

a) 0.05 cm^3 (1 drop) of 0.1 mol dm^{-3} HCl(aq)

b) 0.05 cm^3 (1 drop) of 0.1 mol dm^{-3} NaOH(aq).

(In both instances assume that the volume change on adding 1 drop is insignificant.)

4 Calculate the pH of the solution produced by adding 5 cm^3 of 0.1 mol dm^{-3} NaOH(aq) to 25 cm^3 of a solution of sodium chloride.

5 Show that your answers to questions **1**, **2**, **3** and **4** are consistent with Figure 5.15.

6 What features of the curve plotted in Figure 5.15 are important for the accuracy of acid–base titrations of this kind?

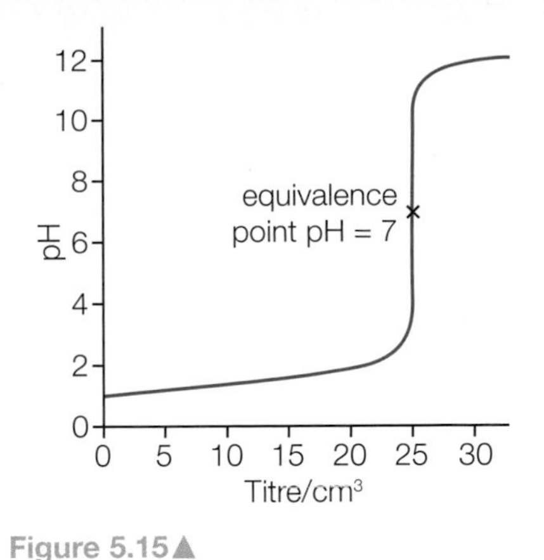

Figure 5.15▲
The pH change on adding 0.1 mol dm^{-3} NaOH(aq) from a burette to 25 cm^3 of a 0.1 mol dm^{-3} solution of HCl(aq).

Titration of a weak acid with a strong base

If the acid in the titration flask is weak, then the equilibrium law applies and the pH curve up to the equivalence point has to be calculated with the help of the expression for K_a.

Consider, for example, the reaction of ethanoic acid with sodium hydroxide during a titration. At the start the flask contains the pure acid.

$$CH_3COOH(aq) \rightleftharpoons H^+(aq) + CH_3CO_2^-(aq)$$

The pH of the pure acid can be calculated from K_a as shown in Section 5.4. However, when some strong alkali runs in from the burette, some of the ethanoic acid reacts to produce sodium ethanoate. Once this has happened $[H^+(aq)] \neq [CH_3CO_2^-(aq)]$ and the method of calculating the pH has to change to account for this.

Practical guidance

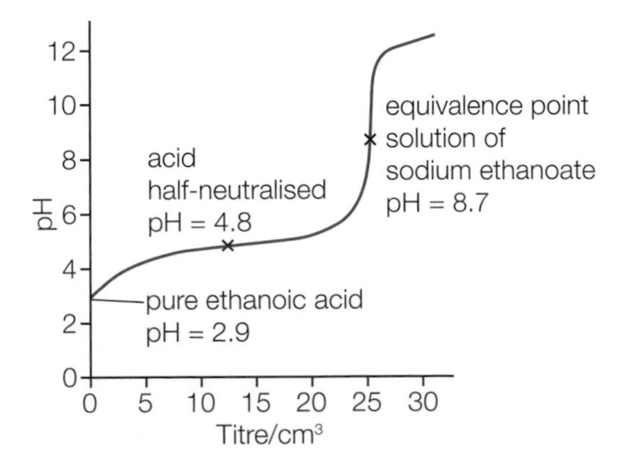

Figure 5.16◀
The pH change on adding 0.1 mol dm^{-3} NaOH(aq) from a burette to 25 cm^3 of a 0.1 mol dm^{-3} solution of CH_3COOH(aq).

Worked example

What is the pH of a mixture formed during a titration after adding 20.0 cm^3 of 0.10 mol dm^{-3} NaOH(aq) to 25.0 cm^3 of a 0.10 mol dm^{-3} solution of $CH_3COOH(aq)$ if its $K_a = 1.7 \times 10^{-5}$ mol dm^{-3}?

Notes on the method

The pH of the mixture can be estimated quite accurately using the equilibrium law by assuming that the concentration of:

- ethanoic acid molecules at equilibrium is determined by the amount of acid which has yet to be neutralised
- ethanoate ions is determined by the amount of acid converted to sodium ethanoate.

Answer

$$K_a = \frac{[H^+(aq)][CH_3CO_2^-(aq)]}{[CH_3COOH\ (aq)]}$$

This rearranges to give: $[H^+(aq)] = \frac{K_a \times [CH_3COOH(aq)]}{[CH_3CO_2^-(aq)]}$

The total volume of the solution = 45.0 cm^3.

5.0 cm^3 of the 0.1 mol dm^{-3} ethanoic acid remains not neutralised. This is now diluted to a total volume of 45 cm^3 solution.

Concentration of ethanoic acid molecules $= \frac{5.0\ cm^3}{45.0\ cm^3} \times 0.10$ mol dm^{-3}

Also the concentration of ethanoate ions $= \frac{20.0\ cm^3}{45.0\ cm^3} \times 0.10$ mol dm^{-3}

So the ratio $\frac{[CH_3COOH(aq)]}{[CH_3CO_2^-(aq)]} = \frac{5.0}{20.0}$

Substituting in the rearranged expression for the equilibrium law gives:

$$[H^+(aq)] = K_a \times \frac{5.0}{20.0} = 1.7 \times 10^{-5}\ mol\ dm^{-3} \times \frac{5.0}{20.0}$$

$[H^+(aq)] = 4.25 \times 10^{-6}$ mol dm^{-3}

$pH = -\log[H^+(aq)] = -\log(4.25 \times 10^{-6}) = 5.4$

Note that halfway to the equivalence point in Figure 5.16, the added alkali converts half of the weak acid to its salt. In this example, at this point:

$$[CH_3COOH(aq)] = [CH_3CO_2^-(aq)]$$

So: $[H^+(aq)] = K_a \times \frac{[CH_3COOH(aq)]}{[CH_3CO_2^-(aq)]} = K_a$

Hence $pH = pK_a$ halfway to the equivalence point.

At the equivalence point itself the solution contains sodium ethanoate. As Figure 5.16 shows, the solution at this point is not neutral. A solution of a salt of a weak acid and a strong base is alkaline (see Section 5.7). Sodium ions have no effect on the pH of a solution, but ethanoate ions are basic. The ethanoate ion is the conjugate base of a weak acid.

Beyond the equivalence point the curve is determined by the excess of strong base and so the shape is the same as for Figure 5.15.

Test yourself

31 Calculate the pH of a 0.1 mol dm^{-3} solution of $CH_3COOH(aq)$.

32 Calculate the pH of a mixture formed during a titration after adding 10.0 cm^3 of 0.10 mol dm^{-3} NaOH(aq) to 25.0 cm^3 of a 0.10 mol dm^{-3} solution of $CH_3COOH(aq)$.

33 Explain why a solution of sodium ethanoate is alkaline.

34 Why is the equivalence point reached at 25 cm^3 in both the titrations illustrated in Figures 5.15 and 5.16?

Titration of a strong acid with a weak base

During the titration of a strong acid with a weak base, the flask contains a strong acid at the start and the titration curve follows the same line as in Figure 5.15. In a titration of hydrochloric acid with ammonia solution, for example, the salt formed at the equivalence point is ammonium chloride.

Since ammonia is a weak base, the ammonium ion is an acid. So a solution of ammonium chloride is acidic and the pH is below 7 at the equivalence point. After the equivalence point (Figure 5.17) the curve rises less far than in Figure 5.15 because the excess alkali is a weak base and is not fully ionised.

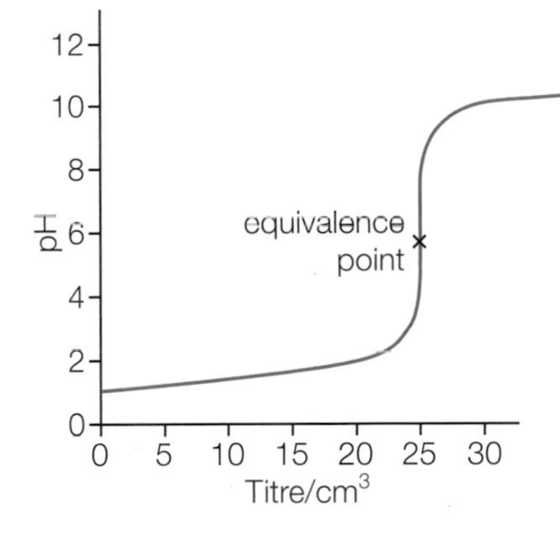

Figure 5.17 The pH change on adding a 0.1 mol dm^{-3} solution of the weak base $NH_3(aq)$ from a burette to 25 cm^3 of a 0.1 mol dm^{-3} solution of the strong acid HCl(aq).

Test yourself

35 Explain why a solution of ammonium chloride is acidic.

36 Write a balanced equation for the neutralisation of ethanoic acid by ammonia solution.

Titration of a weak acid with a weak base

In practice it is not usual to titrate a weak acid with a weak base. As shown in Figure 5.18, the change of pH around the equivalence point is gradual and not very marked if both the acid and base are weak. This means that it is hard to fix the end-point precisely. If the dissociation constants for the weak acid and for the weak base are approximately equal (as is the case for ethanoic acid and ammonia) then the salt formed at the equivalence point is neutral and pH = 7 at this point.

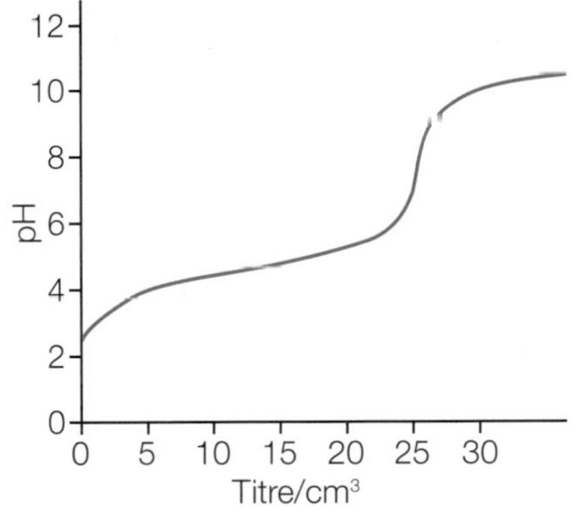

Figure 5.18 The pH change on adding 0.1 mol dm^{-3} $NH_3(aq)$ from a burette to 25 cm^3 of a 0.1 mol dm^{-3} solution of $CH_3COOH(aq)$. Before the end-point the curve is essentially the same as in Figure 5.16, while after the end-point it is as in Figure 5.17. Note the resulting small change of pH around the equivalence point.

Working with logarithms

There can be an advantage to working with a logarithmic form of the equilibrium law when calculating the pH of a mixture of a weak acid and one of its salts during titrations.

In general, for a weak acid HA:

$$HA(aq) \rightleftharpoons H^+(aq) + A^-(aq)$$

$$K_a = \frac{[H^+(aq)][A^-(aq)]}{[HA(aq)]}$$

This rearranges to give $[H^+(aq)] = K_a \times \dfrac{[HA(aq)]}{[A^-(aq)]}$

Taking logs and substituting pH for $-\log\,[H^+(aq)]$ and pK_a for $-\log K_a$, gives:

$$pH = pK_a + \log\left(\frac{[A^-(aq)]}{[HA(aq)]}\right)$$

Note the change of sign and the inversion of the log ratio. This follows because:

$$-\log\left(\frac{[A^-(aq)]}{[HA(aq)]}\right) = +\log\left(\frac{[HA(aq)]}{[A^-(aq)]}\right)$$

In a mixture of a weak acid and one of its salts, the weak acid is only slightly ionised, while the salt is fully ionised, so it is often accurate enough to make the assumption that all the negative ions come from the salt present and all the un-ionised molecules from the acid.

Hence: $$\text{pH} = \text{p}K_a + \log\frac{[\text{salt}]}{[\text{acid}]}$$

This form of the equilibrium law cannot be used to calculate the pH of a solution of a weak acid on its own. However, it can help to explain the properties of acid–base indicators (Section 5.6) and to account for the behaviour of buffer solutions (Section 5.8).

Note

If you want to use the logarithmic form of the equilibrium law, make sure that you can derive it for yourself. Also check that you understand the assumptions made when deriving this form of the law so that you know when it applies.

5.6 Indicators

Acid–base indicators change colour when the pH changes. They signal the end-point of a titration. No one indicator is right for all titrations and the equilibrium law can help chemists to choose the indicator that will give accurate results.

The indicator chosen for a titration must change colour completely in the pH range of the near vertical part of the pH curve (see Figures 5.15, 5.16 and 5.17). This is essential if the visible end-point is to correspond to the equivalence point when exactly equal amounts of acid and base are mixed.

Table 5.2 gives some data for four common indicators. Note that each indicator changes colour over a range of pH values, which differs from one indicator to the next.

Indicator	pK_a	Colour change HIn/In$^-$	pH range over which the colour change occurs
Methyl orange	3.6	Red/yellow	3.2–4.2
Methyl red	5.0	Yellow/red	4.2–6.3
Bromothymol blue	7.1	Yellow/blue	6.0–7.6
Phenolphthalein	9.4	Colourless/red	8.2–10.0

Table 5.2▲

Figure 5.19▲
The flowers of this variety of hydrangea respond to soil pH. In acidic soils they produce blue flowers; elsewhere the flowers are pink.

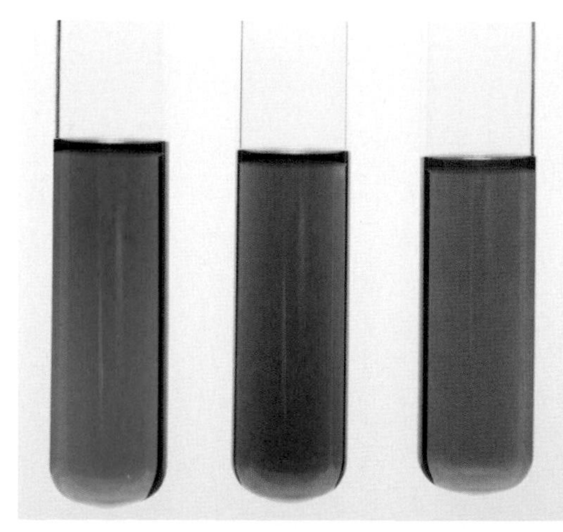

Figure 5.20▲
The colours of screened methyl orange indicator at pH 6 (left), pH 4 (middle) and pH 2 (right). This indicator includes a green dye to make the colour change easier to see.

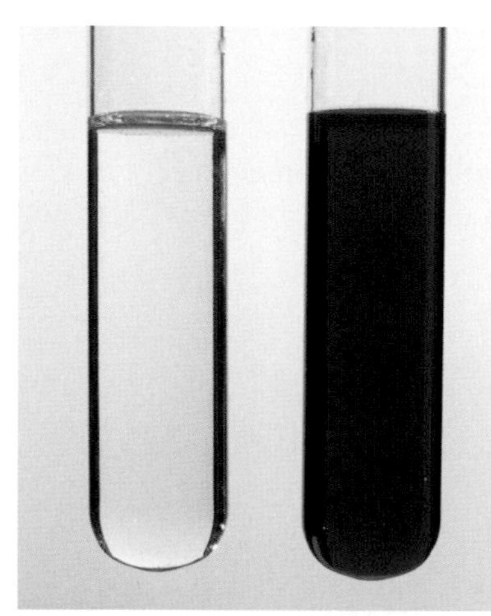

Figure 5.21▲
The colours of phenolphthalein indicator at pH 7 (left) and pH 11 (right).

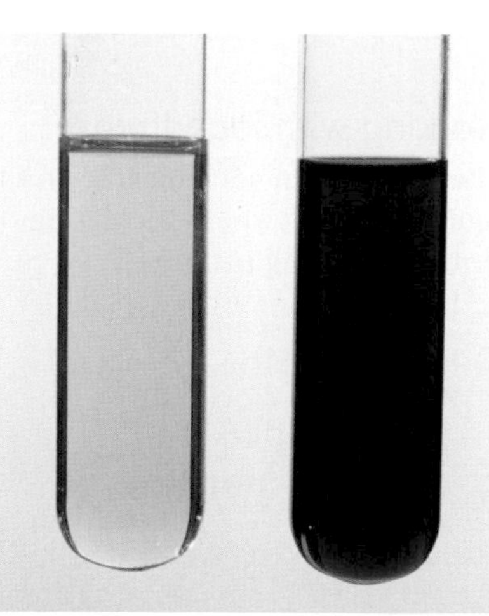

Figure 5.22▲
The colours of bromothymol blue indicator at pH 5 (left) and pH 8 (right).

Test yourself

37 Explain, qualitatively, with the help of le Chatelier's principle, why methyl red is yellow when the pH = 3 but is red at pH = 7.

38 a) Why is methyl orange an unsuitable indicator for the titration illustrated by Figure 5.16?

b) Why is phenolphthalein an unsuitable indicator for the titration illustrated by Figure 5.17?

c) Identify the indicators that can be used to detect the equivalence points of the titrations illustrated by Figures 5.15, 5.16 and 5.17.

d) Explain why it is not possible to use an indicator to give a sharp and accurate end-point for the titration illustrated by Figure 5.18.

Indicators are themselves weak acids or bases which change colour when they lose or gain protons. When added to a solution, an indicator gains or loses protons depending on the pH of the solution. It is conventional to represent a weak acid indicator as HIn, where In is a shorthand for the rest of molecule other than the ionisable hydrogen atom. In water:

$$\underset{\substack{\text{un-ionised indicator}\\ \text{colour 1}}}{HIn(aq)} \rightleftharpoons H^+(aq) + \underset{\substack{\text{indicator after losing a proton}\\ \text{colour 2}}}{In^-(aq)}$$

Note that an analyst only adds a drop or two of indicator during a titration. This means that there is so little indicator that it cannot affect the pH of the mixture. The pH is determined by the titration (as shown in Figures 5.15, 5.16 and 5.17). The position of the equilibrium for the ionisation of the indicator shifts one way or the other as dictated by the pH of the solution in the titration flask.

The pH range over which an indicator changes colour is determined by its strength as the acid. Typically the range is given roughly by $pK_a \pm 1$. The logarithmic form of the equilibrium law derived at the end of Section 5.5 shows why this is so. For an indicator it takes this form:

$$pH = pK_a + \log\left(\frac{[In^-(aq)]}{[HIn(aq)]}\right)$$

When $pH = pK_a$, $[HIn(aq)] = [In^-(aq)]$ and the two different colours of the indicator are present in equal amounts. The indicator is mid-way through its colour change.

Add a few drops of acid and the pH falls. If the two colours of the indicator are equally intense, it turns out that our eyes see the characteristic acid colour of the indicator clearly when $[HIn] = 10 \times [In^-(aq)]$.

At this point $pH = pK_a + \log 0.1 = pK_a - 1$, since $\log 0.1 = -1$

Add a few drops of alkali and the pH rises. Similarly, our eyes see the characteristic alkaline colour of the indicator clearly when $[In^-(aq)] = 10 \times [HIn(aq)]$.

At this point: $pH = pK_a + \log 10 = pK_a + 1$, since $\log 10 = +1$.

Test yourself

39 Suggest an explanation for the fact that the indicators shown in Table 5.2 do not all change colour over a pH range of 2 units.

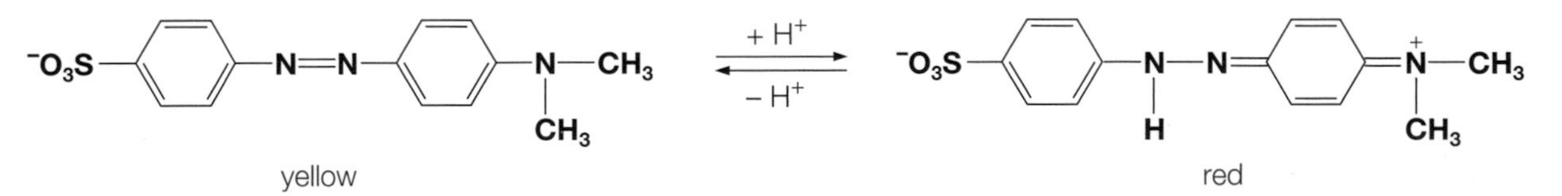

Figure 5.23▲
The structures of methyl orange in acid (right) and alkaline solutions (left). In acid solution the added hydrogen ion (proton) localises two electrons to form a covalent bond. In alkaline solution the removal of the hydrogen ion allows the two electrons to join the other delocalised electrons (see Section 12.3). The change in the number of delocalised electrons causes a shift in the peak of the wavelengths of light absorbed, so the colour changes and the molecule acts as an indicator.

Figure 5.24▲
Limestone etched by rainwater made acidic by dissolved carbon dioxide. Bloody cranes-bill is a plant that can only flourish when growing in soil where the minerals, such as limestone, neutralise acids and keep the soil water alkaline.

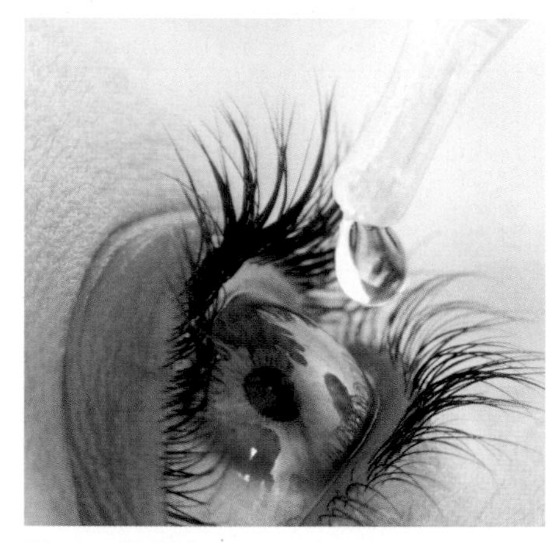

Figure 5.25▲
Eye drops contain a buffer solution to make sure that they do not irritate the sensitive surface of the eye.

Figure 5.26▲
Many shampoos, for animals as well as people, contain a buffer solution. They are marketed as 'pH balanced' shampoos.

5.7 Neutralisation reactions

Chemists use the term neutralisation to describe any reaction in which an acid reacts with a base to form a salt, even when the pH does not equal 7 on mixing equivalent amounts of the acid and the alkali.

The pH of salts

Mixing equal amounts (in moles) of hydrochloric acid with sodium hydroxide produces a neutral solution of sodium chloride. Strong acids, such as hydrochloric acid, and strong bases, such as sodium hydroxide, are fully ionised in solution. So is the salt formed from the reaction of hydrochloric acid and sodium hydroxide, sodium chloride. Writing ionic equations for these examples shows that neutralisation is essentially a reaction between aqueous hydrogen ions and hydroxide ions. This is supported by the values for enthalpies of neutralisation for different acids and bases.

$$H^+(aq) + OH^-(aq) \rightleftharpoons H_2O(l)$$

The surprise is that 'neutralisation reactions' do not always produce neutral solutions. 'Neutralising' a weak acid, such as ethanoic acid, with an equal amount, in moles, of a strong base, such as sodium hydroxide, produces a solution of sodium ethanoate which is alkaline (see Figure 5.16).

'Neutralising' a weak base, such as ammonia, with an equal amount of the strong acid hydrochloric acid produces a solution of ammonium chloride which is acidic (see Figure 5.17).

Where a salt has either a 'parent acid' or a 'parent base' which is weak, it dissolves to give a solution which is not neutral. The 'strong parent' in the partnership 'wins':

- weak acid/strong base – the salt is alkaline in solution
- strong acid/weak base – the salt is acidic in solution.

Test yourself

40 Predict, with the help of a table of K_a or pK_a values, whether the salt formed on mixing equivalent amounts of these acids and alkalis gives a solution with pH = 7, pH above 7 or pH below 7:
  a) nitric acid and potassium hydroxide
  b) chloric(I) acid and sodium hydroxide
  c) hydrobromic acid and ammonia
  d) propanoic acid and sodium hydroxide.

5.8 Buffer solutions

Buffer solutions are mixtures of molecules and ions in solution which help to keep the pH more or less constant. Buffer solutions help to stabilise the pH of blood, medicines, shampoos, swimming pools and of many other solutions in living things, domestic products and in the environment.

Buffer mixtures are important in the food industry because many properties of drinks and foods depend upon their being formulated to the proper pH. The quality of food can be affected significantly if the pH shifts too far from the required value. Maintaining a low pH can help to prevent the growth of bacteria or fungi which spoil food or give rise to food poisoning.

Buffers are also important in living organisms. The pH of blood, for example, is closely controlled by buffers within the narrow range 7.35 to 7.45. Chemists use buffers when they want to investigate chemical reactions at a fixed pH.

A buffer solution cannot prevent pH changes but it evens out the large swings in pH that can happen without a buffer.

Buffers are equilibrium systems which illustrate the practical importance of the equilibrium law. A typical buffer mixture consists of a solution of a weak acid and one of its salts; for example, a mixture of ethanoic acid and sodium ethanoate. There must be plenty of both the acid and its salt.

$CH_3COOH(aq)$ acid molecules are a reservoir of H^+ ions	+ $H_2O(l)$ $\rightleftharpoons$	$CH_3CO_2^-(aq)$ base ions – with the capacity to accept H^+ ions	+ $H_3O^+(aq)$
plenty of weak acid to supply more H^+ ions if alkali is added		plenty of the ions from the salt able to combine with H^+ ions if acid is added	stays roughly constant so the pH hardly changes

Figure 5.27▲
The action of a buffer solution.

Le Chatelier's principle provides a qualitative explanation of the buffering action. Adding a little strong acid temporarily increases the concentration of $H^+(aq)$ so the equilibrium shifts to the left to counteract the change. Adding a little strong alkali temporarily decreases the concentration of $H^+(aq)$ so the equilibrium shifts to the right to counteract the change.

The pH of buffer solutions

By choosing the right weak acid, it is possible to prepare buffers at any pH value throughout the pH scale. If the concentrations of the weak acid and its salt are the same, then the pH of the buffer is equal to pK_a for the acid. The pH of a buffer mixture can be calculated with the help of the equilibrium law.

$$K_a = \frac{[H^+(aq)][A^-(aq)]}{[HA(aq)]}$$

This rearranges to give: $[H^+(aq)] = K_a \times \frac{[HA(aq)]}{[A^-(aq)]}$

So the equilibrium law makes it possible to calculate the pH of a buffer solution made from a mixture of a weak acid and its conjugate base.

In a mixture of a weak acid and its salt, the weak acid is only slightly ionised while the salt is fully ionised. This means that it is often accurate enough to assume that:

- all the molecules HA come from the added acid
- all the negative ions, $A^-(aq)$, come from the added salt.

So the calculation of the hydrogen ion concentration of a buffer solution can be based on the formula:

$$[H^+(aq)] = K_a \times \frac{[acid]}{[salt]}$$

Alternatively, you can use the logarithmic form of this relationship (see Section 5.5).

Diluting a buffer solution with water does not change the ratio of the concentrations of the salt and acid, so the pH does not change, unless the dilution is so great that the assumptions used to arrive at this formula break down.

Buffer solutions form during the titration of a weak acid with a strong base. This is illustrated by Figure 5.28. Once some of the base has been added the pH does not change by much until the titration begins to approach the end-point.

Test yourself

41 Show that the buffer solution in Figure 5.27 consists of an conjugate acid–base pair.

42 Explain why a weak acid on its own cannot make a buffer solution, but a mixture of a weak acid and one of its salts can.

Note

The theory of acid–base indicators and the theory of buffer solutions is essentially the same. The difference is that a large amount of buffer mixture is added to dictate the pH of a solution, whereas the drop or two of an indicator in a titration flask is too little to affect the pH. An indicator follows the pH changes dictated by the mixture of acid and alkali during the titration.

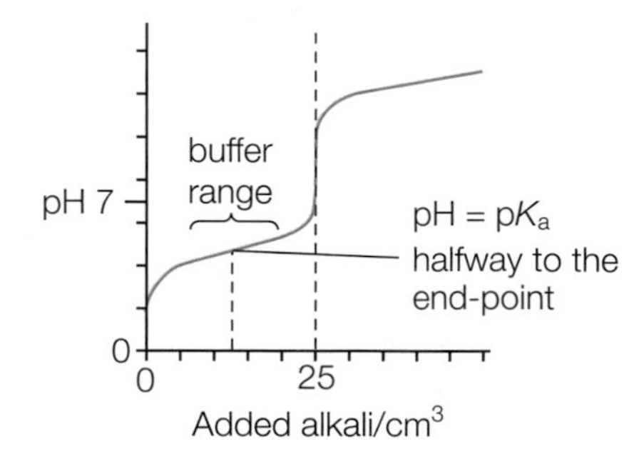

Figure 5.28▲
In the buffering range, the pH changes little on adding substantial volumes of strong alkali. Over this range the flask contains significant amounts of both the acid and the salt formed from the acid.

Test yourself

43 Calculate the pH of these buffer mixtures.

a) A solution containing equal amounts in moles of $H_2PO_4^-(aq)$ and $HPO_4^{2-}(aq)$. K_a for the dihydrogenphosphate(V) ion is 6.3×10^{-7} mol dm^{-3}.

b) A solution containing 12.2 g benzoic acid (C_6H_5COOH) and 7.2 g of sodium benzoate in 250 cm^3 solution. K_a for benzoic acid is 6.3×10^{-5} mol dm^{-3}.

c) A solution containing 12.2 g benzoic acid (C_6H_5COOH) and 7.2 g of sodium benzoate in 1000 cm^3 solution.

44 What must be the ratio of the concentrations of the ethanoate ions and ethanoic acid molecules in a buffer solution with pH = 5.4 if $K_a = 1.7 \times 10^{-5}$ mol dm^{-3} for ethanoic acid?

Worked example

What is the pH of a buffer solution containing 0.40 mol dm^{-3} methanoic acid and 1.00 mol dm^{-3} sodium methanoate?

Notes on the method

Look up the value of K_a in a table of data. K_a of methanoic acid is 1.6×10^{-4} mol dm^{-3}.

Make the assumptions that all the molecules of methanoic acid come from the added acid, and that all the methanoate ions come from the added salt.

Answer

From the information in the question:

$$[\text{acid}] = 0.40 \text{ mol dm}^{-3}$$

$$[\text{salt}] = 1.00 \text{ mol dm}^{-3}$$

Substituting in the formula gives:

$$[H^+(aq)] = K_a \times \frac{[\text{acid}]}{[\text{salt}]}$$

$$= 1.6 \times 10^{-4} \text{ mol dm}^{-3} \times \frac{0.40 \text{ mol dm}^{-3}}{1.00 \text{ mol dm}^{-3}}$$

$$[H^+(aq)] = 6.4 \times 10^{-5} \text{ mol dm}^{-3}$$

$$pH = -\log[H^+(aq)] = -\log[6.4 \times 10^{-5}] = 4.2$$

Data

Tutorial

Activity

Blood buffers

In a healthy person the pH of blood lies within a narrow range (7.35–7.45). Chemical reactions in cells tend to upset the normal pH. Respiration, for example, produces carbon dioxide all the time. The carbon dioxide diffuses into the blood where it is mainly in the form of carbonic acid. However, the blood pH stays constant because it is stabilised by buffer solutions, in particular by the buffer system based on the equilibrium between carbon dioxide, water and hydrogencarbonate ions. This is the carbonic acid–hydrogencarbonate buffer.

$$CO_2(g) + H_2O(l) \rightleftharpoons H^+(aq) + HCO_3^-(aq)$$

Proteins in blood, including haemoglobin, can also contribute to the buffering action of blood pH. This is because the molecules contain both acidic and basic functional groups (see Section 14.3).

Two major organs help to control the total amounts of carbonic acid and hydrogencarbonate ions in the blood. The lungs remove excess carbon dioxide from the blood and the kidneys remove excess hydrogencarbonate ions.

The brain responds to the level of carbon dioxide in the blood. During exercise, for example, the brain speeds up the rate of breathing.

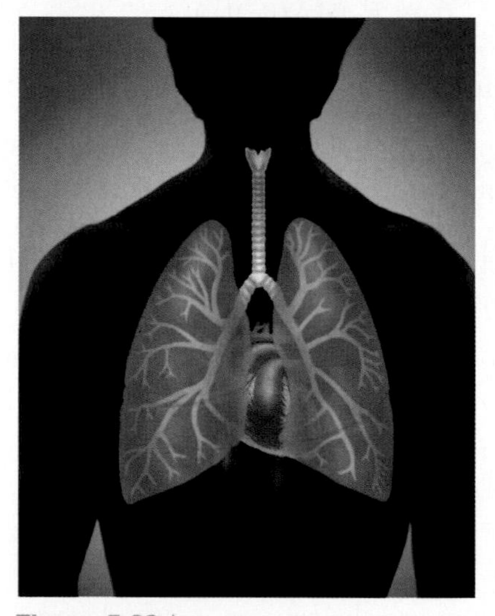

Figure 5.29▲ The lungs have a vital part to play in maintaining the pH of the blood.

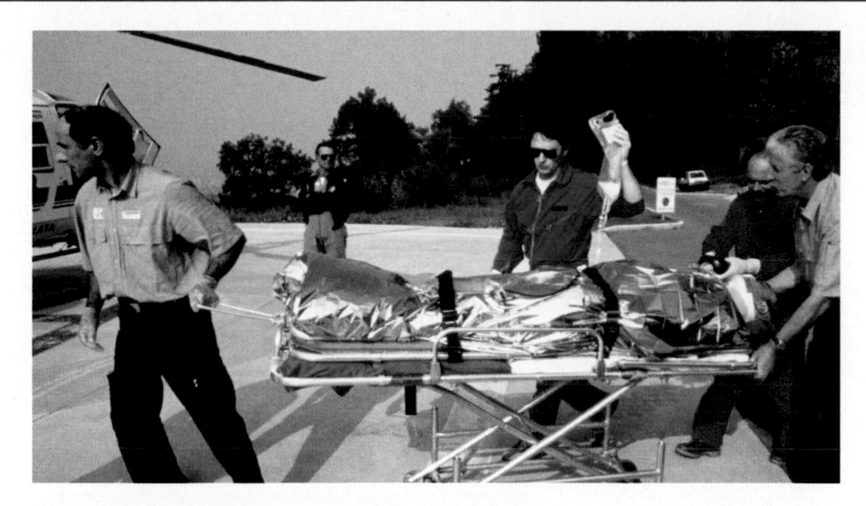

Figure 5.30 ◀ Paramedics transferring a patient to a helicopter ambulance. One of the paramedics is holding an intravenous drip bag.

The consequences can be fatal if the blood pH moves outside the normal range. Patients who have been badly burned or suffered other serious injuries are treated quickly with a drip into a vein. One of the purposes of an intravenous drip is to help maintain the pH of the blood close to its normal value.

1 Write an equation to show aqueous carbon dioxide reacting with water to form hydrogen ions and hydrogencarbonate ions.

2 Explain why breathing faster and more deeply tends to raise the blood pH.

3 a) Suggest two reasons why the blood pH tends to fall during strenuous exercise.

b) Why do people breathe faster and more deeply when running?

4 a) Write the expression for K_a for the equilibrium between carbon dioxide, water, hydrogen ions and hydrogencarbonate ions

b) In a sample of blood, the concentration of hydrogencarbonate ions = 2.5×10^{-2} mol dm^{-3}. The concentration of aqueous carbon dioxide = 1.25×10^{-3} mol dm^{-3}. The value of $K_a = 4.5 \times 10^{-7}$ mol dm^{-3}. Use this information to calculate:

i) the hydrogen ion concentration in the blood

ii) the pH of the blood.

c) What can you conclude about the person who gave the blood sample?

5 Explain why a mixture of carbon dioxide, water and hydrogencarbonate ions can act as a buffer solution.

6 Why are blood buffers on their own unable to maintain the correct blood pH for any length of time?

7 Suggest reasons why people may need treatment to adjust their blood pH if they have been rescued after breathing thick smoke during a fire.

REVIEW QUESTIONS

Extension questions

1 a) Write an equation for the reaction which occurs when a weak acid HX is added to water. (1)

b) Write an expression for the acid dissociation constant of the weak acid HX. (1)

c) The ionisation of HX in aqueous solution is endothermic. Predict the effect, if any, of:

i) an increase in temperature on the value of the acid dissociation constant for HX (1)

ii) an increase in temperature on the pH of an aqueous solution of the weak acid HX (1)

iii) a decrease in concentration of the acid HX on the value of its acid dissociation constant. (1)

2 a) Describe and explain the use of buffer solutions with the help of examples. (6)

b) i) What is the pH of a buffer solution in which the concentration of ethanoic acid is 0.080 mol dm^{-3} and the concentration of sodium ethanoate is 0.040 mol dm^{-3}. K_a for ethanoic acid is 1.7×10^{-5} mol dm^{-3}. (2)

ii) Calculate the new pH value if 0.020 mol of NaOH is dissolved in 1 dm^3 of the buffer solution in i). (2)

iii) Calculate the pH of a solution of 0.020 mol of NaOH in 1 dm^3 of water. (2)

iv) Comment on the effectiveness of the buffer solution. (1)

3 a) Give examples to explain the difference between a strong acid and a weak acid. (4)

b) At 298 K, what is the pH of:

i) 0.01 mol dm^{-3} HNO_3(aq) (1)

ii) 0.01 mol dm^{-3} KOH(aq)? (2)

c) Butanoic acid, C_3H_7COOH, has an acid dissociation constant, K_a, of 1.5×10^{-5} mol dm^{-3} at 298 K.

i) Calculate the pH of a 0.020 mol dm^{-3} solution of butanoic acid at this temperature. (2)

ii) Draw a sketch graph to show the change of pH when 50 cm^3 0.020 mol dm^{-3} NaOH(aq) is added to 25 cm^3 of 0.020 mol dm^{-3} butanoic acid. (4)

iii) Choose from the table the indicator that would be most suitable for detecting the end-point of a titration between butanoic acid and sodium hydroxide. Give your reasons. (2)

Indicator	Colour change acid/alkaline	pH range over which colour change occurs
Thymol blue	Red/yellow	1.2–2.8
Congo red	Violet/red	3.0–5.0
Thymolphthalein	Colourless/blue	8.3–10.6

4 For a solution containing 0.050 mol dm^{-3} chloric(I) acid (HClO) and 0.050 mol dm^{-3} sodium chlorate(I), the pH = 7.43 at 298 K.

a) i) Write an equation for the ionisation of chloric(I) acid. (1)

ii) Write an expression for the acid dissociation constant of chloric(I) acid. (1)

b) Work out the values of K_a and pK_a for chloric(I) acid showing your working. (5)

6 Stereochemistry

Stereochemistry is the study of molecular shapes and the effect of shape on chemical properties. The study of isomerism led chemists to start thinking about molecules in three dimensions. This was an essential step in the development of their understanding of chemical reactions in living things. Smell and taste seem to depend on molecular shape. For example, in one form the hydrocarbon limonene smells of oranges while in another form with a different shape it smells of lemons. Molecular shape can also subtly alter the physiological effects of drugs.

6.1 Isomerism

If two molecules have the same molecular formula, but a different arrangement of their atoms, they are isomers. These isomers are distinct compounds with different physical properties and, in most cases, different chemical properties. Isomers and isomerism occur most commonly with carbon compounds because of the way in which carbon atoms can form chains and rings.

There are two ways in which the atoms can be rearranged to give isomers.

- The atoms are joined together in a different order forming different structures. This is called structural isomerism (Figure 6.2).
- The atoms are joined together in the same order, but they occupy different positions in space. This is called stereoisomerism.

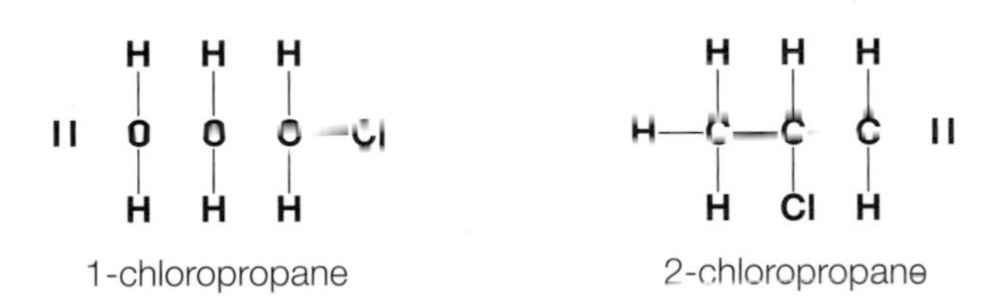

Figure 6.2▲
An example of structural isomerism

Figure 6.3 shows how the two different types of isomerism are further divided. Structural isomerism can be divided into three different types: chain, position and functional group isomerism.

There are two different types of stereoisomerism: *E–Z* (or *cis–trans*) isomerism and optical isomerism. In both these forms of stereoisomerism, the stereoisomers have the same molecular formula and the same structural formula, but different three-dimensional shapes in which their atoms occupy different positions in space.

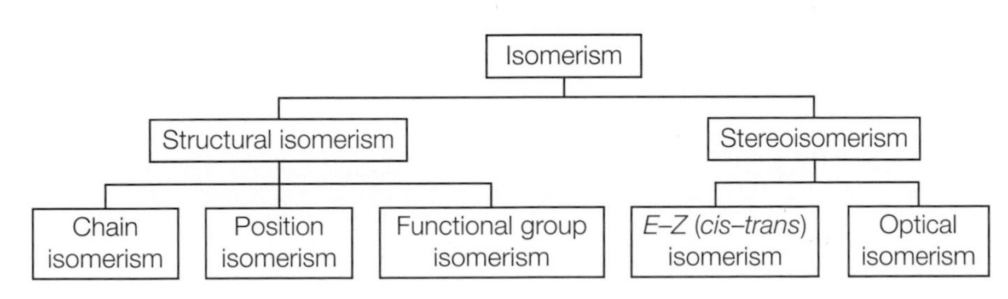

Figure 6.3▲
A 'family tree' showing the relationship between different forms of isomerism.

Figure 6.1▲
a) Caraway seeds and **b)** spearmint leaves. The compound carvone is largely responsible for the different tastes and smells of caraway and spearmint. There are two forms of carvone molecules which have the same formula and structure but subtly different shapes. Chemists describe them as optical isomers.

Test yourself

1. Draw the structures and name the chain isomers of 2,2-dimethylbutane.
2. Draw the structures and name the two position isomers of 1-bromopentane.
3. Draw the structures and name two functional group isomers with the molecular formula C_4H_8O.
4. There are three isomers with the formula C_5H_{12}. Their boiling temperatures are: 10 °C, 28 °C and 36 °C. Draw the structures of the three compounds and match the structures with the boiling temperatures.

Figure 6.4▲
Silk moths (male right and female lower) with eggs on a cocoon.

Note

Z comes from the German word *zusammen* meaning together. *E* comes from the word *entgegen*.

Note

In many cases a compound classed as *trans* in one system is *E* in the other and a compound classed as *cis* is *Z* in the other, but there are examples where this is **not** the case.

6.2 *E–Z* isomerism

The traditional system for naming the isomers of alkenes, in which the same groups are arranged differently, is to name them as *cis* or *trans*. This is illustrated by the sex attractant, bombycol, secreted by female silk moths (Figures 6.4 and 6.5). This messenger molecule strongly attracts male moths of the same species. Analysis shows that two double bonds help to determine the shape of the compound. Chemists are interested in pheromones because they offer an alternative to pesticides for controlling damaging insects. By baiting insect traps with sex attractants it is possible to capture large numbers of insects before they mate.

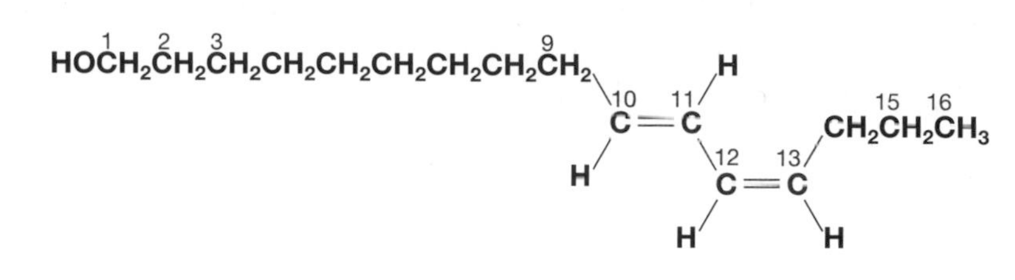

Figure 6.5▲
The sex attractant bombycol which is hexadeca-*trans*,10-*cis*,12-dien-1-ol

However, there are many examples where the *cis–trans* system is not easily applied and the *E–Z* system was developed to name these more complex molecules. The advantage of the *E–Z* system is that it always works, whereas the *cis–trans* system can break down in some cases.

Follow these steps in applying the *E–Z* naming system:

1. Look at the atoms bonded to the first carbon atom in the C=C bond. The atom with the higher atomic number has the higher priority.
2. If two atoms with the same atomic number, but in different groups, are attached to the first carbon atom, then the next bonded atom is taken into account. Thus, CH_3CH_2– has precedence over CH_3–.
3. Similarly, identify the group with the higher priority of the two attached to the second carbon atom in the C=C bond.
4. If the two groups of higher priority are on the same side of the double bond, the isomer is designated *Z*- but if the two groups of highest priority are on opposite sides of the double bond, the isomer is designated *E*- (Figure 6.6).

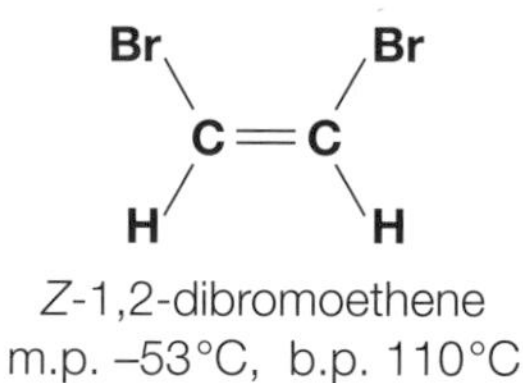

Z-1,2-dibromoethene
m.p. –53°C, b.p. 110°C

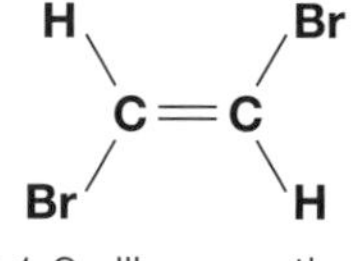

E-1,2-dibromoethene
m.p. –9°C, b.p. 108°C

Figure 6.6▲
The *E* and *Z* isomers of 1,2-dibromoethene are distinct compounds with different melting temperatures and different boiling temperatures. The relative atomic masses of bromine atoms are higher than the relative atomic masses of hydrogen atoms so they have the higher priority.

Test yourself

5. Name these compounds using **i)** the *cis–trans* system **ii)** the *E–Z* system:
   a) pent-2-ene
   b) 2-bromobut-2-ene
   c) 1-chloro-2-methylbut-1-ene.
6. Draw the skeletal formula of:
   a) (1*E*,4*Z*)-1,5-dichlorohexa-1,4-diene
   b) (*E*)-3-methyl-4-propyloct-3-ene.

6.2 Chirality and optical isomerism

Mirror image molecules

Every molecule has a mirror image. Generally, the mirror image of a molecule can be turned around to show that it is identical to the original molecule. Sometimes, however, a molecule and its mirror image are not quite the same. The molecule and its mirror image cannot be superimposed.

A molecule is chiral if, like one of your hands, it cannot be superimposed on its mirror image. The word 'chiral' (pronounced kiral) comes from the Greek for 'hand'.

Chiral molecules are asymmetric. This means that they have mirror image forms which are not identical. The commonest chiral compounds are organic molecules in which there is a carbon atom attached to four different atoms or groups.

Figure 6.7 shows the two structures of lactic acid, which forms when milk turns sour. The two molecules each have the same four atoms or groups attached to their central carbon atom: a CH_3– group, an –OH group, a –COOH group and a H– atom. However, it is impossible to superimpose the mirror images of lactic acid. No matter how the molecules are rotated, it is not possible to get the two to look identical with groups and atoms in the same positions in space.

Note

Chemists have conventions for drawing three-dimensional molecules on paper.

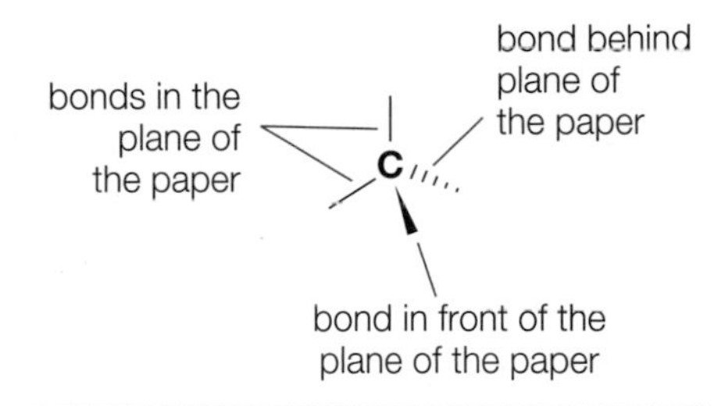

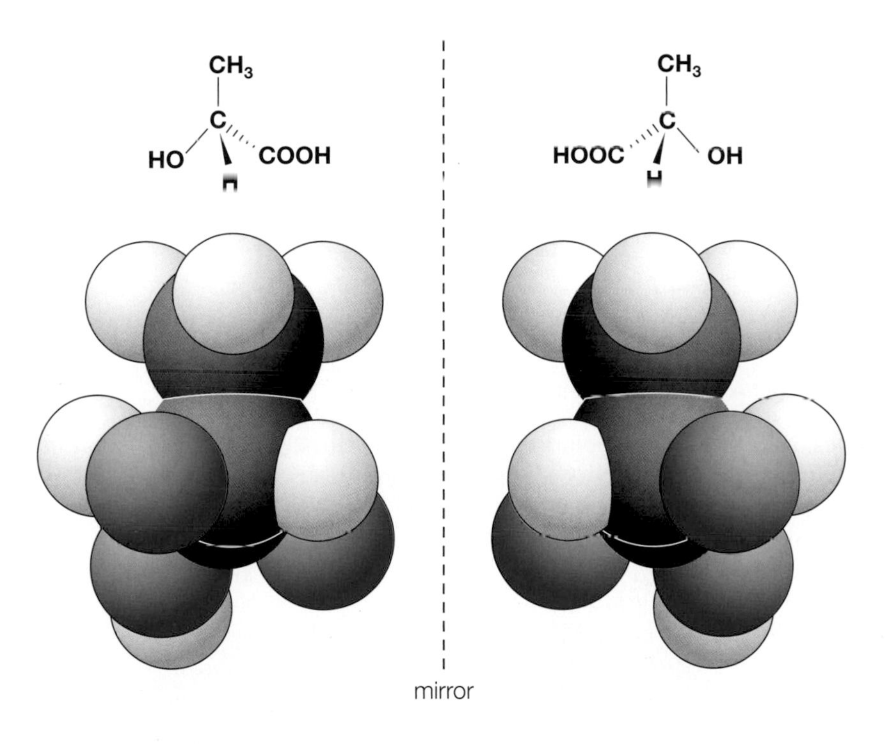

Figure 6.7▲
Molecules of lactic acid (2-hydroxypropanoic acid) are chiral. It is not possible to superimpose the two mirror image molecules.

The two forms of lactic acid behave identically in all their chemical reactions and all their physical properties except for their effect on polarised light. This optical property is the only way of telling the two forms of lactic acid apart. So, chemists call them optical isomers. The word 'enantiomers' is also used to describe mirror image molecules that are optical isomers. The word 'enantiomer' comes from a Greek word meaning 'opposite'.

Definition

Asymmetric molecules are molecules with no centre, axis or plane of symmetry. Asymmetric molecules are chiral and exist in mirror image forms. Any carbon atom with four different groups or atoms attached to it is asymmetric and chiral.

Test yourself

7 Identify the chiral objects in Figure 6.8.

8 With the help of molecular models, decide which of the following molecules are chiral:
NH_3, CH_2Cl_2, CH_2ClBr, $CH_3CHClBr$, $CH_3CHOHCOOH$.

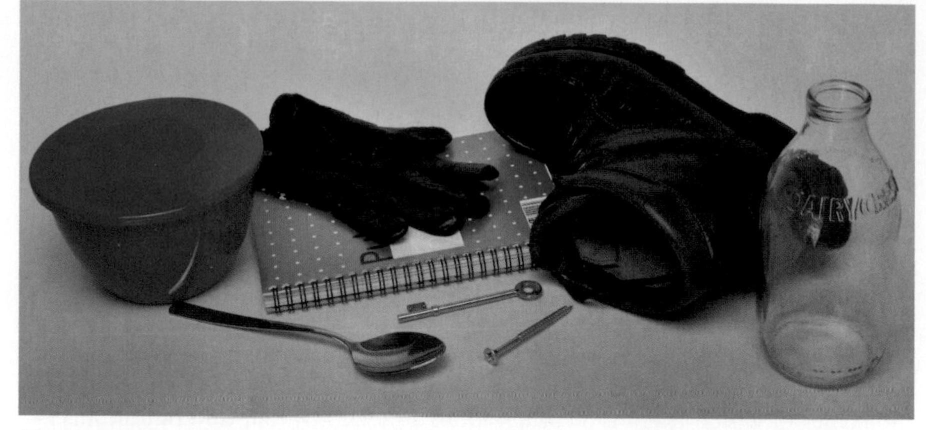

Figure 6.8▲

Optical isomerism and polarised light

A light beam becomes polarised after passing through a sheet of Polaroid, the material used to make some sunglasses. The Polaroid prevents vibrations of the light waves in all but one plane. So, in polarised light, all the waves are vibrating in the same plane (Figure 6.9).

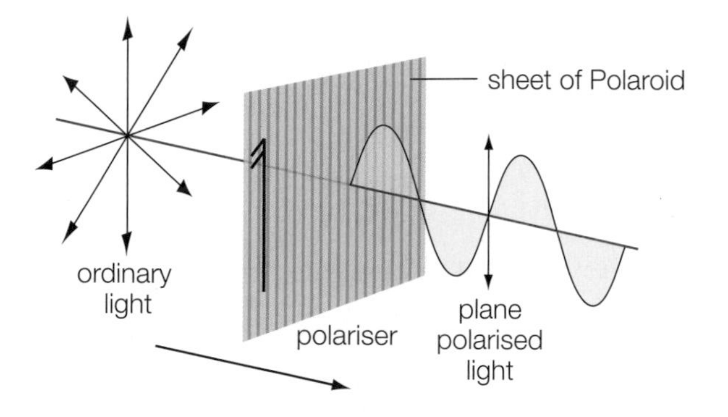

Figure 6.9▲
Ordinary light and polarised light.

Light is said to be plane polarised after passing through a sheet of Polaroid. If the polarised light is then directed at a second sheet of Polaroid, all the polarised beam passes through if the second sheet of Polaroid is aligned in the same way as the first (Figure 6.10a). However, no light gets through if the second sheet is rotated through 90° relative to the first sheet (Figure 6.10b).

Figure 6.10▶
The effect of a second sheet of Polaroid on polarised light.

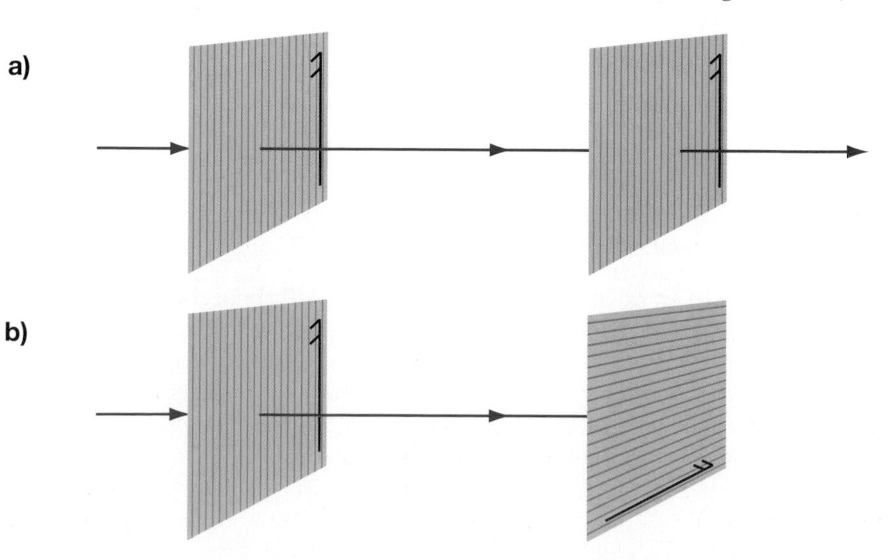

When polarised light passes through a solution of just one form of a chiral compound, it rotates the plane of polarisation. One isomer rotates the plane of plane-polarised light clockwise. This is named the + isomer. The other isomer rotates the plane of plane-polarised light anticlockwise and this is the – isomer. For accurate results, chemists measure the rotations with monochromatic light (light of one colour or frequency) in an instrument called a polarimeter (Figure 6.11).

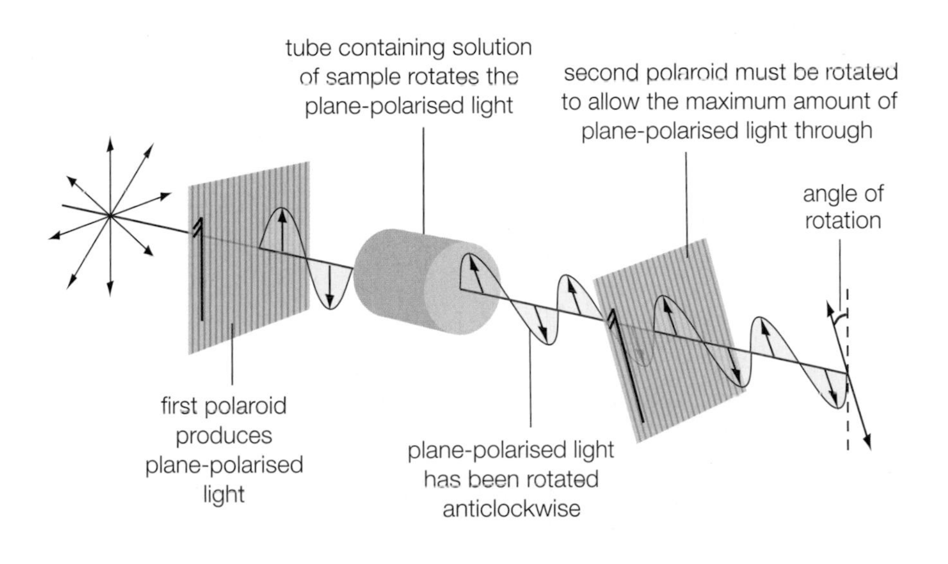

Figure 6.11 ▲
The effect of passing plane-polarised light through a solution of a chiral compound.

Definitions

Optical isomers are mirror-image forms of a chiral compound. One mirror-image form rotates the plane of plane-polarised light clockwise. The other form rotates the plane of plane-polarised light anticlockwise.

A **racemic mixture** is a mixture of equal amounts of the two mirror-image forms of a chiral compound. The mixture does not rotate polarised light because the two optical isomers have equal and opposite effects so they cancel each other out.

Test yourself

9 Which of these alcohols can exist as optical isomers: butan-1-ol, butan-2-ol, pentan-1-ol, pentan-2-ol, pentan-3-ol?

6.3 Optical isomerism and reaction mechanisms

The optical activity of the reactants and products of organic reactions can help chemists to determine the mechanisms of reactions. This is illustrated by the outcomes of the S_N1 and S_N2 mechanisms in nucleophilic substitution reactions.

During the one-step S_N2 mechanism, the three groups that remain attached to the central carbon atom are turned inside out. The molecule is inverted like an umbrella in a high wind (Figure 6.12).

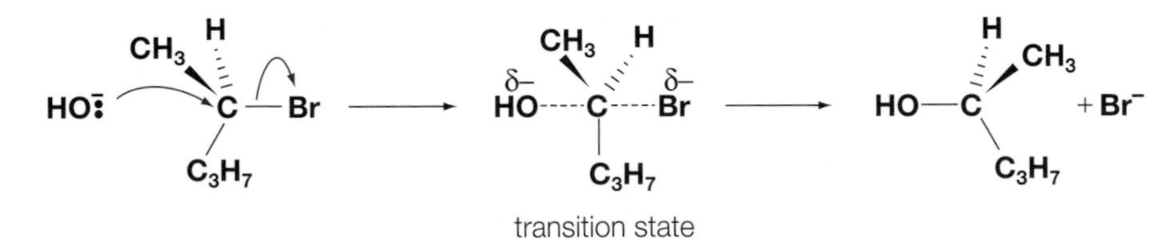

Figure 6.12 ▲
Nucleophilic substitution by the S_N2 mechanism with a molecule that is chiral.

This means that an optically active halogenoalkane gives rise to an optically active alcohol if substitution takes place by the S_N2 mechanism.

In the two-step S_N1 mechanism, a planar intermediate is formed after the first step. However, attack by the nucleophile during the second step can happen from either side of the planar intermediate. The result is that starting

Note

There no simple relationship between the three-dimensional shape of a chiral compound and the direction that it rotates polarised light. What this means is that molecular inversion of the + isomer of a halogenoalkane during an S_N2 reaction does give an optical isomer of the alcohol, but it is not possible to predict whether it will be the + or the – isomer.

with one optical isomer of a halogenoalkane leads to a product which is a racemic mixture of the two forms of the chiral alcohol. This means that the product is optically inactive (Figure 6.13).

Test yourself

10 Suggest explanations to account for the fact that the reaction of iodide ions with an optically active isomer of 3-chloro-3-methylhexane gives:
  a) a mixture of the two optical isomers of 3-iodo-3-methylhexane
  b) a product mixture which is slightly optically active.

11 Account for the fact that the reaction of 2-bromooctane with sodium hydroxide in an aqueous solvent is stereospecific: (+) 2-bromooctane reacts to form (–) octan-2-ol while (–) 2-bromooctane gives (+) octan-2-ol

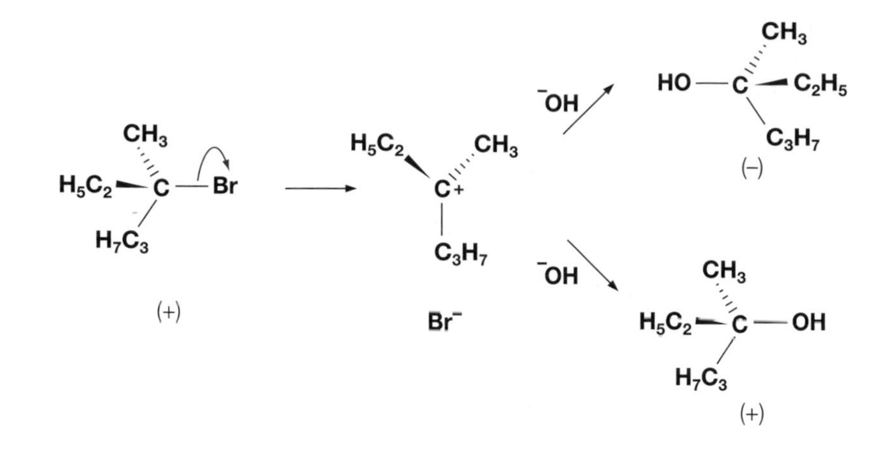

Figure 6.13▲
Hydrolysis of an optically active halogenoalkane by the S_N1 mechanism. The product is a mixture of the two optical isomers of the alcohol. This is a racemic mixture.

Activity

Chirality and living things

Human senses are sensitive to molecular shape. The optical isomers of some molecules have different tastes and smells.

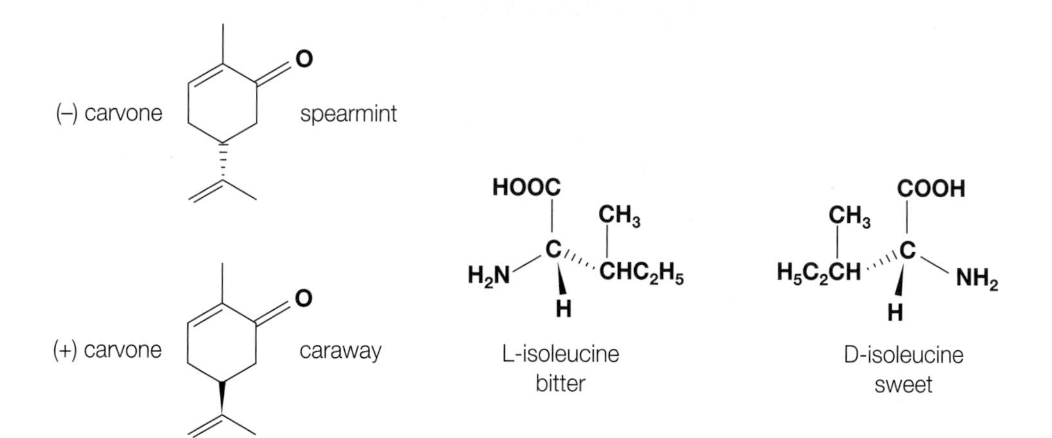

Figure 6.14▲
Optical isomers with differing tastes and smells.

What is true of the sensitive cells in the nose and on the tongue is also true of most of the rest of the human body. Living cells are full of messenger and carrier molecules which interact selectively with the active sites and receptors in other molecules such as enzymes. These messenger and carrier molecules are all chiral and the body works with only one of the mirror-image forms. This is particularly true of amino acids and proteins (see Section 14.1).

The chemists who synthesise and test new drugs have to pay close attention to chirality. Dextropropoxyphene, for example, is a painkiller. The molecule has two asymmetric carbon atoms. Its mirror image is no use for treating pain but is a useful ingredient of cough mixtures.

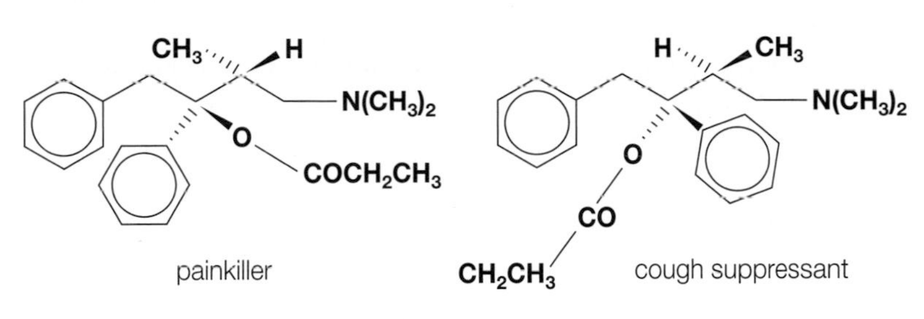

Figure 6.15 ◀ Dextropropoxyphene and its mirror image.

1 Identify the functional groups in:

a) carvone **b)** isoleucine.

2 Identify the chiral centres in:

a) isoleucine **b)** carvone

c) dextropropoxyphene.

3 Explain why the amino acid alanine, CH_3CHNH_2COOH, has optical isomers while the amino acid glycine, CH_2NH_2COOH, does not.

4 Suggest reasons why chemists working on new drugs need to develop effective methods to separate optical isomers of new compounds or methods to synthesise selectively each of the pairs of isomers.

REVIEW QUESTIONS

1 Write the displayed formulae of all the isomers of the following, stating which types of isomerism are involved:

a) C_3H_7Cl (2)

b) $C_2HFClBr$ (6)

c) $C_2H_3Cl_2Br$ (2)

2 a) What is meant by a chiral centre in a molecule? (1)

b) Explain why a chiral centre in a molecule gives rise to optical isomers. (2)

c) Which of these compounds can exist as optical isomers? (1)

$CH_2(NH_2)COOH$

CH_2OHCH_2COOH

$CH_3CHOHCOOH$

d) Explain why a racemic mixture of optical isomers has no effect on the plane of plane-polarised light. (2)

e) Salbutamol is a drug used to relieve the symptoms of asthma.

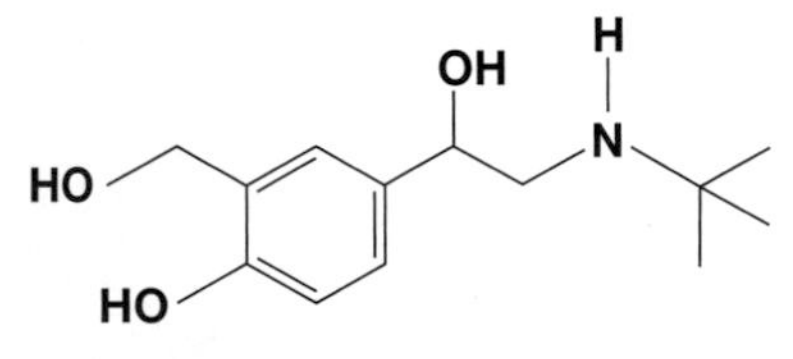

i) What is the molecular formula of salbutamol? (1)

ii) Identify any chiral centres in the salbutamol molecule. (1)

3 Explain the term stereoisomerism and describe the different types of stereoisomerism, illustrating your answer with suitable examples. (15)

7 Carbonyl compounds

The two types of carbonyl compounds are the aldehydes and the ketones. The reactive carbonyl group plays an important part in the chemistry of living things, in laboratory chemistry and in industry. As expected for a double bond, the characteristic reactions are addition reactions. The C=O bond, is polar because oxygen is highly electronegative. As a result the mechanism of addition to carbonyl compounds is not the same as the mechanism for addition to alkenes.

7.1 Aldehydes

Names and structures

Aldehydes are carbonyl compounds in which a carbonyl group (C=O) is attached to two hydrogen atoms or to a hydrocarbon group and a hydrogen atom. So the carbonyl group is at the end of a carbon chain. The names are based on the alkane with the same carbon skeleton with the ending changed from –**ane** to –**anal**.

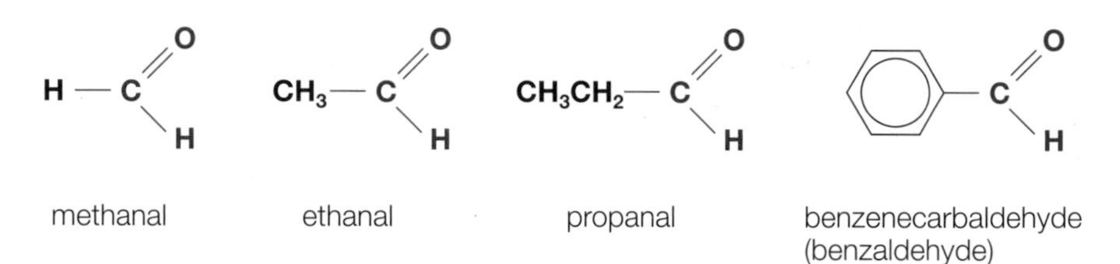

methanal ethanal propanal benzenecarbaldehyde (benzaldehyde)

Figure 7.1▲
Structures and names of aldehydes. The –CHO group is the functional group which gives aldehydes their characteristic reactions.

Note

Always write the aldehyde group as –CHO. Writing –COH is unconventional and easily leads to confusion with alcohols.

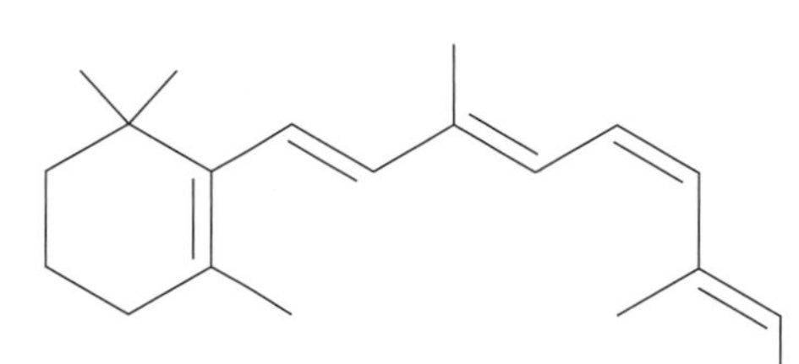

Figure 7.2▲
The skeletal formula of retinal, which is a naturally occurring aldehyde. Combined with a protein it forms the light sensitive part of the visual pigment in the rod cells of the retina. When light falls on a rod cell the retinal molecule changes shape. This sets off a series of changes that lead to a signal being sent to the brain.

Occurrence and uses

Biologists use a solution of methanal to preserve specimens. Since the beginning of the twentieth century it has been the main ingredient of the fluids used by embalmers. Methanal is also an important industrial chemical because it is a raw material for the manufacture for a range of thermosetting plastics (see Section 12.12).

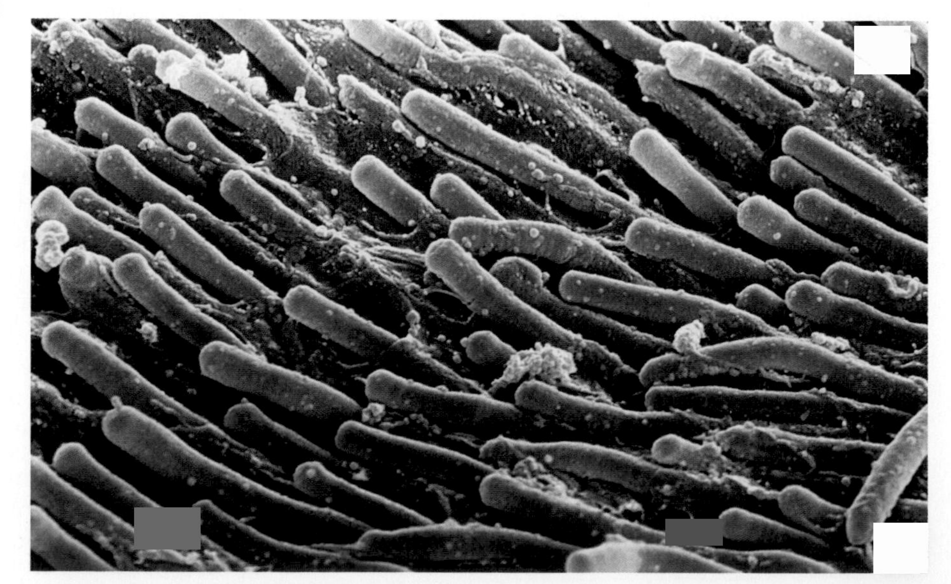

Figure 7.3▶
A coloured scanning electron micrograph of rod cells in the retina of an eye. The cells are magnified about 3000 times. Rod cells contain a visual pigment that can respond to dim light but cannot distinguish colours.

Formation

Oxidation of primary alcohols by heating with a mixture of dilute sulfuric acid and potassium dichromate(VI) produces aldehydes if the conditions allow the aldehyde to distil off as it forms. Unlike ketones, aldehydes can easily be oxidised further to carboxylic acids by longer heating with an excess of the reagent.

Test yourself

1 Write the structure of 2-methylbutanal.
2 An aldehyde can be made by adding propan-1-ol a few drops at a time to a mixture of sodium dichromate(VI) and dilute sulfuric acid. The product is then distilled from the reaction mixture.
 a) Write an equation for the reaction. (Represent the oxygen from the oxidising agent as [O].)
 b) Suggest a reason for adding the propan-1-ol a few drops at a time.
 c) Explain why the reaction mixture is not heated in a flask fitted with a reflux condenser before distilling off the product.

7.2 Ketones

Names and structures

In ketones the carbonyl group is attached to two hydrocarbon groups. Chemists name ketones after the alkane with the same carbon skeleton by changing the ending **–ane** to **–anone**. Where necessary a number in the name shows the position of the carbonyl group.

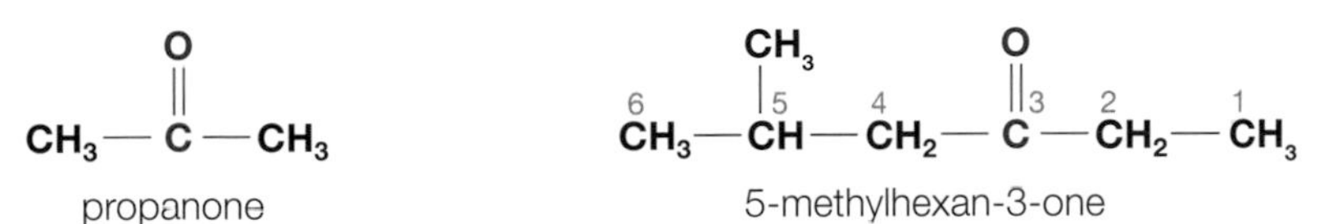

Figure 7.4▲
Structures and names of two ketones.

Figure 7.5▲
A bottle of peppermint oil and leaves of the peppermint plant. The oil is used in aromatherapy. Peppermint oil contains menthone together with a range of chemicals which include menthol, methyl ethanoate and volatile oils.

Occurrence and uses

The most widely used ketone is propanone, which is a common solvent. It has a low boiling temperature and evaporates quickly, making it suitable for cleaning and drying parts of precision equipment. Propanone is also the starting point for producing the monomer of the glass-like addition polymer in display signs, plastic baths and the cover of car lights. Propanone, and other ketones, form during normal metabolism, especially at night and during fasting when the levels of propanone and other ketones in the blood rise.

Formation

Oxidation of secondary alcohols with hot, acidified potassium dichromate(VI) produces ketones which, unlike aldehydes, are not easily oxidised further.

Test yourself

3 Write the structure of 4,4-dimethylpentan-2-one.
4 What is the molecular formula of menthone?
5 Show that propanone and propanal are functional group isomers.
6 Write an equation for the oxidation of butan-2-ol to butanone. (Represent the oxygen from the oxidising agent as [O].)

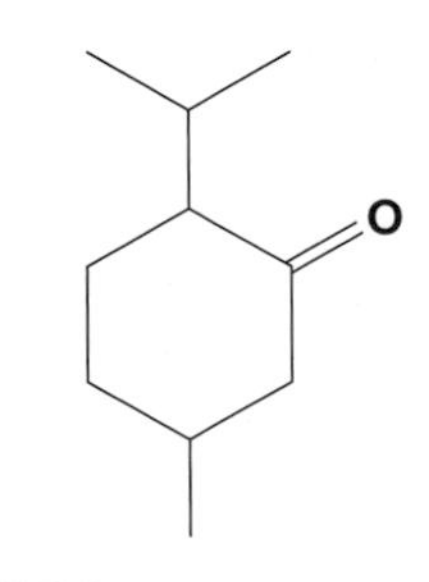

Figure 7.6▲
The skeletal formula of the naturally occurring ketone called menthone, which is found in the oils extracted from some plants.

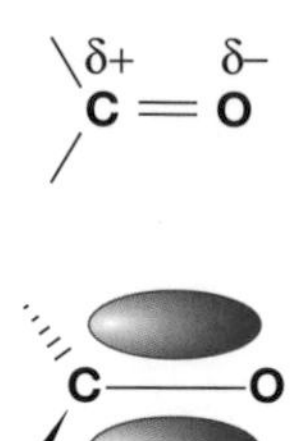

Figure 7.7▲
Representations of the carbonyl group in aldehydes and ketones. The bond is polar because the oxygen atom is more electronegative than carbon.

7.3 Physical properties of carbonyl compounds

The C=O bond in carbonyl compounds is polar (Figure 7.7). As a result the intermolecular forces include dipole–dipole attractions as well as London forces. There are no hydrogen atoms bonded to oxygen in carbonyl compounds so hydrogen bonding cannot occur between the molecules of aldehydes or ketones. However, hydrogen bonding is possible between oxygen atoms in carbonyl groups and the –OH group in water.

Methanal is a gas at room temperature. Ethanal boils at 21 °C so it may be a liquid or gas at room temperature depending on the conditions. Other common aldehydes are all liquids too. Similarly, the common ketones are liquids with boiling temperatures similar to those of the corresponding aldehydes.

The simpler aldehydes such, as methanal and ethanal, are freely soluble in water. The simplest ketone, propanone, mixes freely with water.

Test yourself

www Data

7 a) Show that the boiling temperatures of aldehydes are higher than those of alkanes with similar relative molecular masses, but lower than those of the corresponding alcohols.
b) Account for the values of the boiling temperatures of aldehydes relative to those of alkanes and alcohols in terms of intermolecular forces.

8 Explain why propanone is freely soluble in water.

9 Why does an aldehyde such as ethanal mix freely with water while benzaldehyde, C_6H_5CHO, is much less soluble?

Activity

Chemicals in perfumes

The perfume 'Chanel N° 5' was innovative when produced for the first time in 1921. As well as natural extracts from flowers, the scent includes a high proportion of synthetic aldehydes such as dodecanal. This produces a highly original perfume.

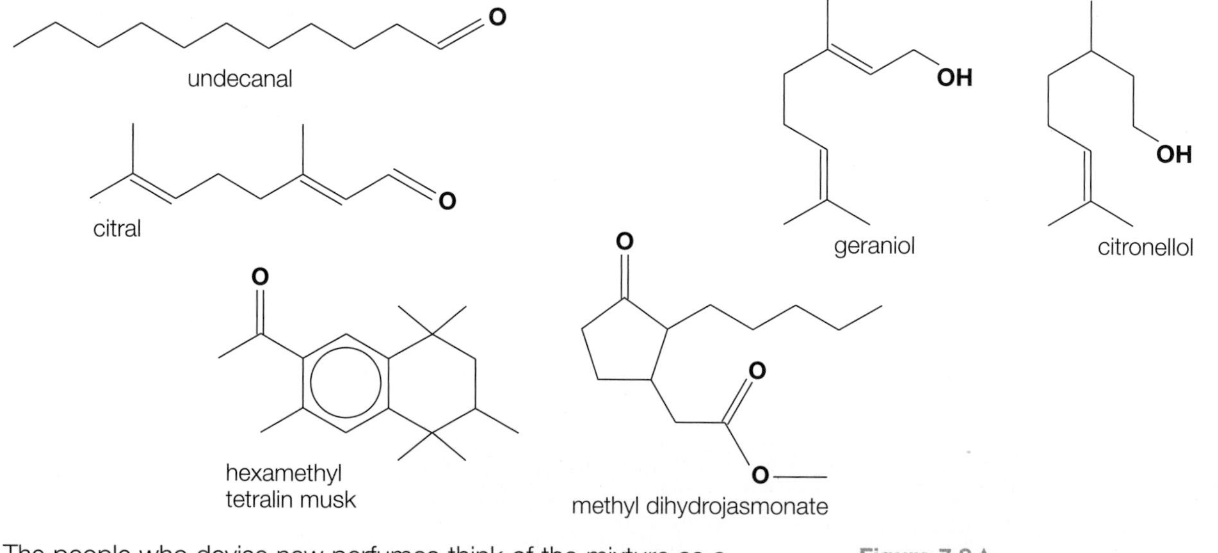

Figure 7.8▲
Skeletal formulae of some perfume chemicals.

The people who devise new perfumes think of the mixture as a sequence of 'notes'. You first smell the 'top notes', but the main effect depends on the 'middle notes', while the more lasting elements of the perfume are the 'end notes'. The overall balance of the three is critical. This means that the volatility of perfume chemicals is of great importance to the perfumer.

Note	Natural chemicals	Synthetic chemicals	Boiling temperature or melting temperature
Top	Citrus oils Lavender	Octanal (citrus) Undecanal (green)	(boils) 168 °C (boils) 117 °C
Middle	Rose Violet	Geraniol (floral) Citronellol (rosy)	(boils) 146 °C (boils) 224 °C
End	Balsam Musk	Indane (musk) Hexamethyl tetralin (musk)	(melts) 53 °C (melts) 55 °C

Table 7.1◄
Natural and synthetic chemicals used to make perfumes.

1 Draw the skeletal formula of octanal.

2 Suggest two advantages for the perfumer of using synthetic chemicals instead of chemicals extracted from living things.

3 Identify the carbonyl compounds among the compounds shown in Figure 7.8. In each case state whether the compound includes the functional group of an aldehyde or of a ketone.

4 Like many perfume chemicals, geraniol is a terpene. Terpene molecules are built from units derived from 2-methylbuta-1,3-diene.

a) What is the structure of 2-methylbuta-1,3-diene?

b) How many 2-methylbuta-1,3-diene units are needed to make up the hydrocarbon skeleton of geraniol?

5 Use your knowledge of intermolecular forces to explain why:

a) aldehydes are useful as top notes while alcohols are more commonly used as middle notes

b) the musks used as 'end notes' also help to 'fix', or retain, the more volatile components of a perfume

c) geraniol is more soluble in water than undecanal.

7.4 Reactions of aldehydes and ketones

Oxidation

Oxidising agents easily convert aldehydes to carboxylic acids. It is much harder to oxidise ketones. Oxidation of ketones is only possible with powerful oxidising agents which break up the molecules. Chemical tests to distinguish aldehydes and ketones are based on the difference in the ease of oxidation (Section 7.5).

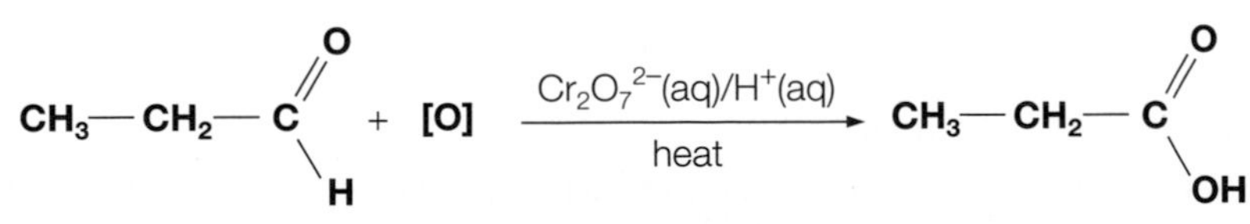

Figure 7.9▲
Oxidation of propanal to propanoic acid.

Acidified potassium dichromate(VI) is orange and contains $Cr_2O_7^{2-}$ ions. After oxidising an aldehyde to a carboxylic acid the reagent is green, containing a solution of green Cr^{3+} ions.

Test yourself

10 Identify the oxidation states of chromium before and after the oxidation of an aldehyde by a solution of potassium dichromate(VI).

11 a) Outline the reagents and conditions for converting butanal to butanoic acid.
b) What apparatus is used i) to carry out the reaction and ii) to separate the product from the reaction mixture?

Reduction

Metal hydrides can reduce carbonyl compounds to alcohols. Lithium tetrahydridoaluminate(III), $LiAlH_4$, is a powerful reducing agent that converts aldehydes to primary alcohols, and ketones to secondary alcohols. $LiAlH_4$ is easily hydrolysed so the reagent is dissolved in dry ether (ethoxyethane).

Test yourself

12 Name the products of reducing butanal and butanone. Which is a primary alcohol and which a secondary alcohol?

13 Show that reduction of an aldehyde or ketone with $LiAlH_4$ has the effect of adding hydrogen to the double bond.

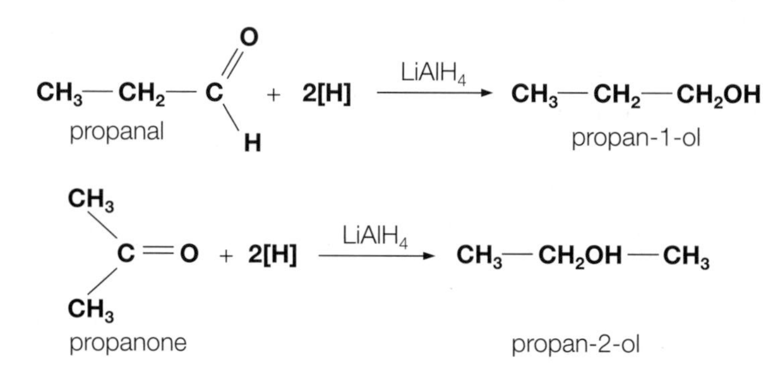

Figure 7.10▲
Reduction of propanal and propanone. The 2[H] comes from the reducing agent. This is a shorthand way of balancing a complex equation involving reduction.

Addition of hydrogen cyanide

Hydrogen cyanide rapidly adds to carbonyl compounds at room temperature. Hydrogen cyanide is a highly toxic gas which is formed in the reaction mixture by adding potassium cyanide and dilute sulfuric acid. The potassium cyanide must be in excess to ensure that there are free cyanide ions ready to start the reaction.

Definition

A **nitrile** is a compound with a functional group –C≡N. Note that the carbon atom in the –CN group counts as part of the carbon chain when naming these compounds.

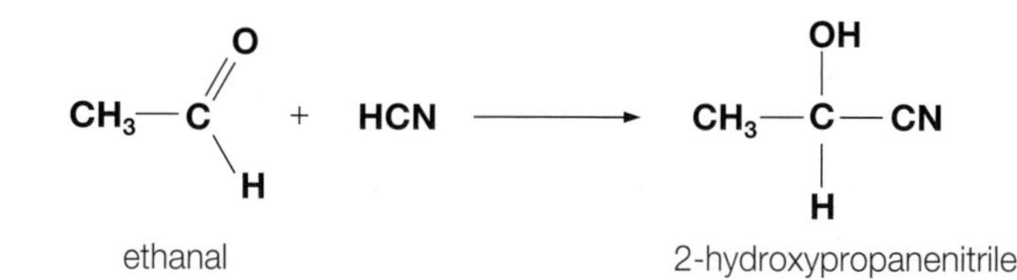

Figure 7.11▲
Addition of hydrogen cyanide to ethanal. The product is a hydroxynitrile.

This reaction is a useful step in synthetic routes to valuable compounds because the –CN group can be hydrolysed to a carboxylic acid (see Section 8.2). The reaction adds a carbon atom to the carbon skeleton of the original molecule.

Nucleophilic addition

The reduction of a carbonyl compound with $LiAlH_4$ and its reaction with hydrogen cyanide are both examples of nucleophilic addition. The carbon atom in a carbonyl group is open to attack by nucleophiles because the electronegative oxygen draws electrons away from it.

The incoming nucleophile uses its lone pair to form a new bond with the carbon atom. This displaces one pair of electrons in the double bond onto oxygen. Oxygen has thus gained one electron from carbon and now has a negative charge.

www
Tutorial

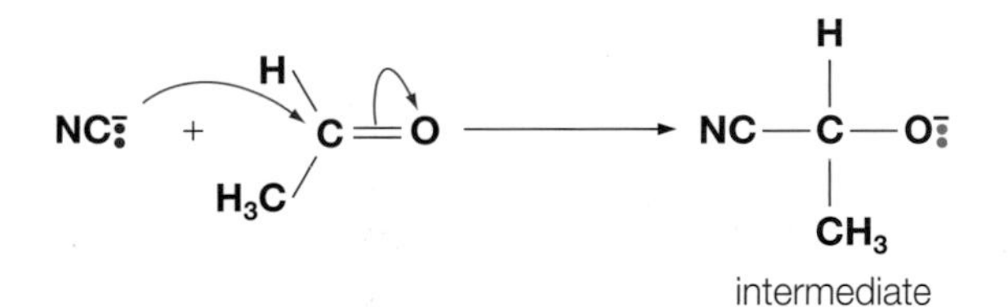

Figure 7.12▲
First step of nucleophilic addition of hydrogen cyanide to ethanal.

To complete the reaction, the negatively charged oxygen acts as a base and gains a proton.

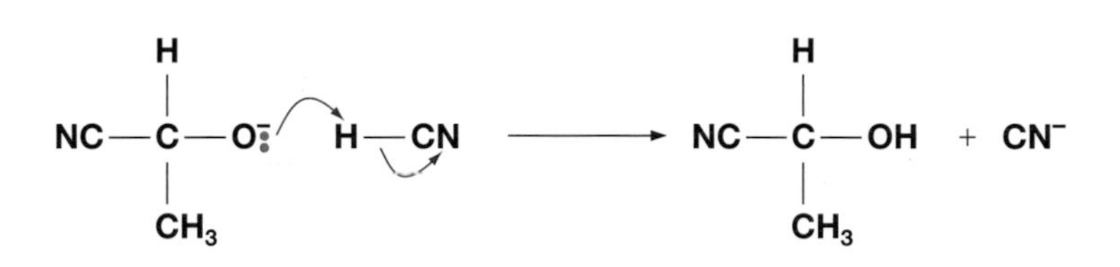

Figure 7.13▲
Second step of nucleophilic addition of hydrogen cyanide to ethanal. Note that taking a proton from HCN produces another cyanide ion.

This mechanism helps to account for the fact that if the product of addition is chiral, the outcome is a racemic mixture of the two optical isomers. This is illustrated by Figure 7.14, which shows the addition of hydrogen cyanide to ethanal in three dimensions. The atoms around the carbon atom of a carbonyl group lie in a plane. The attacking nucleophile has an equal chance of bonding to the carbon atom from either side of this plane.

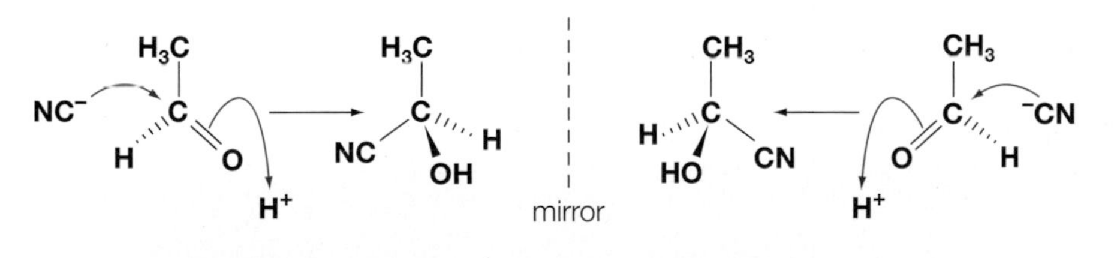

Figure 7.14◄
Formation of two optical isomers by addition of HCN to ethanal. Nucleophiles attack randomly from either side of the ethanal molecules, giving rise to equal numbers of the two isomer molecules.

Test yourself

14 Answer these questions about the nucleophilic addition of hydrogen cyanide by reaction of a carbonyl compound with HCN.
- **a)** What features of the cyanide ion mean that it is a nucleophile?
- **b)** What type of bond breaking takes place in each step?
- **c)** Explain why there is a negative charge on the oxygen atom at the end of step 1.
- **d)** Which molecule acts as an acid in step 2?

15 Show that the hydroxynitrile formed from ethanal and HCN is chiral.

16 Write the equation for the formation of 2-hydroxy-2-methylpropanenitrile from a ketone.

17 a) Write equations to show the mechanisms of the reduction of ethanal by $LiAlH_4$. The nucleophile is the hydride ion, H^-. A water molecule is involved in the second step of the process.
- **b)** How can the mechanism account for the fact that $LiAlH_4$ reduces the double bond in carbonyl compounds but not the double bond in alkenes?

7.5 Tests for aldehydes and ketones

Definition

Chemists can use **derivatives** to identify unknown organic compounds. Converting a compound to a crystalline derivative produces a product that can be purified by recrystallisation and then identified by measuring its melting temperature.

Recognising carbonyl compounds

Today chemists can identify aldehydes and ketones with the help of instrumental techniques such as mass spectrometry and infrared spectroscopy (Topic 9). Traditionally chemists characterised these compounds by combining them with a reagent that could convert them to a solid product. The solid, a so-called crystalline derivative, is a chemical which can be purified by recrystallisation and then identified by measuring its melting temperature.

The reagent 2,4-dinitrophenylhydrazine reacts with carbonyl compounds forming 2,4-dinitrophenylhydrazone derivatives, which are solid at room temperature and bright yellow or orange. The solid derivative can be filtered off, recrystallised and identified by measuring its melting temperature. Put together with the boiling temperature of the original aldehyde or ketone, this makes it possible to identify the carbonyl compound.

Figure 7.15▲
A bright orange 2,4-dinitrophenylhydrazone derivative.

Distinguishing aldehydes and ketones

Fehling's solution, Benedict's solution and Tollens' reagent are three mild oxidising agents which are used to distinguish aldehydes from ketones. Aldehydes are easily oxidised but ketones are not.

Fehling's reagent does not keep so it is made when required by mixing two solutions. One solution is copper(II) sulfate in water. The other is a solution of 2,3-dihydroxybutanedioate (tartrate) ions in strong alkali. The 2,3-dihydroxybutanedioate salt forms a complex with copper(II) ions so that they do not precipitate as copper(II) hydroxide with the alkali.

Benedict's solution is similar to Fehling's solution but is more stable. It is less strongly alkaline and does not react so reliably with all aldehydes.

Aldehydes reduce the copper(II) ions in Fehling's, or Benedict's, solution to copper(I) which then precipitates in the alkaline conditions to give an orange-brown precipitate of copper(I) oxide.

Tollens' reagent (ammoniacal silver nitrate) similarly distinguishes aldehydes from ketones. Tollens' reagent consists of an alkaline solution of diamminesilver(I) ions, $[Ag(NH_3)_2]^+$. Forming a complex with ammonia keeps the silver(I) ions in solution under alkaline conditions. Aldehydes reduce the silver ions to metallic silver.

Figure 7.16▲
Fehling's reagent is used to test for aldehydes. The reagent has a blue colour as it contains copper(II) ions. The test tube in the middle contains Fehling's reagent that has been reduced by an aldehyde, to form an orange-brown precipitate of copper(I) oxide. The test tubes on the left and right contain Fehling's reagent and ketones. Ketones are very similar to aldehydes but do not react with Fehling's reagent, hence the colour is unchanged.

Figure 7.17▲
Warming Tollens' reagent with an aldehyde produces a precipitate of silver which coats clean glass with a shiny layer of silver so that it acts like a mirror (left). There is no reaction with a ketone (right).

Test yourself

18 Write an ionic equation for the reaction of copper(II) ions with an alkali in the absence of 2,3-dihydroxybutanedioate ions.

19 a) Write an equation for the reaction of Tollens' reagent with propanal. Use the symbol [O] to represent the reagent.

b) Use the oxidation numbers of the metal ions and atoms to show that propanal reduces Tollens' reagent.

20 Hydrolysis of A, C_4H_9Cl, with hot, aqueous sodium hydroxide produces B, $C_4H_{10}O$. Heating B with an acidic solution of potassium dichromate(VI) and distilling off the product as it forms gives C, C_4H_8O. C gives a yellow precipitate with 2,4-dinitrophenylhydrazine and forms a silver mirror when warmed with Tollens' reagent. Give the names and structures of these compounds.

Activity

Identifying an unknown carbonyl compound

Figure 7.18 shows stages in making, purifying and identifying a carbonyl compound.

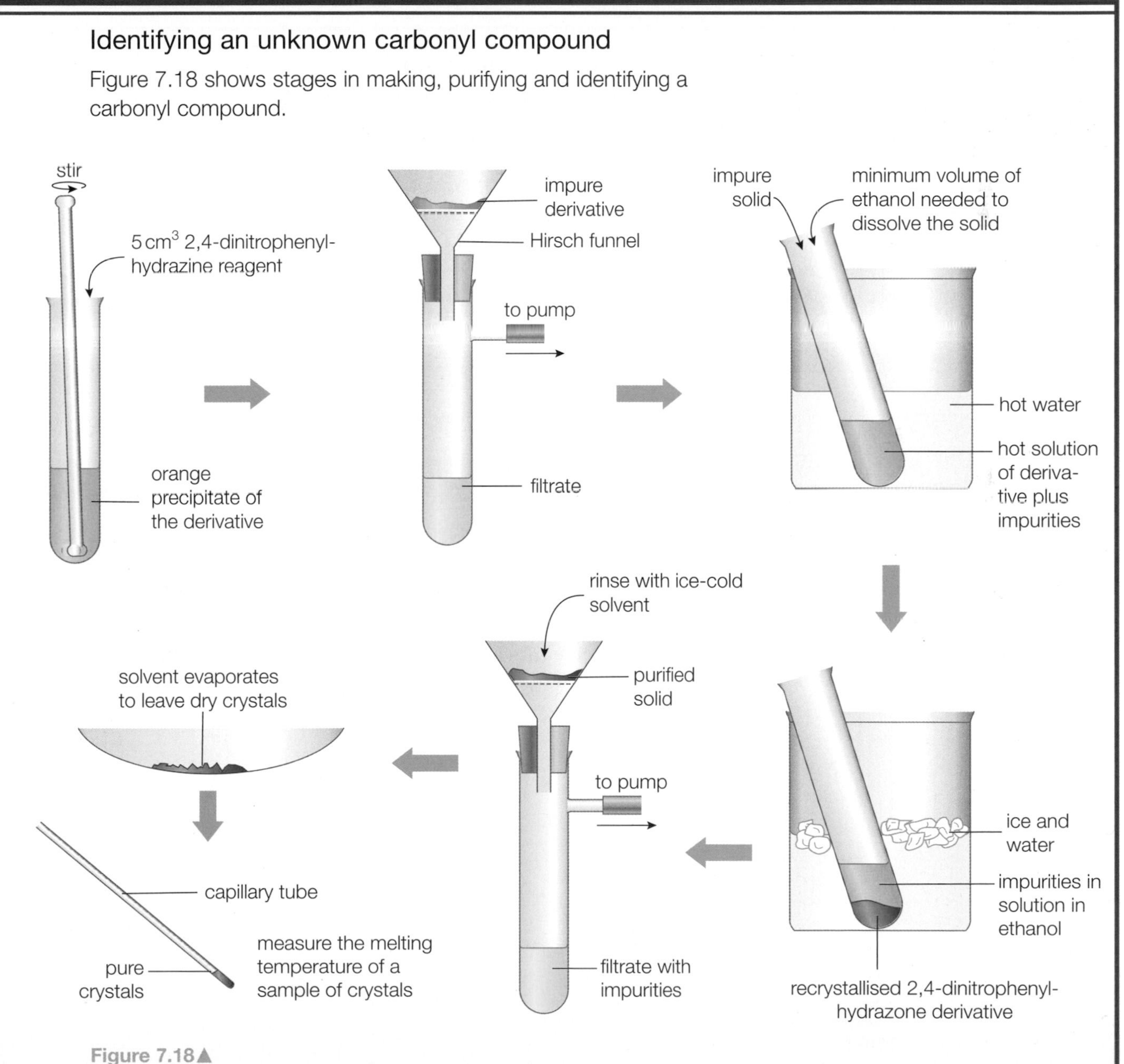

Figure 7.18▲
Making a pure crystalline derivative of a carbonyl compound.

1 Why is it necessary to purify the derivative before measuring its melting temperature?

2 Explain how the procedure illustrated in Figure 7.18 removes impurities from the derivative.

3 In this instance, ethanol is the solvent used for recrystallising the derivative. What determines the choice of solvent?

4 When measuring the melting temperature, what are the signs that the derivative is pure?

5 Identify the carbonyl compound that forms a 2,4-dinitrophenylhydrazone that melts at 115 °C. The carbonyl compound boils at 80 °C. The compound does not give an orange precipitate with Fehling's solution.

Note

An alternative reagent for the iodoform reaction is a mixture of potassium iodide and sodium chlorate(I).

The triiodomethane reaction

A compound containing a CH_3CO- group produces a yellow precipitate of triiodomethane (iodoform) when warmed with a mixture of iodine and sodium hydroxide. The test shows the presence of a methyl group next to a carbonyl group in an organic molecule. The reaction takes place in two steps.

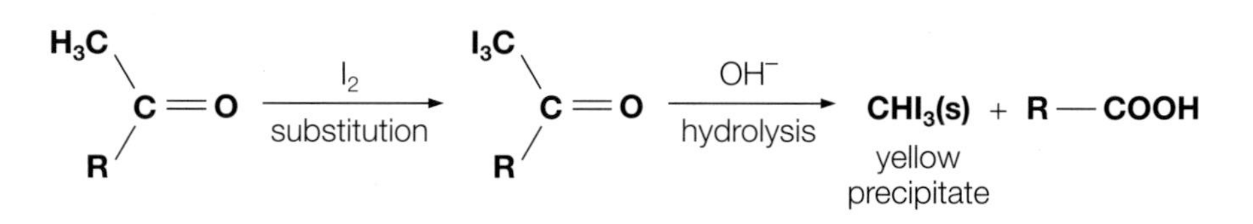

Figure 7.19▲
The equation for the triiodomethane reaction.

Test yourself

21 a) Explain why alcohols with this group CH_3CHOH- also give a positive result with the triiodomethane reaction.
b) Explain why a mixture of iodine with sodium hydroxide solution is chemically equivalent to a mixture of potassium iodide and sodium chlorate(I) solutions.

22 Name and write out the displayed formulae of the two isomers of $C_5H_{10}O$ that form a yellow precipitate when they react with iodine in the presence of alkali.

23 Which is the only aldehyde to undergo the triiodomethane reaction?

REVIEW QUESTIONS

Extension questions

1 Copy and complete the table to show three different reactions of propanal.

Reactant	Reagent	Organic product	
		Name	Displayed formula
CH_3CH_2CHO	Tollens' reagent		
CH_3CH_2CHO		Propanoic acid	
CH_3CH_2CHO	$LiAlH_4$		

(6)

2 Citral and β-ionone are perfume chemicals. β-ionone is one of the chemicals in the oil extracted from violets.

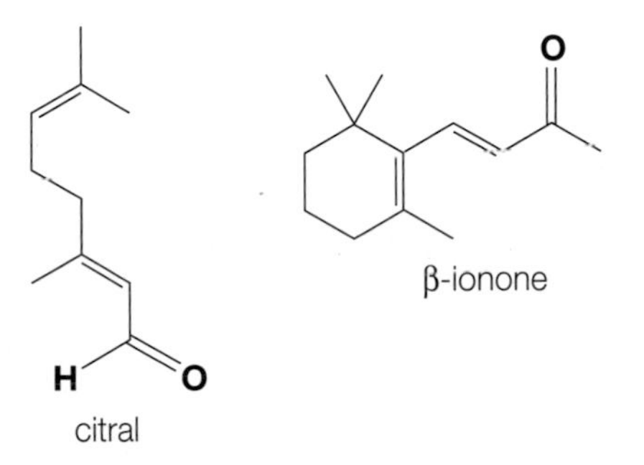

a) Work out the molecular formula of citral. (1)

b) i) Name the two functional groups in citral. (2)

ii) Name the functional group which is present in β-ionone but not in citral. (1)

c) Describe the observations you would expect with each of the two compounds on:

i) warming them with Fehling's solution (2)

ii) mixing them with a solution of 2,4-dinitrophenylhydrazine (2)

iii) warming them with a mixture of iodine and sodium hydroxide. (2)

d) i) Draw the skeletal formula of the product of treating β-ionone with $LiAlH_4$ in dry ether. (1)

ii) Draw the skeletal formula of the compound formed when citral is warmed with an acidic solution of potassium dichromate(VI). (1)

e) i) Which of the two compounds have *E–Z* isomers? (1)

ii) Draw an *E–Z* isomer of one of the two compounds. (2)

iii) Which of the two compounds has optical isomers? (1)

3 The molar mass of a hydrocarbon W is 56 g mol^{-1} and it contains 85.7% carbon. W reacts with hydrogen bromide to form X. Heating X under reflux with aqueous sodium hydroxide produces Y. Heating Y with an acidified solution of potassium dichromate(VI) converts it to Z. Z gives a yellow precipitate with 2,4-dinitrophenylhydrazine but it does not give a precipitate with Benedict's reagent.

a) Identify W, X, Y and Z and give your reasoning. (8)

b) Write equations for the reactions mentioned of W, X and Y. (3)

4 Describe the mechanisms for the reactions of propene with bromine, and propanone with HCN. Identify similarities and differences between the two mechanisms with reference to the nature of the bonds and bond breaking, the reagents involved, the formation of intermediates and the overall effects of the change. (10)

8 Carboxylic acids and their derivatives

Many organic acids are instantly recognisable by their odours. Ethanoic acid, for example, gives vinegar its taste and smell. Butanoic acid is responsible for the foul smell of rancid butter, while the body odour of goats is a blend of the three unbranched organic acids with 6, 8 and 10 carbon atoms. Organic acids play a vital part in the biochemistry of life because of the great variety of their reactions. The acids can be converted to a wide range of derivatives that are important in the laboratory and in industrial chemistry. These derivatives help to give rise to a variety of new materials, especially synthetic fibres and plastics.

8.1 Carboxylic acids

Occurrence

Carboxylic acids are compounds with the formula R–COOH where R represents an alkyl group, aryl group or a hydrogen atom. The carboxylic acid group –COOH is the functional group which gives the acids their characteristic properties.

Figure 8.1▲
The traditional names for organic acids were based on their natural origins. The original name for methanoic acid was formic acid because it was first obtained from red ants and the Latin name for 'ant' is formica. This red wood ant can spray attackers with methanoic acid (magnification ×10).

Figure 8.2◀
Many vegetables contain ethanedioic acid, which is commonly called oxalic acid. The level of the acid in rhubarb leaves is high enough for it to be dangerous to eat the leaves. The acid kills by lowering the concentration of calcium ions in blood to a dangerously low level.

Names and structures

The carboxylic acid group can be regarded as a carbonyl group, C=O, attached to an –OH group but is better seen as a single functional group with distinctive properties.

Chemists name carboxylic acids by changing the ending of the corresponding alk**ane** to –**oic** acid. So ethane becomes ethanoic acid.

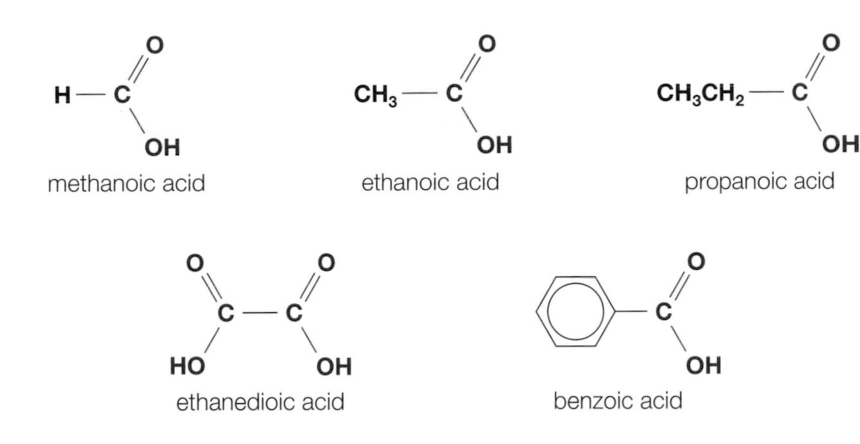

Figure 8.3▲
Names and structures of carboxylic acids.

Figure 8.4▲
A ball-and-stick model of a carboxylic acid.

Carboxylic acids form a wide range of derivatives, each with their own characteristics. This is illustrated by the derivatives of ethanoic acid in Figure 8.5.

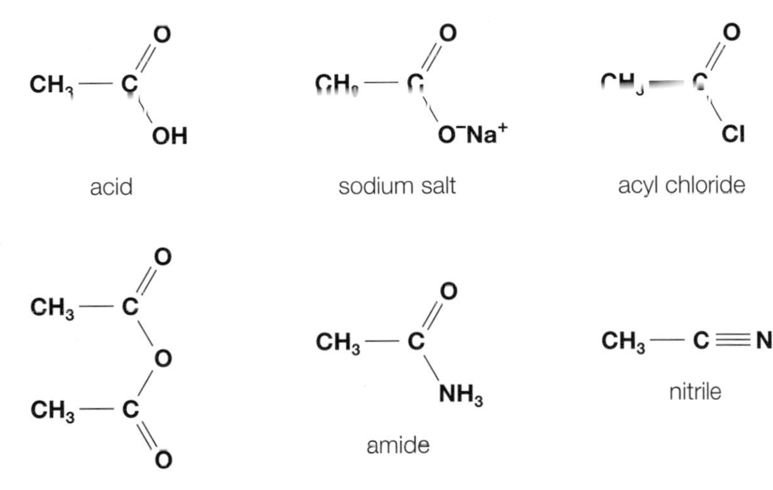

Figure 8.5▲
Compounds related to carboxylic acids.

Test yourself

1 Write out the structural formulae and give the systematic names of the three carboxylic acids which were traditionally derived from the Latin word 'caper', meaning goat: caproic acid (6C), caprylic acid (8C) and capric acid (10C).

2 Give the molecular formula, skeletal formula and name of the acid shown in Figure 8.4.

Physical properties

Even the simplest acids such as methanoic acid and ethanoic acid are liquids at room temperature because of hydrogen bonding between the carboxylic acid groups. Carboxylic acids with more than eight carbon atoms in the chain are solids. Benzoic acid is also a solid at room temperature.

Carbon–oxygen bonds are polar. There is also the possibility of hydrogen bonding between water molecules and the –OH groups and oxygen atoms in carboxylic acid molecules. This means that the acids are soluble in water.

Data

Test yourself

3 Draw a diagram to show hydrogen bonding between ethanoic acid molecules and water molecules.

4 In a non-polar solvent, ethanoic acid molecules dimerise through hydrogen bonding.
   a) Suggest a reason why the acid dimerises in a non-polar solvent but not in water.
   b) Draw a diagram to show an ethanoic acid dimer with two hydrogen bonds between the molecules.

Test yourself

5 Give the name and structure of the carboxylic acid formed on:
   a) heating pentan-1-ol under reflux with an acidic solution of potassium dichromate(VI) and then distilling off the product
   b) heating butanal under reflux with an acidic solution of potassium dichromate(VI) and then distilling off the product
   c) heating propanenitrile under reflux with aqueous sodium hydroxide, acidifying the mixture and then distilling off the product.

6 Write the equation for the hydrolysis of ethanenitrile with an excess of:
   a) dilute hydrochloric acid
   b) aqueous sodium hydroxide.

7 After hydrolysis of a nitrile with aqueous alkali, why must the solution be acidified before distilling off a carboxylic acid from the mixture of products?

Figure 8.8▲
Citrus fruits including lemons, grapefruits, limes, kumquats, clementines and oranges. Citric acid is found in the juice of all citrus fruits. Citric acid contains three carboxylic acid functional groups and has a molecular formula of $C_6H_8O_7$.

8.2 Preparation of carboxylic acids

In the laboratory, carboxylic acids are normally made by oxidising primary alcohols or aldehydes (Section 7.4). The usual oxidising agent is an acidic solution of potassium dichromate(VI).

Carboxylic acids can also be made by hydrolysing nitriles. The reagent for speeding up the hydrolysis can either be a solution of a strong acid or a solution of a strong base.

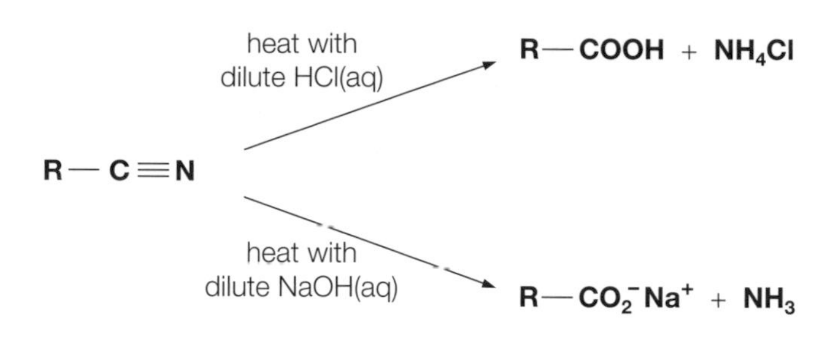

Figure 8.6▲
Hydrolysis of nitriles produces carboxylic acids or their salts.

8.3 Reactions of carboxylic acids

Reactions as acids

Carboxylic acids are weak acids (see Section 5.4). They are only slightly ionised when they dissolve in water.

$$CH_3COOH(aq) \rightleftharpoons CH_3CO_2^-(aq) + H^+(aq)$$

The aqueous hydrogen ions in the solutions of these compounds mean that they show the characteristic reactions of acids with metals, bases and carbonates.

Carboxylic acids are sufficiently acidic to produce carbon dioxide when added to a solution of sodium carbonate or sodium hydrogencarbonate. This reaction distinguishes carboxylic acids from weaker acids such as phenols (see Section 12.10).

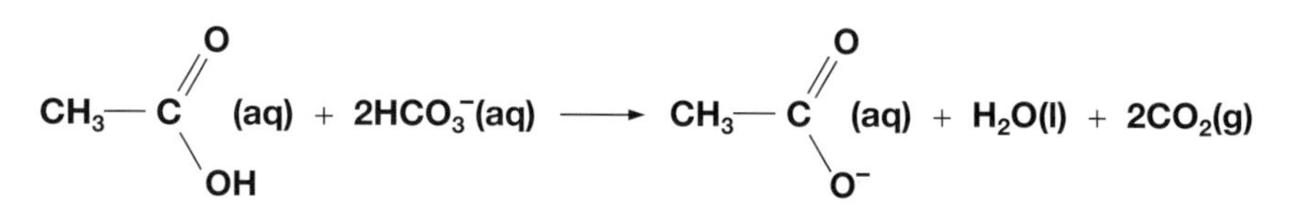

Figure 8.7▲
The reaction of ethanoic acid with hydrogencarbonate ions.

Test yourself

8 For each of these pairs of chemicals describe what you would observe when they react, write equations for the reactions and name the organic products:
   a) ethanoic acid and potassium hydroxide
   b) propanoic acid and sodium carbonate
   c) butanoic acid and ammonia.

9 Why is sodium ethanoate a solid at room temperature while ethanoic acid is a liquid?

10 a) Write an equation for the complete neutralisation of citric acid by sodium hydroxide.
   b) Describe in outline how the reaction of sodium hydroxide with citric acid could be used to estimate the concentration of citric acid in fruit juices, indicating reasons why the result might not be accurate.

Reduction

Carboxylic acids are much harder to reduce than carbonyl compounds. However, they can be reduced to primary alcohols by the powerful reducing agent lithium tetrahydridoaluminate(III). The reagent is suspended in dry ether (ethoxyethane). Adding water after the reaction is complete destroys any excess reducing agent.

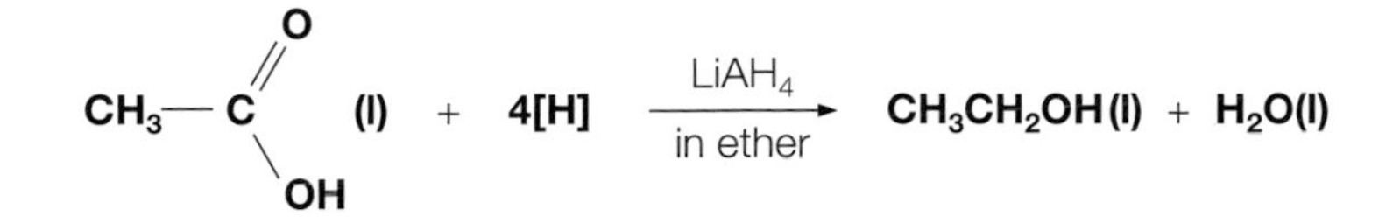

Figure 8.9▲
The reduction of a carboxylic acid with $LiAlH_4$.

Reaction with phosphorus(V) chloride

Phosphorus(V) chloride, PCl_5, reacts vigorously with carboxylic acids at room temperature. The reaction replaces the –OH group with a chlorine atom, forming an acyl chloride. The other product is hydrogen chloride gas which fumes in moist air.

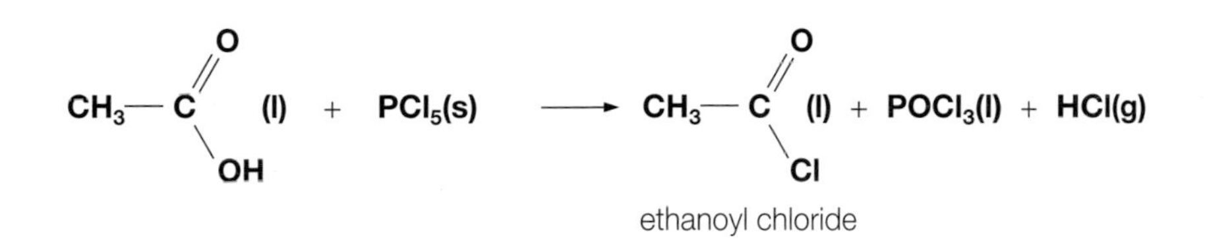

Figure 8.10▲
The reaction of PCl_5 with ethanoic acid.

Test yourself

11 Write the structural formula and give the name of the product of the reaction of:
   a) butanoic acid with $LiAlH_4$
   b) propanoic acid with PCl_5.

12 a) How can the reaction of carboxylic acids with PCl_5 be used as a test tube test and what does the test show?
   b) Alcohols and carboxylic acids both give a positive result when tested with PCl_5. How can they be distinguished?

Esterification

Carboxylic acids react with alcohols to form esters (see Section 8.5). The two organic compounds are mixed and heated under reflux in the presence of a small amount of a strong acid catalyst such as concentrated sulfuric acid.

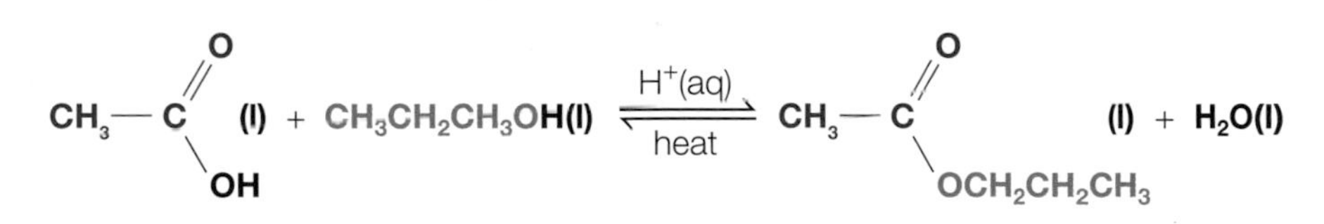

Figure 8.11▲
The formation of an ester from ethanoic acid and propan-1-ol.

This reaction is reversible. The conditions for reaction have to be arranged to increase the yield of the ester. One possibility is to use an excess of either the acid or the alcohol, depending on which is the more available or cheaper. Using more concentrated sulfuric acid than needed for its catalytic effect can help too, because the acid reacts with the water formed. In some esterification reactions it is possible to distil off either the ester or the water as they form, which encourages the reaction to go to completion.

Test yourself

13 Use le Chatelier's principle to explain the methods used to increase the yield of an ester formed from an acid and an alcohol.

Activity

Preparation of an ester

This sequence of diagrams shows the procedure for preparing a small sample of an ester.

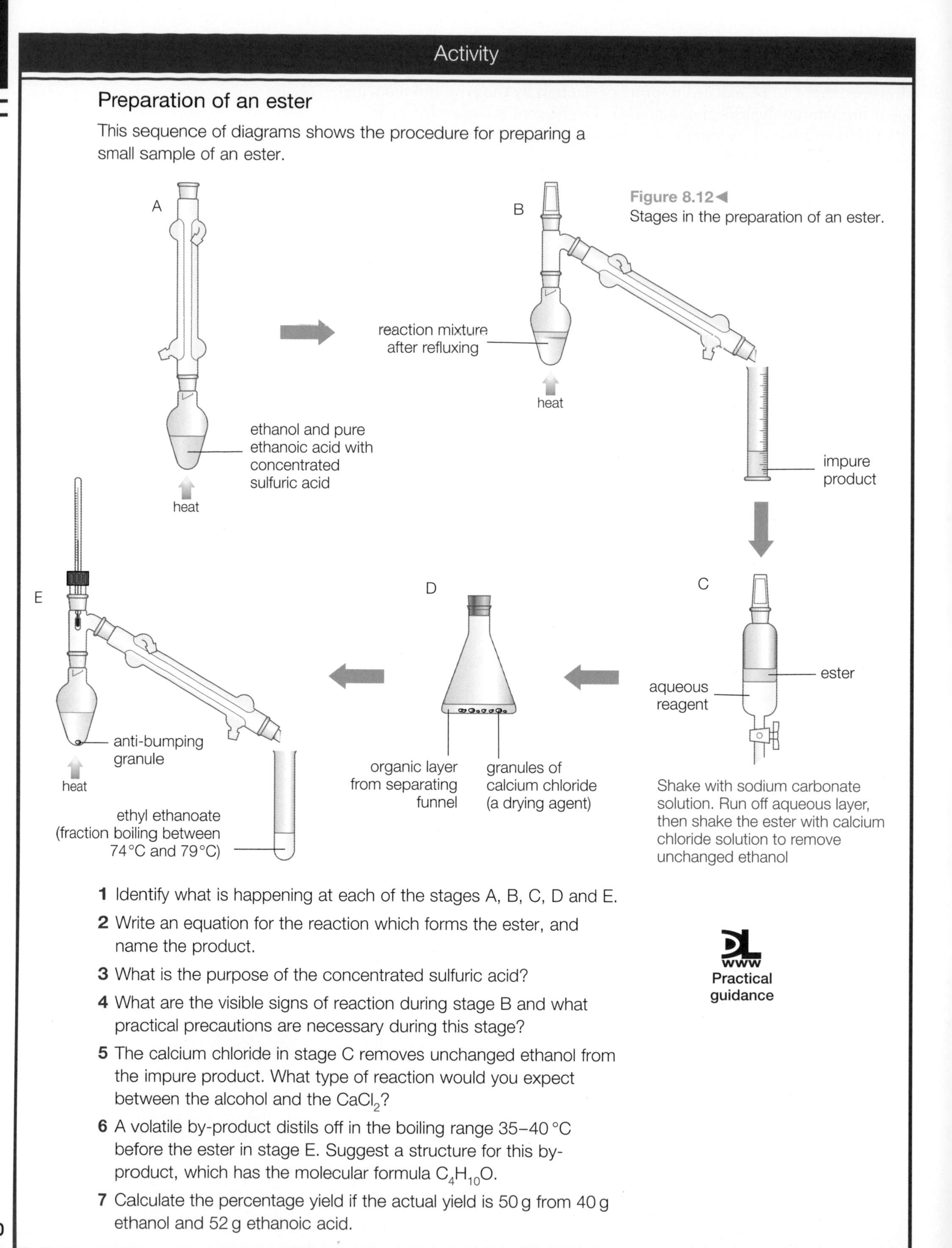

Figure 8.12◀ Stages in the preparation of an ester.

1 Identify what is happening at each of the stages A, B, C, D and E.

2 Write an equation for the reaction which forms the ester, and name the product.

3 What is the purpose of the concentrated sulfuric acid?

4 What are the visible signs of reaction during stage B and what practical precautions are necessary during this stage?

5 The calcium chloride in stage C removes unchanged ethanol from the impure product. What type of reaction would you expect between the alcohol and the $CaCl_2$?

6 A volatile by-product distils off in the boiling range 35–40 °C before the ester in stage E. Suggest a structure for this by-product, which has the molecular formula $C_4H_{10}O$.

7 Calculate the percentage yield if the actual yield is 50 g from 40 g ethanol and 52 g ethanoic acid.

www **Practical guidance**

8.4 Acyl chlorides

Chemists value acyl chlorides as reactive compounds for synthesis both on a small laboratory scale and on a large scale in industry. These reagents often provide the easiest way to make important products such as esters. They are acylating agents.

Acyl chlorides are very reactive. Ethanoyl chloride, for example, is a colourless liquid which fumes as it reacts with moisture in the air. The reaction between ethanoyl chloride and water is violent at room temperature.

Ethanoyl chloride and other acyl chlorides, also react rapidly at room temperature with alcohols to form esters.

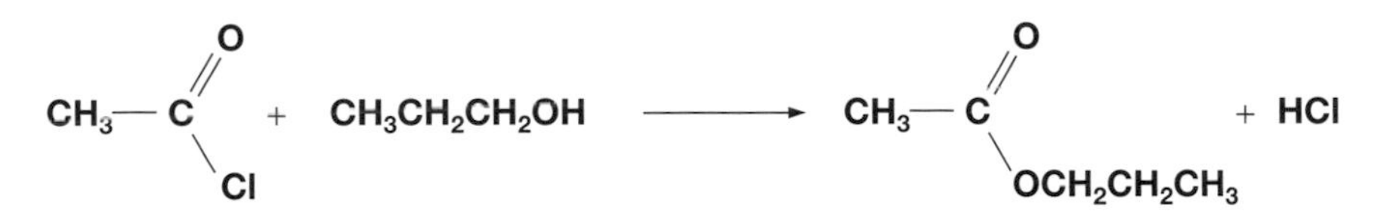

Figure 8.13▲
The formation of an ester from ethanoyl chloride and propan-1-ol.

Acid chlorides such as ethanoyl chloride also react with ammonia and amines to form amides (see Section 13.7).

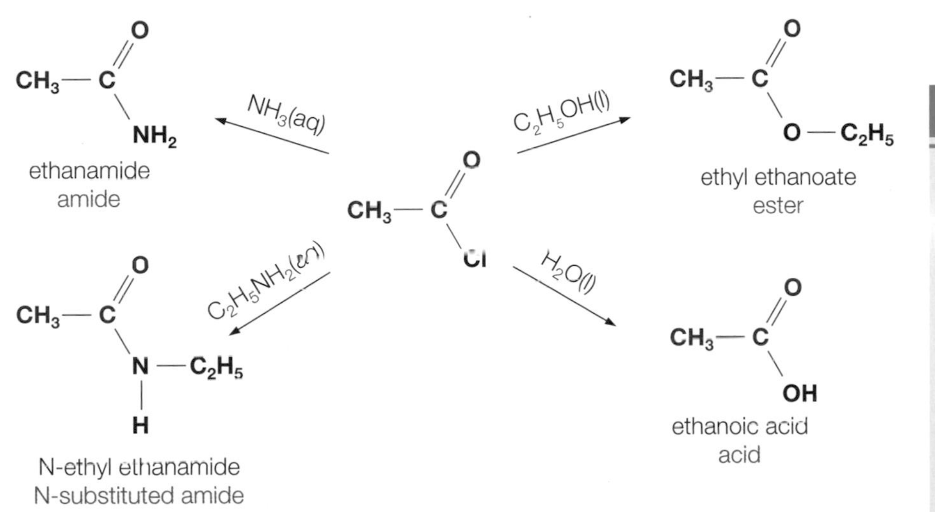

Figure 8.14▲
A summary of the reactions of ethanoyl chloride. All the reactions happen quickly at room temperature. The ethanoyl group in the products is shown in red.

Definitions

An **acyl group** consists of all the parts of a carboxylic acid except the –OH group. The ethanoyl group, CH_3CO–, is an example of an acyl group.

Acylation is a reaction which substitutes an acyl group for a hydrogen atom. The H atom may be part of an –OH group, an $-NH_2$ group or a benzene ring.

Test yourself

14 Write an equation for the reaction between ethanoyl chloride and water. Show that this is an example of hydrolysis.
15 Write an equation for the formation of ethanamide from ethanoyl chloride to show why two moles of ammonia are required for reaction with one mole of the acyl chloride.
16 Draw the structure and name the product of the reaction of propanoyl chloride and 1-aminobutane.

8.5 Esters

Occurrence and uses

Many of the sweet smelling compounds found in perfumes and fruit flavours are esters. Some drugs used in medicine are esters, including aspirin, paracetamol and the local anaesthetics novocaine and benzocaine. The insecticides malathion and pyrethrin are also esters. Compounds with more than one ester link include fats and oils as well as polyester fibres. Other esters are important as solvents and plasticisers.

Definition

Essential oils are the oils that chemists extract from plants as a source of chemicals for perfumes, food flavourings and other uses. They often include esters.

Figure 8.15►
Natural fruit flavours are complex mixtures. Some simpler esters on their own have odours which resemble fruit flavours. Examples are propyl ethanoate (pear), ethyl butanoate (pineapple), octyl ethanoate (orange), 2-methylpropyl ethanoate (apple).

Names and structures

The general formula for an ester is RCOOR', where R and R' are alkyl or aryl groups.

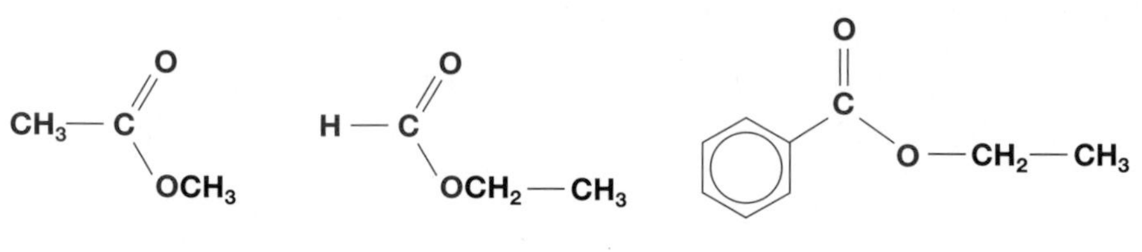

Figure 8.16▲
The names and structures of some esters.

Physical properties

Common esters such as ethyl ethanoate are volatile liquids and only slightly soluble in water.

Hydrolysis reactions

Hydrolysis splits an ester into an acid and an alcohol.

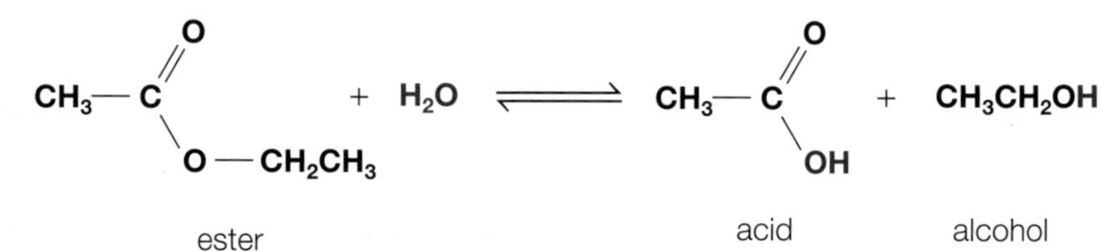

Figure 8.17▲
Hydrolysis of an ester. These are the products when an ester is heated with an excess of dilute acid, such as hydrochloric acid. This reaction is reversible.

Acids or bases can catalyse the hydrolysis. Hydrolysis catalysed by an acid is a reversible reaction. It is the reverse of the reaction used to synthesise esters from carboxylic acids (see Section 8.3).

Base catalysis is generally more efficient because it is not reversible. This is because the acid formed loses its proton by reacting with excess alkali. This turns it into a negative ion which does not react with the alcohol.

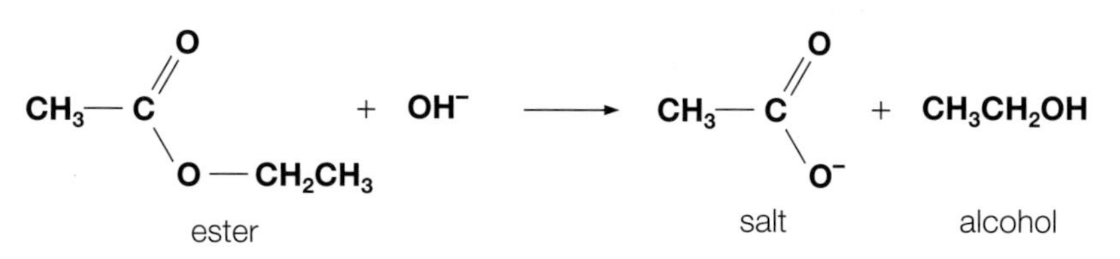

Figure 8.18▲
The result of hydrolysing ethyl ethanoate by heating it with an aqueous alkali such as sodium hydroxide. The salt and alcohol produced do not react with each other so this reaction is not reversible.

Test yourself

17 Give the name and displayed formulae of the esters formed when:
- a) butanoic acid reacts with propan-1-ol
- b) ethanoic acid reacts with methanol
- c) ethanoic acid reacts with butan-1-ol.

Data

Test yourself

18 Explain, in terms of intermolecular forces, why the boiling temperature of ethyl ethanoate is similar to that of ethanol and lower than that of ethanoic acid.

Test yourself

19 Identify the products of heating:
- a) propyl butanoate with dilute hydrochloric acid
- b) ethyl methanoate with aqueous sodium hydroxide.

20 Under acid conditions the reaction of ethyl ethanoate with water is reversible.
- a) What conditions favour the hydrolysis of the ester?
- b) How do these conditions compare with those for the synthesis of the ester?

8.6 Triglycerides

Fats, oils and fatty acids

Fats and vegetable oils are esters of long-chain carboxylic acids and the alcohol propane-1,2,3-triol, better known as glycerol. There are three –OH groups in a glycerol molecule so the alcohol can form three ester links with carboxylic acids, giving rise to triglycerides.

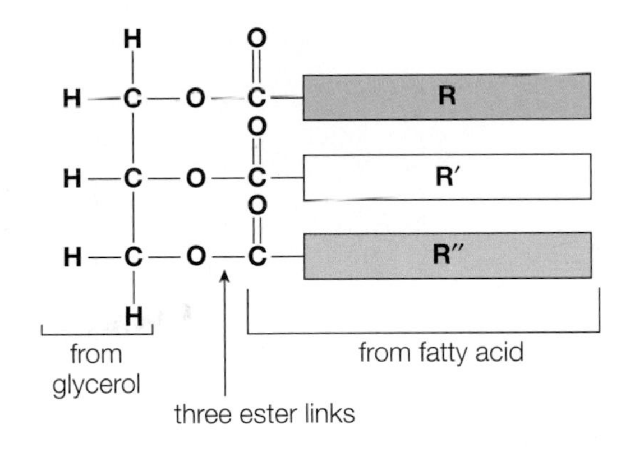

Figure 8.19▲
The general structure of a triglyceride. In natural fats and vegetable oils the hydrocarbon chains may all be the same or they may be different.

Definitions

A **triglyceride** is an ester of glycerol (propane-1,2,3-triol) with three carboxylic acid molecules. The carboxylic acids may or may not all be the same.

Fatty acids are the naturally occurring carboxylic acids in triglycerides. The acids have long hydrocarbon chains.

Saturated fatty acids do not have double bonds in the hydrocarbon chain. There is one, or more, C=C double bond in a molecule of an **unsaturated fatty acid**.

The carboxylic acids in fats are usually referred to as fatty acids. Saturated fatty acids do not have double bonds in the hydrocarbon chain but there are double bonds in the molecules of unsaturated fatty acids.

Fatty acid	Chemical name	Formula
Palmitic	Hexadecanoic	$CH_3(CH_2)_{14}COOH$
Stearic	Octadecanoic	$CH_3(CH_2)_{16}COOH$
Oleic	*cis*-octadec-9-enoic	$CH_3(CH_2)_7CH{=}CH(CH_2)_7COOH$
Linoleic	*cis*, *cis*-octadec-9,12-dienoic	$CH_3(CH_2)_4CH{=}CHCH_2CH{=}CH(CH_2)_7COOH$

Table 8.1◀
Examples of fatty acids. The *cis–trans* system for naming geometric isomers is still generally used for fatty acids.

Fats are solid at around room temperature (below 20 °C). Fats contain a high proportion of saturated fatty acids. Solid triglycerides are generally found in animals. In lard, for example, the main fatty acids are palmitic acid (28%), stearic acid (8%) and only 56% oleic acid, which is a much lower percentage than in a vegetable oil such as olive oil.

Test yourself

21 Classify the acids in Table 8.1 as saturated or unsaturated.

22 Draw the skeletal formula of *cis,cis,cis*-octadec-9,12,15-trienoic acid.

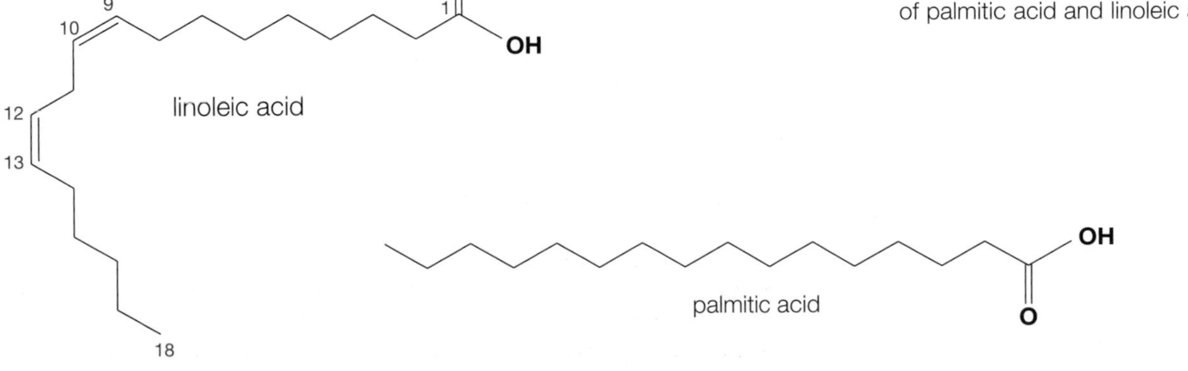

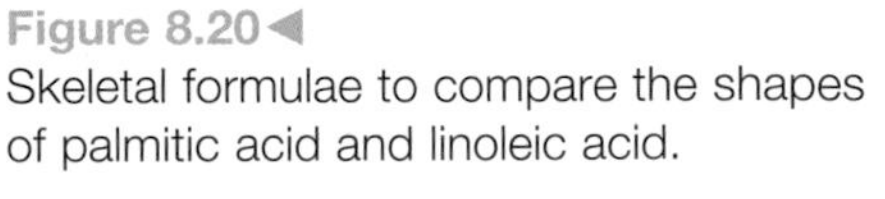
Figure 8.20◀
Skeletal formulae to compare the shapes of palmitic acid and linoleic acid.

Figure 8.21▲
A farmer inspecting sunflowers in a field. He is growing the flowers for their seeds which will be compressed to extract oil for cooking.

The triglycerides with unsaturated fatty acids have a less regular structure than saturated fats (see Figure 8.20). The molecules do not pack together so easily to make solids so they have lower melting temperatures and have to be cooler before they solidify. Triglycerides of this kind occur in plants and are liquids at around room temperature. In a vegetable oil such as olive oil the main fatty acids are oleic acid (80%) and linoleic acid (10%).

Hydrolysis

Hydrolysis of triglycerides produces soaps and glycerol. Soaps are the sodium or potassium salts of fatty acids.

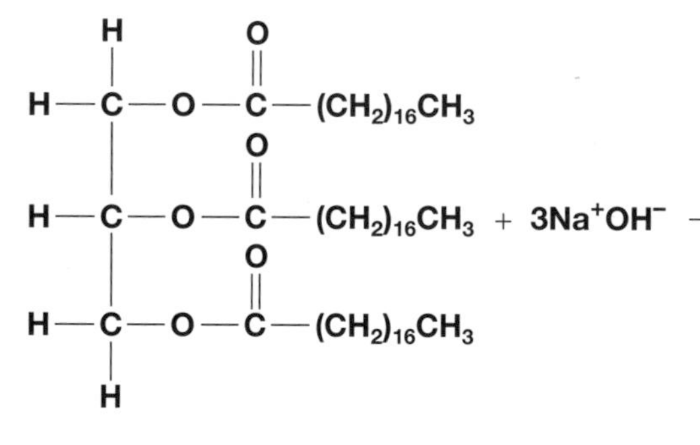

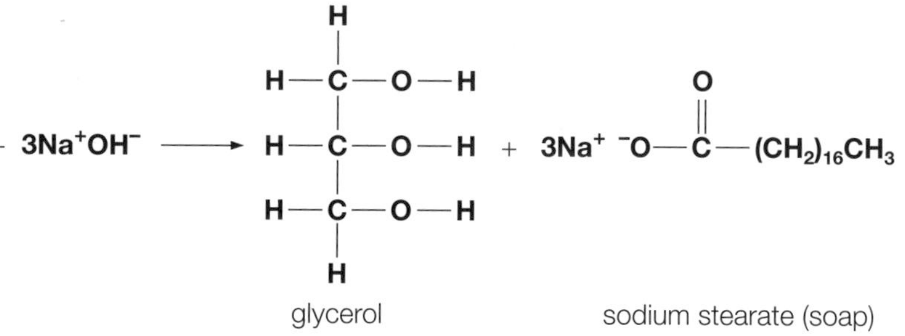

Figure 8.22▲
The hydrolysis of a fat or vegetable oil (triglyceride) with alkali to make soap. This process is also called saponification.

Soaps help to remove greasy dirt because they have an ionic (water-loving) head and a long hydrocarbon (water-hating) tail. Chemists call them surface active agents, or surfactants. Soaps first help to separate greasy dirt from surfaces. Then they keep the dirt dispersed in water so that it rinses away.

Most toilet soaps are made from a mixture of animal fat and coconut palm oil. Soaps from animal fat are less soluble and longer lasting. Soaps from palm oils are more soluble so that they lather quickly but wash away more quickly. A bar of soap also contains a dye and perfume together with an antioxidant to stop the soap and air combining to make irritant chemicals.

Test yourself

23 Why is hydrolysis with alkali preferred to acid hydrolysis when making soaps from fats and oils?

Hydrogenation

Margarine was invented as a cheap substitute for butter. The production of margarine originally depended on adding hydrogen to the double bonds in vegetable oils to turn them into saturated fats. This process turns a liquid oil into a solid fat so it is sometimes called 'hardening'.

The catalyst for hydrogenating a vegetable oil is finely divided nickel which is suspended in the oil during hydrogenation and then recovered by filtration. It is not necessary to hydrogenate all the double bonds in an oil for it to be sufficiently hardened for use in foods.

Margarine, and other non-dairy spreads, are blends of vegetable oils. Some of them contain partly hardened fats to make the product a spreadable solid. Fats made by hardening vegetable oils are also used to manufacture products such as cakes and biscuits.

Naturally occurring fatty acids exist as the *cis* isomers. In the presence of the nickel catalyst used for hydrogenation, some of the *cis* isomers become *trans*. This means that there is a proportion of *trans*-fats in products made using partially hardened oils. There is growing concern that *trans*-fats may be bad for health.

The formation of *trans*-fats contributes to the hardening process because a molecule of a *trans*-fat is straight, rather than kinked. This means that its shape is very like the straight chain of a fully saturated fat.

Definitions

Polyunsaturated fats contain fatty acids with two or more double bonds in the chain.

Some fats or oils provide fatty acids that are essential in the diet. Two **essential fatty acids** are linoleic and linolenic acids.

Test yourself

24 Name the fatty acid formed by completely hydrogenating linoleic acid (see Table 8.1).

25 a) Draw skeletal formulae for *cis*-octadec-9-enoic acid (oleic acid) and *trans*-octadec-9-enoic acid (elaidic acid).

b) What has to happen during hydrogenation for the *cis* isomer to turn into the *trans* isomer?

c) Suggest a reason why a triglyceride with three elaidic acid molecules has a higher melting temperature than a triglyceride with oleic acid molecules.

Rearranging esters in food

Food scientists have now found ways to turn vegetable oils into solid spreads without hydrogenation. Instead they use a process which allows triglycerides to swap fatty acid molecules. The result of this process is a product with a higher melting temperature. This happens because the process works in a way that allows the molecules to pack together more neatly to make a denser material. The process is sometimes called 'trans-esterification'.

Trans-esterification does not produce *trans*-fats but it does produce triglycerides that are not found naturally. Catalysts for trans-esterification include sodium methoxide and the enzyme lipase.

Note

The term trans-esterification is sometimes used to describe the swapping of fatty acids between triglycerides, but strictly this process should be called inter-esterification (see also page 107).

Test yourself

26 Draw a simple diagram to represent the change to triglycerides when they undergo trans-esterification.

27 Sodium methoxide is used as a catalyst for trans-esterification. Suggest reasons why:

a) this can be hazardous

b) the catalyst is destroyed if there is water present in the vegetable oil.

Activity

Health risks and fats

The Food Standards Agency advises the public about diet and health. The Agency explains on its website that it is important to have some fat in our diet because fat helps the body to absorb some vitamins, it is a good source of energy and a source of the essential fatty acids that the body cannot make itself. However, the agency also warns that having a lot of fat makes it easy to have more energy than we need, which means we might be more likely to put on weight.

We are told that we should be cutting down on food that is high in saturated fat or *trans*-fats, or replacing these foods with ones that are high in unsaturated fat instead.

Figure 8.23▲
Milk, full-fat yogurt, cheeses and eggs are rich in saturated fats. Having too much saturated fat can increase the amount of cholesterol in the blood, which increases the chance of developing heart disease.

Figure 8.24◀
Foods containing hydrogenated vegetable oil. Hydrogenated vegetable oil is used because it is both cheap and extends the shelf life of foods. Hydrogenated vegetable oil contains *trans*-fats which have a similar effect on blood cholesterol to saturated fats. They raise the type of cholesterol in the blood that increases the risk of heart disease.

Unsaturated fats can be a healthy choice. These types of fats can actually reduce cholesterol levels and provide us with the essential fatty acids that the body needs. They include the unsaturated fats found in oily fish, which may help to prevent heart disease.

Someone, who has never studied advanced chemistry, has looked at the Food Standards Agency website but wants to know more about fats in the diet. Give answers to these questions using words and diagrams suitable for a non-specialist:

1 What is the difference between a fat and a fatty acid?

2 Why are some fatty acids essential in the diet while others are not?

3 What is the difference between a saturated and an unsaturated fat?

4 Why are *trans*-fats unnatural and how do they get into the food we eat?

5 Why do some spreads include *trans*-fats while others do not?

Definition

Cholesterol plays an important part in metabolism. It makes up part of cell membranes and is converted in the body to steroid hormones, such as the sex hormones testosterone and progesterone. High levels of cholesterol in the blood may lead to deposits building up in arteries, resulting in heart disease.

Esters for biodiesel

Biodiesel is a renewable fuel. At the moment most biodiesel is made from triglycerides in vegetable oils or animal fats. The triglyceride is heated with methanol or ethanol in the presence of a base to act as catalyst. If the base catalyst is sodium hydroxide it can be dissolved in the alcohol before being added to the triglyceride. When the reaction is complete the fatty acids are converted to methyl or ethyl esters and glycerol separates as a by-product. Chemists call this 'trans-esterification'.

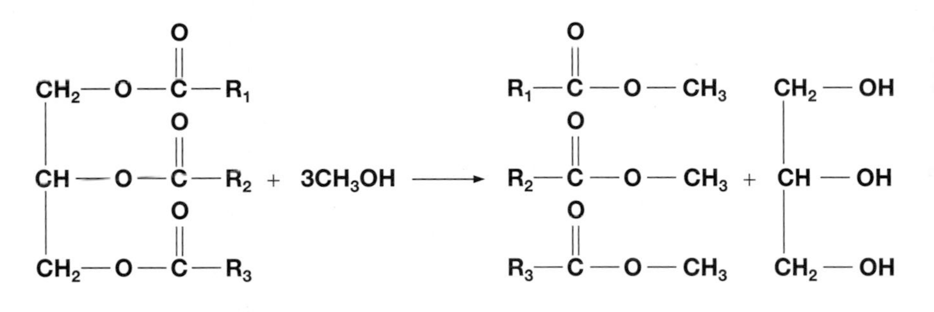

Figure 8.25▲
Trans-esterification of fatty acids in a triglyceride with methanol. R_1, R_2 and R_3 represent the hydrocarbon chains of the fatty acids.

The use of biodiesel is increasing because of the urgent need to reduce carbon dioxide emissions to the atmosphere. The carbon dioxide released when biodiesel burns is equivalent to the carbon dioxide taken in by photosynthesis during growth. However, across the whole life cycle of these fuels they are not carbon neutral because of the energy resources used during the production of the fuel, including the manufacture of fertilisers, plant cultivation, harvesting, extraction of oil and processing of the oil into fuel.

Some studies have investigated the carbon dioxide emissions resulting from preparing rainforests, peatlands and grasslands to grow palms or soya beans for biodiesel. Clearing the land can release between 17 and 420 times more carbon than the annual savings from replacing fossil fuels.

The first biofuels have relied on the direct use of food crops to produce biodiesel from oils extracted from palm nuts or rapeseed and bioethanol from corn starch. New research is developing alternatives derived from either the waste parts of food crops (such as straw from corn and wheat) or based on crops grown on marginal land that cannot be used for food production.

Definition

Trans-esterification involves the reaction of an ester, RCO–OR' with an alcohol, R''–OH to give a new ester with a different alkoxy group, RCO– R''.

Figure 8.26▲
A tube of biofuel made from soy oil extracted from soybeans at a biodiesel plant in Argentina.

Test yourself

28 Suggest a reason why the product of reacting a vegetable oil with methanol produces a much better fuel than the original oil itself.

29 Suggest a reason why the alcohol and the triglyceride must be very dry to ensure a good yield of biodiesel.

30 Explain the difference between trans-esterification to make biodiesel and trans-esterification (strictly, inter-esterification) to raise the melting temperature of vegetable oils.

31 a) Explain why the production and use of biodiesel could in theory be carbon neutral.

b) Why, in practice, is the production of biodiesel not carbon neutral?

32 Why is it increasingly important to develop new raw materials for making biofuels in order to avoid using vegetable oils or corn starch?

8.7 Polyesters

Polyesters are polymers in which the monomers are linked together by ester groups. The commonest polyester is PET. The initials are short for the traditional name for the polymer which is polyethylene terephthalate. It is a polymer formed when ethane-1,2-diol (ethylene glycol) reacts with benzene-1,4-dicarboxylic acid (terephthalic acid).

Definitions

A **condensation reaction** is a reaction in which molecules join together by splitting off a small molecule such as water.

Condensation polymers are produced by a series of condensation reactions between the functional groups of the monomers.

Figure 8.27▲
Bottles for sparkling drinks are made of the polyester, PET. After heat treatment, this polymer is impermeable to gases.

The formation of polyesters involves a series of condensation reactions splitting off water between the monomer molecules to create ester links (see Section 14.7).

REVIEW QUESTIONS

Extension questions

1 Account for the fact that butane, propan-1-ol, propanal and ethanoic acid all have about the same relative molecular mass, but their boiling temperatures are 273 K, 371 K, 322 K and 391 K, respectively. (4)

2 Predict the names and structures of the main organic products of each of these reactions:

a) propanoic acid and phosphorus(V) chloride (1)

b) butanoic acid and $LiAlH_4$ in ethoxyethane (1)

c) pentyl ethanoate and hot, aqueous sodium hydroxide (2)

d) ethanoic acid and aqueous calcium hydroxide (1)

e) ethanoyl chloride and propan-2-ol. (1)

3 Consider this reaction sequence:

$$C_3H_7CN \xrightarrow{\text{step 1}} C_3H_7COOH \xrightarrow{\text{step 2}} X$$

$$\xrightarrow[NH_3]{\text{step 3}} CH_3CH_2CH_2CONH_2$$

a) Give the reagents and conditions for steps 1 and 2 (2)

b) Give the name and structure of X. (2)

c) Give the displayed formula and name of the product of step 3. (2)

4 a) Describe two examples of test tube reactions that you could use to show the similarities and differences between ethanoic acid and hydrochloric acid. (3)

b) Explain, with the help of equations, the observations you have described in a). (6)

5 Ibuprofen is a painkiller with this skeletal formula:

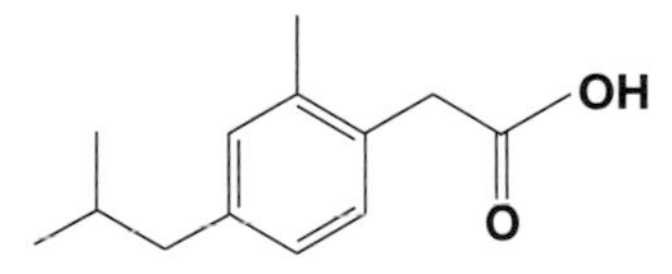

a) What is the molecular formula of ibuprofen? (1)

b) Suggest whether or not ibuprofen is soluble in water. Explain your answer in terms of intermolecular forces. (3)

c) Draw the displayed formula of the organic products formed when ibuprofen reacts with:

i) dilute sodium hydroxide solution (1)

ii) ethanol and a little sulfuric acid on warming. (1)

6 Octadecanoic acid and *cis*-octadec-9-enoic acid are fatty acids present in the triglycerides that make up fats and vegetable oils.

a) Suggest an explanation for the fact that octadecanoic acid is a solid while *cis*-octadec-9-enoic acid is a liquid at room temperature. (3)

b) Describe a chemical test that could be used to distinguish the two fatty acids. (3)

c) Explain what is meant by a triglyceride. (2)

d) Reaction with hydrogen in the presence of a nickel catalyst is used to harden vegetable oils for the food industry. During partial hydrogenation, some of the fatty acids change from the *cis* to the *trans* form.

i) What happens to *cis*-octadec-9-enoic acid when it reacts with hydrogen? (1)

ii) What is the difference between *cis*-octadec-9-enoic acid and *trans*-octadec-9-enoic acid? (2)

iii) Why should people be concerned by the use of partial hydrogenation to prepare ingredients for foodstuffs? (2)

7 a) Explain the meaning of the term condensation polymerisation. (2)

b) Give the structures of a pair of compounds that can react to form a polyester and name the functional groups in the two compounds. (2)

c) State one important use of polyester polymers and state the properties of the polymer on which the use depends. (2)

9 Spectroscopy and chromatography

In a modern laboratory, organic analysis is based on a range of automated and instrumental techniques. These include chromatographic methods to separate and purify chemicals and a range of spectroscopic techniques for analysis.

Figure 9.1 ▲
Using a mass spectrometer in a forensic laboratory to detect drugs in a urine sample.

9.1 Radiation and molecules

The radiation we can see is the very narrow band of the electromagnetic spectrum that our eyes detect as visible light. We see because the photons of visible light have enough energy to bring about chemical changes in the cells of our retinas. These chemical changes then set off nerve signals to the brain. Other types of electromagnetic radiation with different wavelengths can also change atoms and molecules but our eyes cannot detect them.

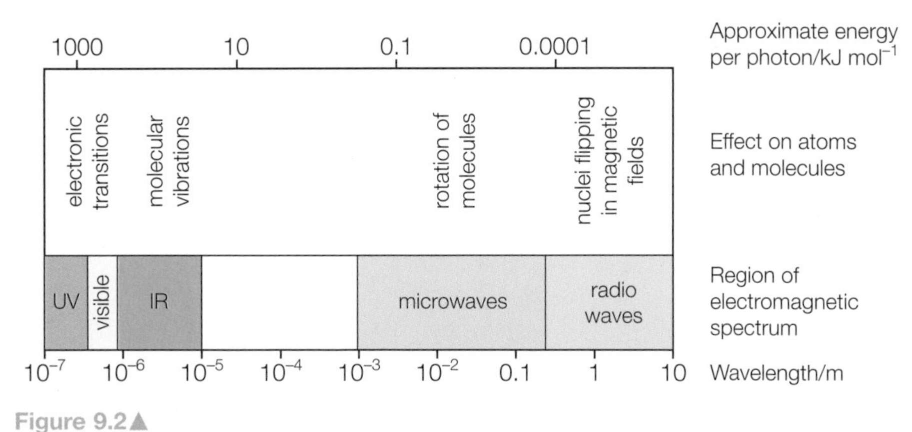

Figure 9.2 ▲
The electromagnetic spectrum and its chemical effects.

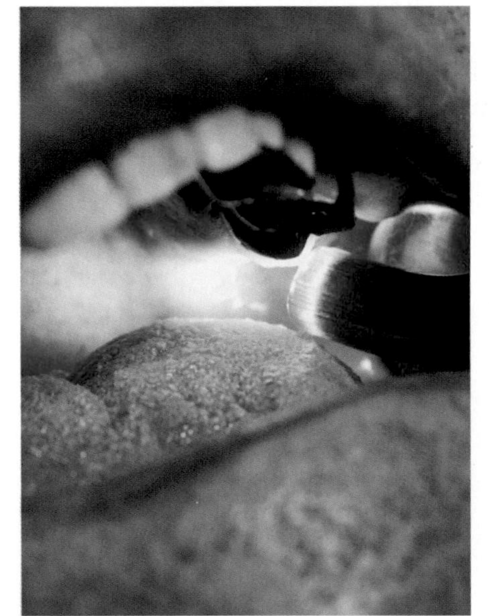

Figure 9.3 ▲
A dentist using an ultraviolet lamp to set the cement being used for the implants in a patient's teeth. The blue light is a guide to the dentist, who cannot see the ultraviolet light. This technique is also used for some types of dental filling.

The shorter the wavelength of radiation, the higher its frequency and the greater the energy per photon. The energy of photons in ultraviolet (UV) radiation, for example, is large enough to break chemical bonds. This means that UV radiation can start chemical reactions, including the formation and destruction of ozone in the stratosphere and the initiation of the reactions of alkanes with chlorine.

Photons with less energy than those in UV light interact with vibrating molecules. This is what happens when chemicals absorb infrared (IR) radiation. The lower IR frequencies correspond to the natural frequencies of vibration of bonds in molecules as they stretch and bend. However, it is only molecules which change their polarity as they vibrate that can absorb IR radiation. This is the basis of IR spectroscopy (see Section 9.3).

At even lower frequencies, the energy of microwaves is absorbed by polar molecules as they rotate. Internal friction resulting from intermolecular forces spreads the energy between molecules, so that the chemical gets hotter as its molecules rotate and vibrate faster. This type of heating can speed up some reactions enormously because the energy is absorbed directly by the molecules. Research and industrial laboratories are making increasing use of microwaves for heating. The greater control that is possible with the help of microwaves can give rise to cleaner and more efficient processes.

The photons of radio waves have low energies but these correspond to the tiny energy jumps as the nuclei of some atoms change their alignment in a magnetic field. This is the basis of nuclear magnetic resonance spectroscopy (Section 9.2).

Test yourself

1 Write an equation to show the effect of UV light on chlorine molecules and explain the significance of this change for the reaction of alkanes with chlorine.
2 Use a simple diagram to show the vibrations of a molecule that can absorb IR radiation.
3 Identify reasons why heating with microwaves is more efficient than heating with an electric mantle or Bunsen flame.

9.2 Nuclear magnetic resonance spectroscopy (nmr)

Nuclear magnetic resonance spectroscopy (nmr) is a powerful analytical technique for finding the structures of carbon compounds. The technique is used to identify unknown compounds, to check for impurities and to study the shapes of molecules.

This type of spectroscopy studies the behaviour of the nuclei of atoms in magnetic fields. It is limited to those nuclei that behave like tiny magnets because they have a property called spin. In common organic compounds the only nuclei to do this are those of carbon-13, ^{13}C, and hydrogen-1, ^{1}H. The nuclei of the much commoner carbon-12, oxygen-16 and nitrogen-14 atoms do not show up in nmr spectra.

When placed in a very strong magnetic field, magnetic nuclei line up in either the same direction as, or the opposite direction to, an external magnetic field. There is a small energy jump when the protons flip from one alignment to the other. The size of the energy jump corresponds to the energies of photons of radio waves.

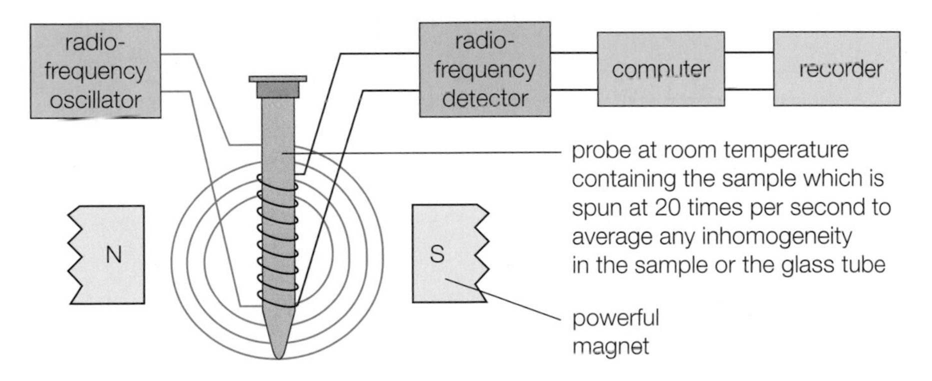

Figure 9.4 ◀ A diagram of an nmr spectrometer showing the key features of the technique.

A tube with the sample is supported in a strong magnetic field (Figure 9.4). The operator turns on a source of radiation at radio frequencies. The radio-frequency detector records the intensity of the signal from the sample as the oscillator emits pulses of radiation across a range of wavelengths.

The sample is dissolved in a solvent (see page 113). Also in the solution is some tetramethylsilane (TMS), $Si(CH_3)_4$, which is a standard reference compound producing a single, sharp absorption peak well away from the peaks produced by samples for analysis.

Definition

The horizontal scale of an nmr spectrum shows the **chemical shifts** of the peaks measured in parts per million (ppm). The symbol δ stands for the 'chemical shift' relative to the zero on the scale which is given by the signal from tetramethylsilane.

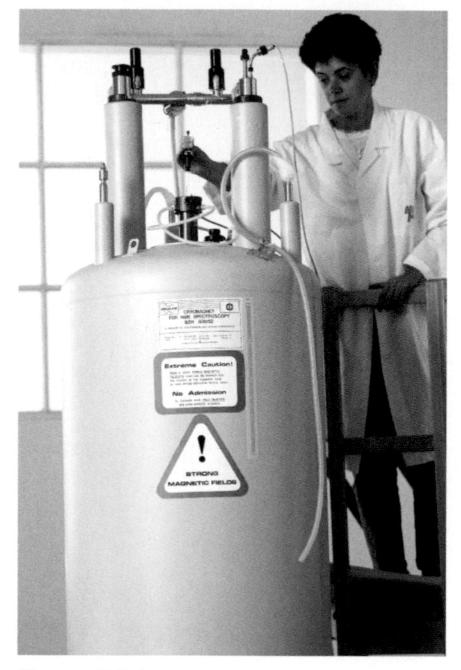

Figure 9.5▲
A researcher adding a sample to a nuclear magnetic resonance spectrometer.

Each peak corresponds to one or more magnetic atoms in a particular chemical environment. Nuclei in different parts of a molecule experience slightly different magnetic fields in an nmr machine. This is because the tiny magnetic fields associated with the electrons of neighbouring bonds and atoms modify the field applied by the spectrometer. These effects are very small and bring about changes in the magnetic field measured in parts per million.

The recorder prints out a spectrum which has been analysed by computer to show peaks wherever the sample absorbs radiation strongly. The zero on the scale is fixed by the absorption of magnetic atoms in the reference chemical. The distances of the sample peaks from this zero are called their 'chemical shifts'. The symbol used for chemical shift is δ.

Proton nmr

Proton nmr relies on the magnetic properties of the hydrogen-1 isotope. The number of main peaks in the spectrum shows how many different chemical environments there are for hydrogen atoms. Also, the values for the chemical shifts are a useful indication of the types of chemical environment of the hydrogen atoms in each peak.

It is possible to work out the number of hydrogen atoms in each environment from the spectrum. This is possible because in a proton nmr spectrum the area under a peak is proportional to the number of nuclei.

An nmr instrument is set up to follow the curve and work out the ratios of the areas below the peaks. Sometimes the results of the calculation are shown by an integration trace, such as the blue line shown in the spectrum for methyl ethanoate in Figure 9.6. Alternatively, the instrument's computer prints a number below each peak which is a measure of the relative area under the curve, as shown for the spectrum in Figure 9.7.

Figure 9.6 shows the proton nmr spectrum for methyl ethanoate. This is a low-resolution spectrum which shows the main peaks but no fine detail.

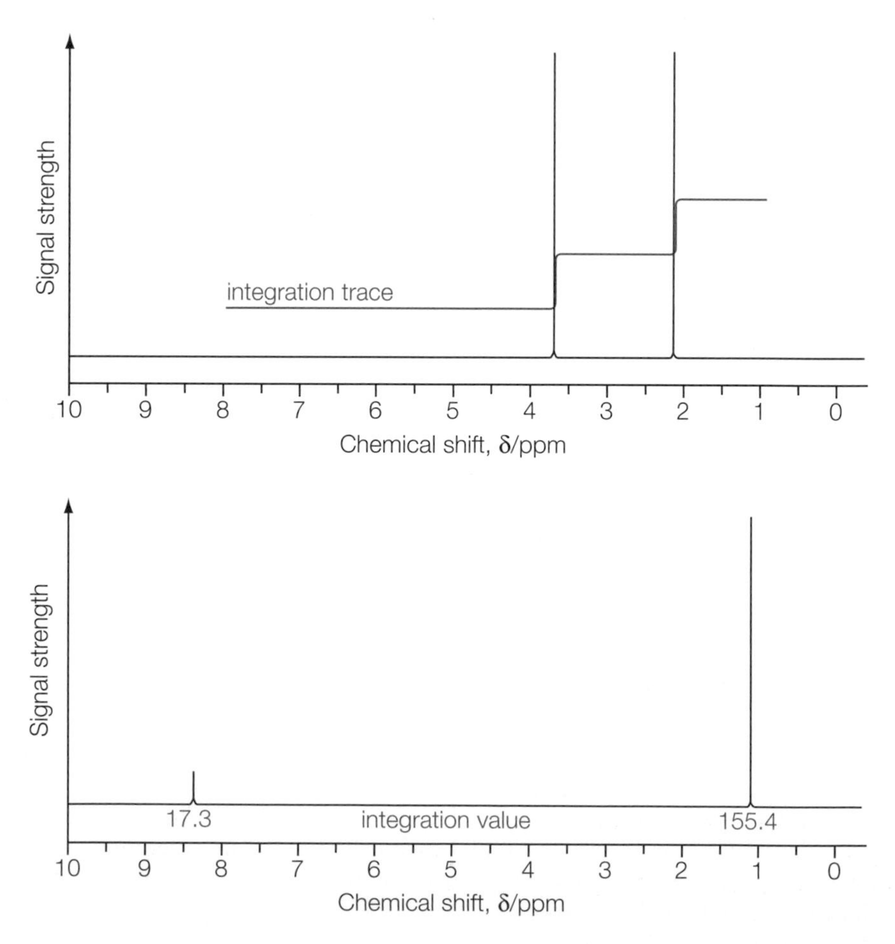

Figure 9.6►
A proton nmr spectrum of methyl ethanoate in $CDCl_3$.

Figure 9.7►
A proton nmr spectrum of a compound $C_5H_{10}O$ in $CDCl_3$.

Proton nmr spectra are usually recorded with the sample in solution. It is important that the solvent does not contain hydrogen atoms that would give peaks with chemical shifts similar to those in the sample. One possibility is to use a solvent that contains no hydrogen atoms, such as tetrachloromethane, CCl_4. The other is to use a solvent in which atoms of the hydrogen-1 isotope have been replaced by deuterium atoms, D or ${}^2_1H$. A compound that is often used is $CDCl_3$. Deuterium atoms produce peaks in regions of the spectrum well away from the chemical shifts for proton nmr.

The reference chemical for proton nmr is tetramethylsilane (TMS). There are twelve hydrogen atoms in a molecule of TMS and they all have the same chemical environment. TMS provides a single strong peak which marks the zero on the scale of chemical shifts.

Coupling

At high resolution it is possible to produce proton nmr spectra with more detail which provide even more information about molecular structures. The spins of protons connected to neighbouring carbon atoms interact with each other. Chemists call this interaction 'spin–spin coupling' and they find that the effect is to split the peaks into a number of lines.

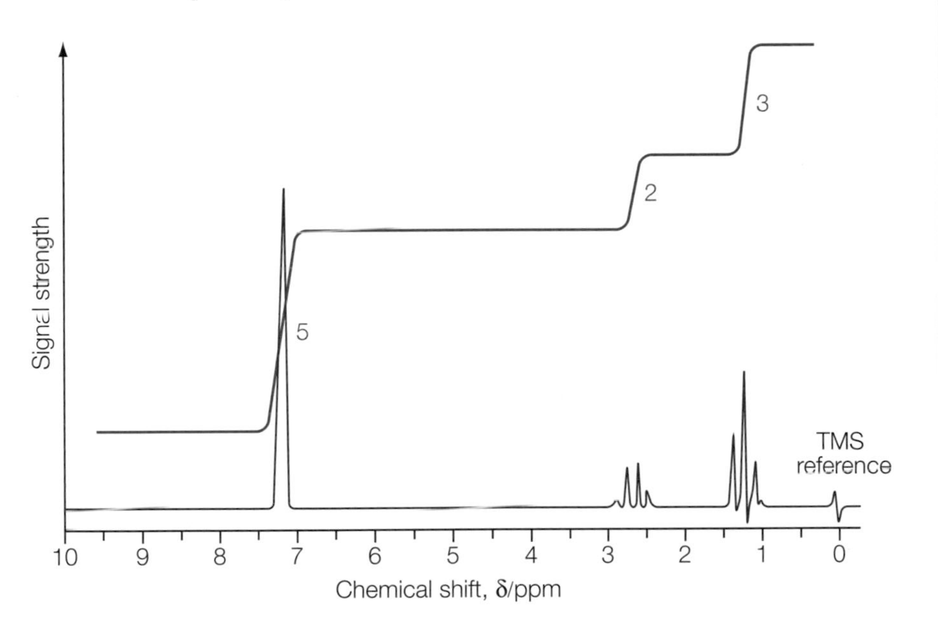

Figure 9.8▲
A high-resolution proton nmr spectrum for a hydrocarbon with a benzene ring (Topic 12). Note the extra peaks compared with a low-resolution spectrum.

A peak from protons bonded to an atom which is next to an atom with two protons splits into three lines with the central line being twice as large as the other two. This happens because there are three energy states available to the two protons, depending on whether each proton is aligned with or against the magnetic field:

- both aligned with the field
- one aligned with the field and one against the field (with two possible combinations)
- both aligned against the field.

For similar reasons, a peak from protons bonded to an atom which is next to an atom with three protons splits into four lines. In general the 'n + 1' rule makes it possible to work out the numbers of coupled protons where n is the number of protons on the adjacent atom.

Test yourself

4 Nmr based on hydrogen nuclei is sometimes called ^1H nmr and sometimes proton nmr. Show that these are equivalent names.

5 Refer to the proton nmr spectrum in Figure 9.6 and the Edexcel Data booklet for chemical shifts.
   a) Explain why there are two peaks in the spectrum.
   b) Use the chemical shift values to decide which hydrogen atoms give rise to each peak.
   c) Show that the integration trace is as expected for the molecule.

6 Refer to the proton nmr spectrum in Figure 9.7 and the Edexcel Data booklet for chemical shifts.
   a) How many different chemical environments are there for hydrogen atoms in the molecule?
   b) Use the integration values under each peak to work out the ratios of hydrogen atoms in each environment.
   c) Use your answers to **a)** and **b)** and chemical shift values to suggest a structure for the compound.
   d) Describe two chemical tests that could be used to confirm the presence of the main functional group in the molecule. State what you would do and what you would expect to observe.

7 With the help of the table of chemical shifts in the Edexcel Data booklet, sketch the low-resolution nmr spectrum you would expect for:
   a) butanone
   b) 2-methylpropan-2-ol.

Note

Protons in the same chemical environment, with the same chemical shift, do not couple with each other.

number of equivalent protons causing splitting	splitting pattern and relative intensity of the peaks
1	1 1
2	1 2 1
3	1 3 3 1
4	1 4 6 4 1

Figure 9.9▲
Pascal's triangle predicts the pattern of peaks and the relative peak heights.

Test yourself

8 Show that when a proton can couple with three protons in a methyl group, there are four alignments of the methyl protons that can give rise to peaks with relative intensities 1 : 3 : 3 : 1.
9 Use the chart of chemical shifts in the Edexcel Data booklet to suggest a structure for the hydrocarbon with a benzene ring which has the proton nmr spectrum shown in Figure 9.8.
10 Sketch the high-resolution nmr spectrum you would expect to observe with:
   a) propane
   b) ethoxyethane.

Labile protons

Hydrogen bonding affects the properties of compounds with hydrogen atoms attached to highly electronegative atoms such as oxygen or nitrogen. These molecules can rapidly exchange protons as they move from one electronegative atom to another. Chemists describe these protons as labile.

Labile protons do not couple with the protons linked to neighbouring atoms. This means that the nmr peak for a proton in an –OH group appears as a single peak in a high-resolution spectrum.

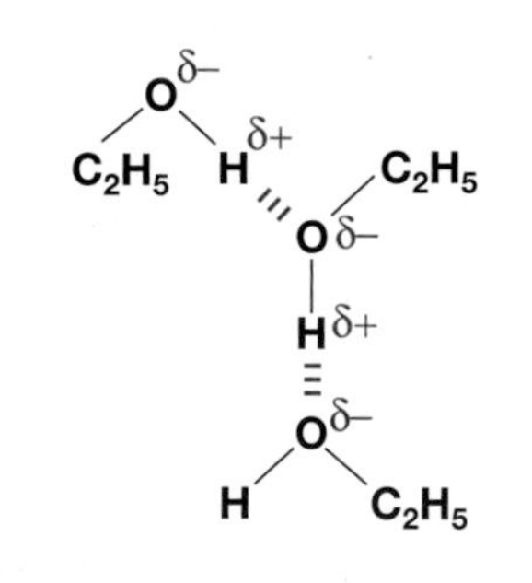

Figure 9.10▲
Hydrogen bonding in ethanol.

Definition

An atom is **labile** if it quickly and easily reacts or moves from one molecule to another.

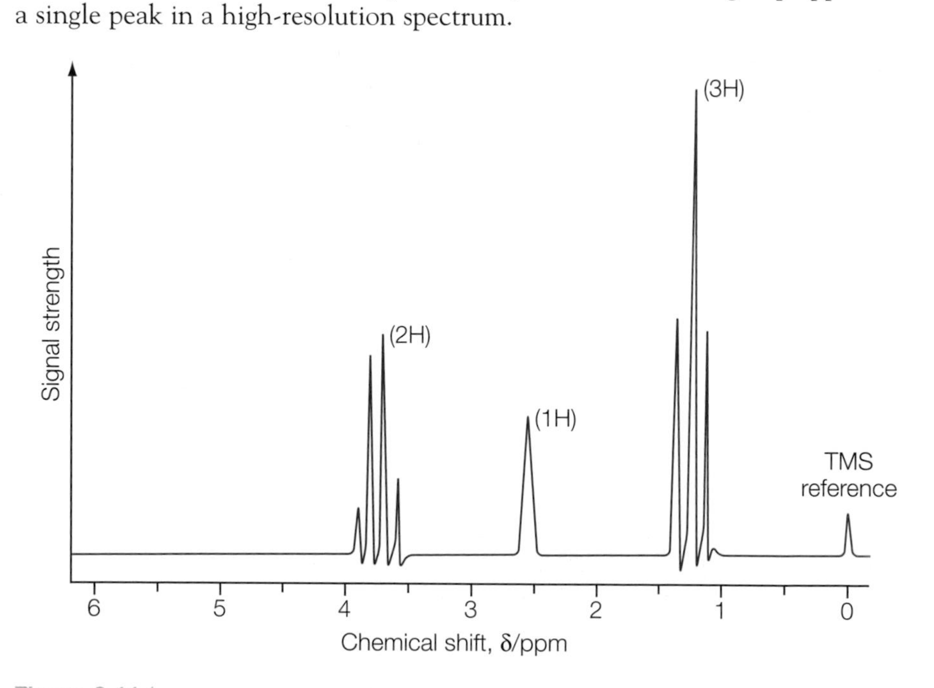

Figure 9.11▲
The high-resolution spectrum of ethanol. Note that there is no coupling between the proton in the –OH group and the protons in the next door $–CH_2$ group.

A useful technique for detecting labile protons is to measure the nmr spectrum in the presence of deuterium oxide (heavy water), D_2O. Deuterium nuclei can exchange rapidly with labile protons. Deuterium nuclei do not show up in the proton nmr region of the spectrum and so the peaks of any labile protons disappear.

Test yourself

11 Write an equation to show the reversible exchange of deuterium and hydrogen nuclei between ethanol and deuterium oxide.
12 Account for the splitting pattern shown for the peaks in the proton nmr spectrum of ethanol in Figure 9.11.
13 Table 9.1 shows the main features of the high-resolution nmr spectrum of a compound containing carbon, hydrogen and oxygen. The peak with a chemical shift of 11.7 disappears in the presence of D_2O. Deduce the structure of the compound.

Chemical shift/ppm	Number of lines	Integration ratio
1.2	Triplet	3
2.4	Quartet	2
11.7	Singlet	1

Table 9.1 ▲

Medical benefits from nmr

In medicine, magnetic resonance imaging uses nmr to detect the hydrogen nuclei in the human body, especially in water and lipids. A computer translates the information from a body scan into 3D images of the soft tissue and internal organs which are normally transparent to X-rays.

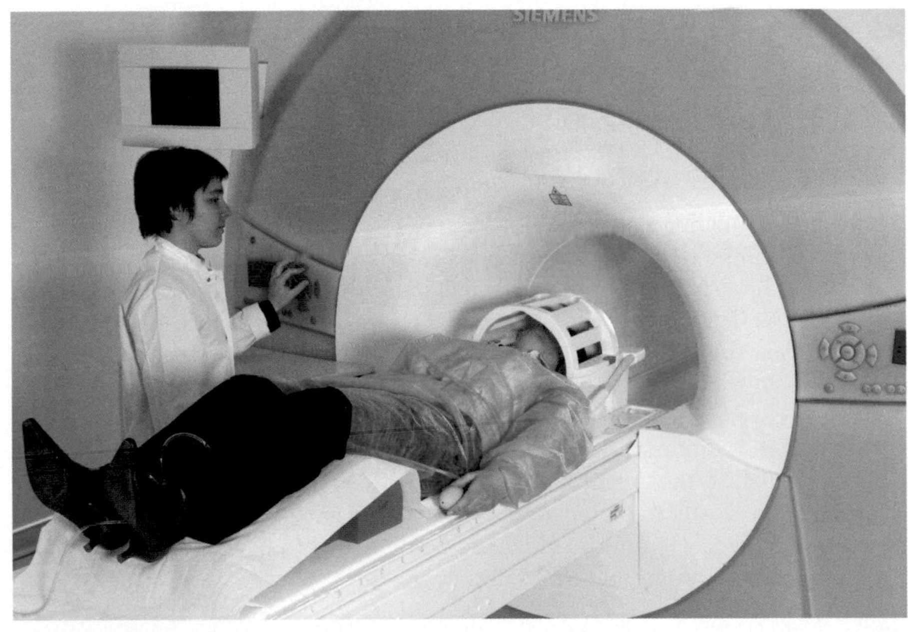

Figure 9.12 ▲
A researcher preparing a test subject for scanning in an advanced magnetic resonance imaging (MRI) machine.

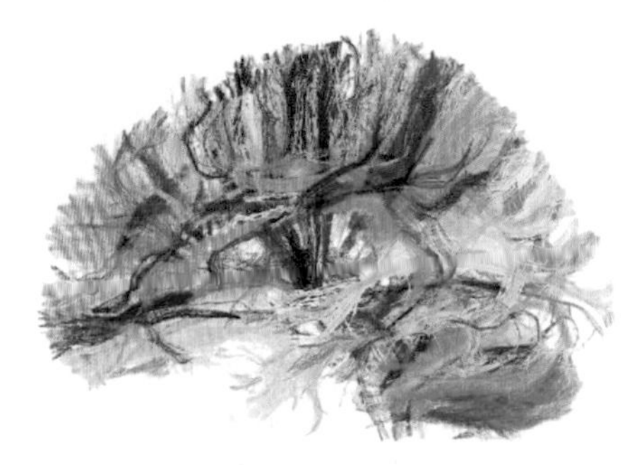

Figure 9.13 ▲
A coloured 3D MRI scan of a normal human brain. The front of the brain is on the left. The coloured lines indicate the many neural pathways that exist in the brain.

Test yourself

14 Suggest a reason why doctors and radiographers refer to MRI scanning rather than to nmr imaging, even though the technologies used in medicine and chemical research are essentially the same.

9.3 Combined techniques

Analytical chemists use a combination of techniques to identify organic compounds and determine their structures.

- Mass spectrometry gives the relative molecular mass of a compound and suggests a likely structure for a compound.
- Infrared spectroscopy shows the presence of particular functional groups by detecting their characteristic vibration frequencies.
- Nuclear magnetic resonance techniques help to detect groups with hydrogen atoms in particular environments in molecules.

Definitions

The instruments used for analysis are variously called **spectroscopes** (emphasising the use of the techniques for making observations) or **spectrometers** (emphasising the importance of measurements).

Mass spectrometry

Mass spectrometry is used to determine the relative molecular masses and molecular structures of organic compounds. In this way it can be used to identify unknown compounds.

The combination of gas chromatography (GC) with mass spectrometry is of great importance in modern chemical analysis (Section 9.6). First, gas chromatography separates the chemicals in an unknown mixture, such as a sample of polluted water; then mass spectrometry detects and identifies the components (Figure 9.15).

All mass spectrometers have the components shown in Figure 9.14.

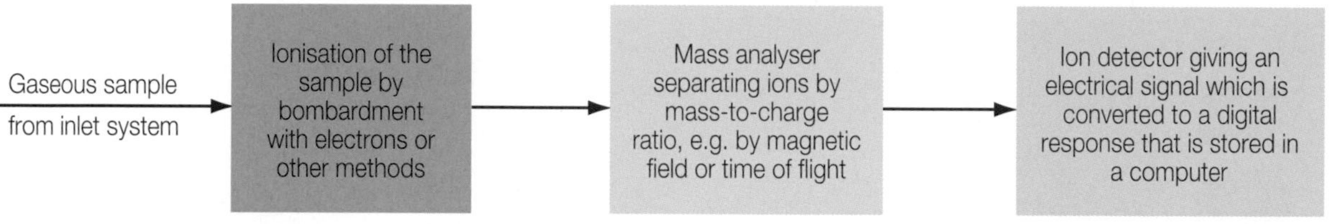

Figure 9.14▲
A schematic diagram to show the key features of a mass spectrometer.

There are five main types of mass spectrometer. They differ in the method used to separate ions with different mass-to-charge ratios. One type uses an electric field to accelerate ions into a magnetic field which then deflects the ions onto the detector. A second type accelerates the ions and then separates them by their flight time through a field-free region. A third type, the so-called transmission quadrupole instrument, is now much the most common because it is very reliable, compact and easy to use. It varies the fields in the instrument in a subtle way to allow ions with a particular mass-to-charge ratio to pass through to the detector at any one time.

Inside a mass spectrometer there is a high vacuum. This allows ionised atoms and molecules from the chemical being tested to be studied without interference from atoms and molecules in the air.

Definition

The **mass-to-charge ratio** (*m/e*) is the ratio of the relative mass, *m*, of an ion to its charge, *e*, where *e* is the number of charges (1, 2 and so on). In all the examples that you will meet $e = 1$.

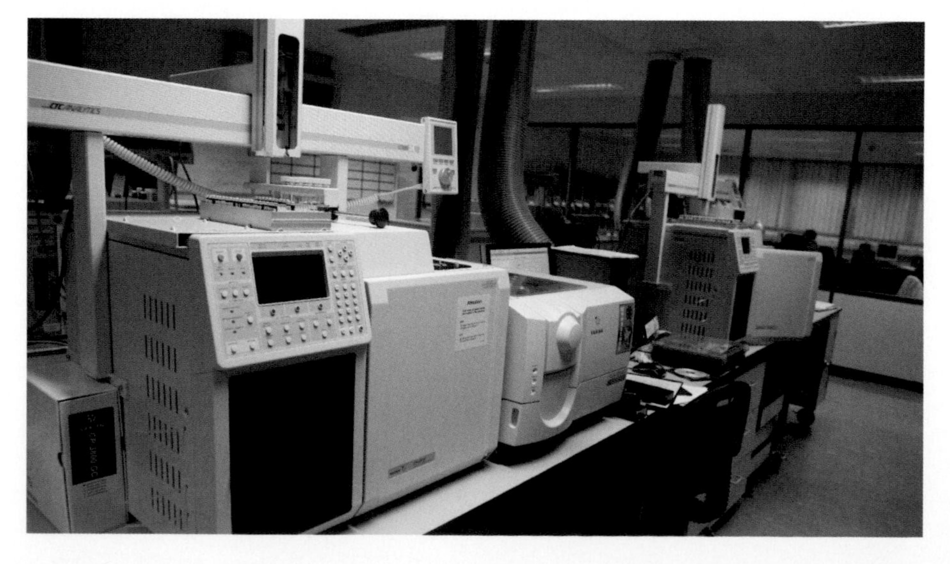

Figure 9.15►
The smaller instrument on the right is a mass spectrometer being used to analyse samples from the gas chromatography instrument on the left.

In a mass spectrometer, a beam of high-energy electrons bombards the molecules of the sample. This turns them into ions by knocking out one or more electrons.

Bombarding molecules with high-energy electrons not only ionises them but usually splits them into fragments. As a result the mass spectrum consists of a 'fragmentation pattern'.

Molecules break up more readily at weak bonds or at bonds which give rise to stable fragments. The highest peaks correspond to positive ions which are relatively more stable, such as tertiary carbocations or ions such as RCO^+ (the acylium ion) or the fragment $C_6H_5^+$ from aromatic compounds related to benzene, C_6H_6 (Topic 12).

After ionisation and fragmentation, the charged species are separated to produce the mass spectrum which distinguishes the fragments on the basis of their ratios of masses to charges. When analysing molecular compounds, the peak of the ion with the highest mass is usually the whole molecule ionised. So the mass of this 'parent ion' or 'molecular ion', M^+, is the relative molecular mass of the compound.

Chemists study mass spectra with these ideas in mind and as a result can gain insight into the structure of new molecules. They identify the fragments from their masses and then piece together likely structures with the help of evidence from other methods of analysis such as infrared spectroscopy and nmr spectroscopy.

Chemists have also built up a very large database of mass spectra of known compounds for use in analysis. They regard the spectra in databases as 'fingerprints' for identifying chemicals during analysis. The computer of a mass spectrometer is programmed to search its database to find a good match between the spectrum of a compound being analysed and a spectrum in the database.

Definition

A **carbocation** is an unstable intermediate in which a carbon atom carries a positive charge. A carbocation with the charge on a tertiary carbon atom, is more stable than one with the charge on a secondary carbon atom, which in turn is more stable than one with the charge on a primary carbon atom. This order of stability arises because of the slight electron-donating effect of alkyl groups which can partly stabilise the positive charge of a carbocation.

Test yourself

15 An organic molecule M can be represented as a combination of two parts: $m_1.m_2$. Draw a diagram to represent the ionisation and then fragmentation of the molecule and explain why only one of the two fragments shows up in the mass spectrum.

Analysing mass spectra

When analysing molecular compounds, the peak of the ion with the highest mass is usually the whole molecule ionised. So the mass of this 'parent ion', M^+, is the relative molecular mass of the compound.

Test yourself

16 Suggest the identify of the peaks labelled in the mass spectrum shown in Figure 9.16.

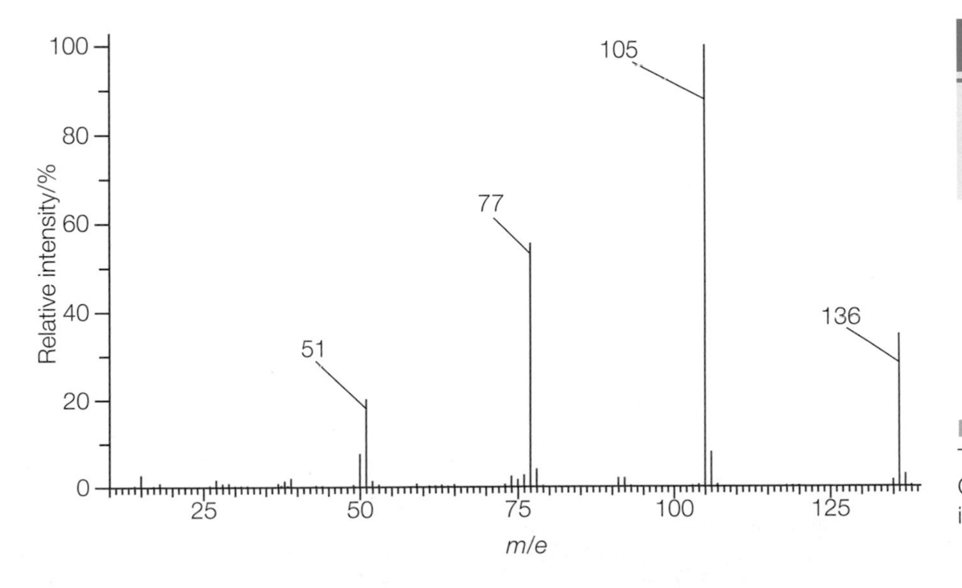

Figure 9.16 ◀ The mass spectrum of methyl benzoate, $C_6H_5COOCH_3$. The pattern of fragments is characteristic of this compound.

Test yourself

17 Account for these facts about the mass spectrum of dichloroethene:
a) it includes three peaks at *m/e* values of 96, 98 and 100 with intensities in the ratio 9 : 6 : 1,
b) it includes two peaks at *m/e* values 61 and 63 with intensities in the ratio of 3 : 1.

The presence of isotopes shows up in spectra of organic compounds containing chlorine or bromine atoms. Chlorine has two isotopes, ^{35}Cl and ^{37}Cl. Chlorine-35 is three times more abundant than chlorine-37. If a molecule contains one chlorine atom its molecular ion appears as two peaks separated by two mass units. The peak with the lower value of *m/e* is three times higher than the peak with the higher value.

Bromine consists largely of two isotopes, ^{79}Br and ^{81}Br, in roughly equal proportions. If a molecule contains one bromine atom the molecular ion shows up as two peaks of roughly equal intensity separated by two mass units.

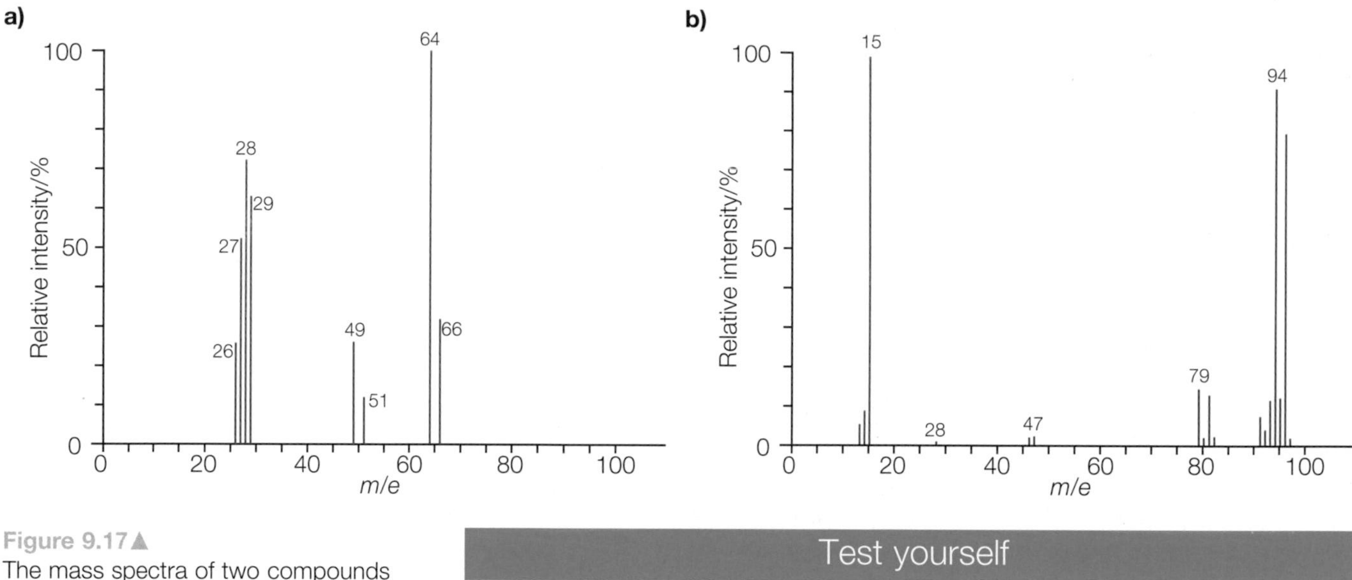

Figure 9.17▲
The mass spectra of two compounds containing halogen atoms.

Test yourself

18 One of the mass spectra in Figure 9.17 is bromomethane and the other is chloroethane. Match the spectra to the compounds and identify as many fragments in the spectra as you can.

Infrared spectroscopy

Spectroscopists have found that it is possible to correlate absorptions in the region 4000–1500 cm^{-1} with the stretching or bending vibrations of particular bonds. As a result the infrared spectrum gives valuable clues to the presence of functional groups in organic molecules.

Definition

An **absorption spectrum** is a plot showing how strongly a sample absorbs radiation over a range of frequencies. Absorption spectra from infrared spectroscopy give chemists valuable information about the bonding and structure of chemicals.

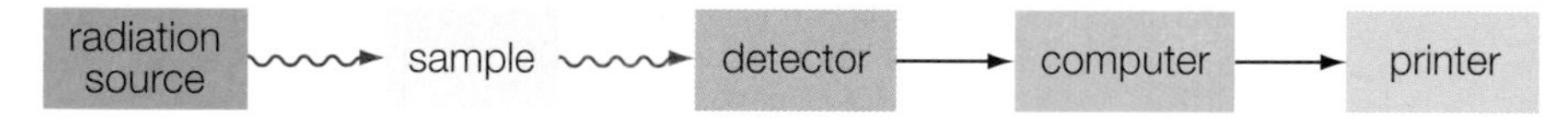

Figure 9.18▲
The essential features of a modern single-beam IR spectrometer.

Note

Most organic molecules contain C–H bonds. As a result most organic compounds have a peak at around 3000 cm^{-1} in their IR spectrum.

The important correlations between different bonds and observed absorptions are shown in Figure 9.19. Hydrogen bonding broadens the absorption peaks of –OH groups in alcohols and even more so in carboxylic acids.

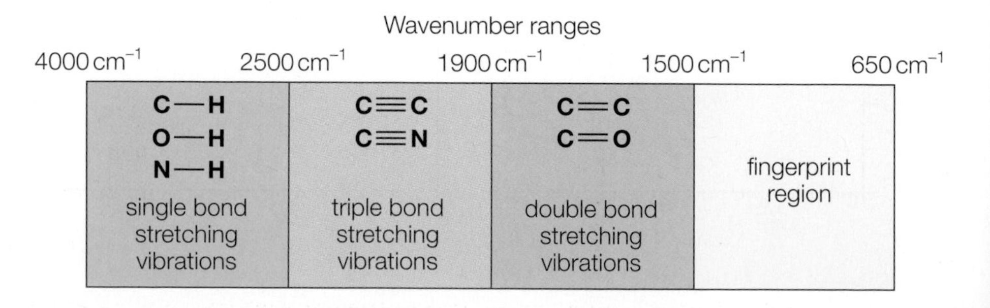

Figure 9.19►
A chart to show the main regions of the infrared spectrum and important correlations between bonds and observed absorptions.

Molecules with several atoms can vibrate in many ways because the vibrations of one bond affect others close to it. The complex pattern of vibrations can be used as a 'fingerprint' to be matched against the recorded IR spectrum in a database.

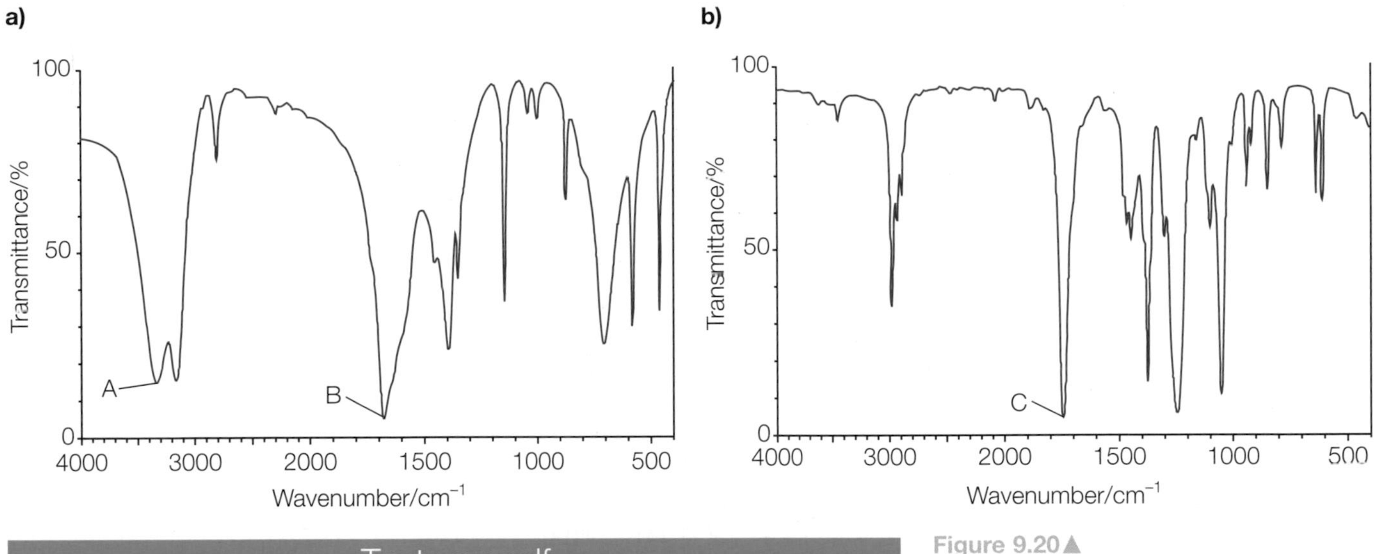

Figure 9.20 ▲ Two IR spectra

Test yourself

19 Figure 9.20 shows the infrared spectra of ethyl ethanoate and ethanamide. Use the Edexcel Data booklet to work out:
- **a)** which bonds give rise to the peaks marked with letters
- **b)** which spectrum belongs to which compound.

20 A compound P is a liquid which does not mix with water; its molecular formula is C_7H_6O and it has an infrared spectrum with strong, sharp peaks at 2800 cm^{-1}, 2720 cm^{-1} and 1700 cm^{-1} with a weaker absorption peak between 3000 and 3100 cm^{-1}. Oxidation of P gives a white crystalline solid Q with a strong broad IR absorption band in the region 2500–3300 cm^{-1} and another strong absorption at 1680–1750 cm^{-1}.
- **a)** Suggest possible structures for P and Q.
- **b)** What chemical tests could you use to check your suggestions?

Data

Activity

Analysing a perfume chemical

Jasmine blossom is a source of chemicals used in perfumes. Analysis of an extract from the blossom by gas chromatography shows that it can contain over 200 compounds. Two of the compounds in the mixture are mainly responsible for the smell of the blossom. One of these two chemicals is jasmone, which has the empirical formula $C_{11}H_{16}O$.

The structure of jasmone has been studied by mass spectrometry, IR spectroscopy and nmr, with the results shown in Figures 9.22 and 9.23.

Figure 9.21 ▲ A harvester gathering jasmine flowers for the French perfume industry.

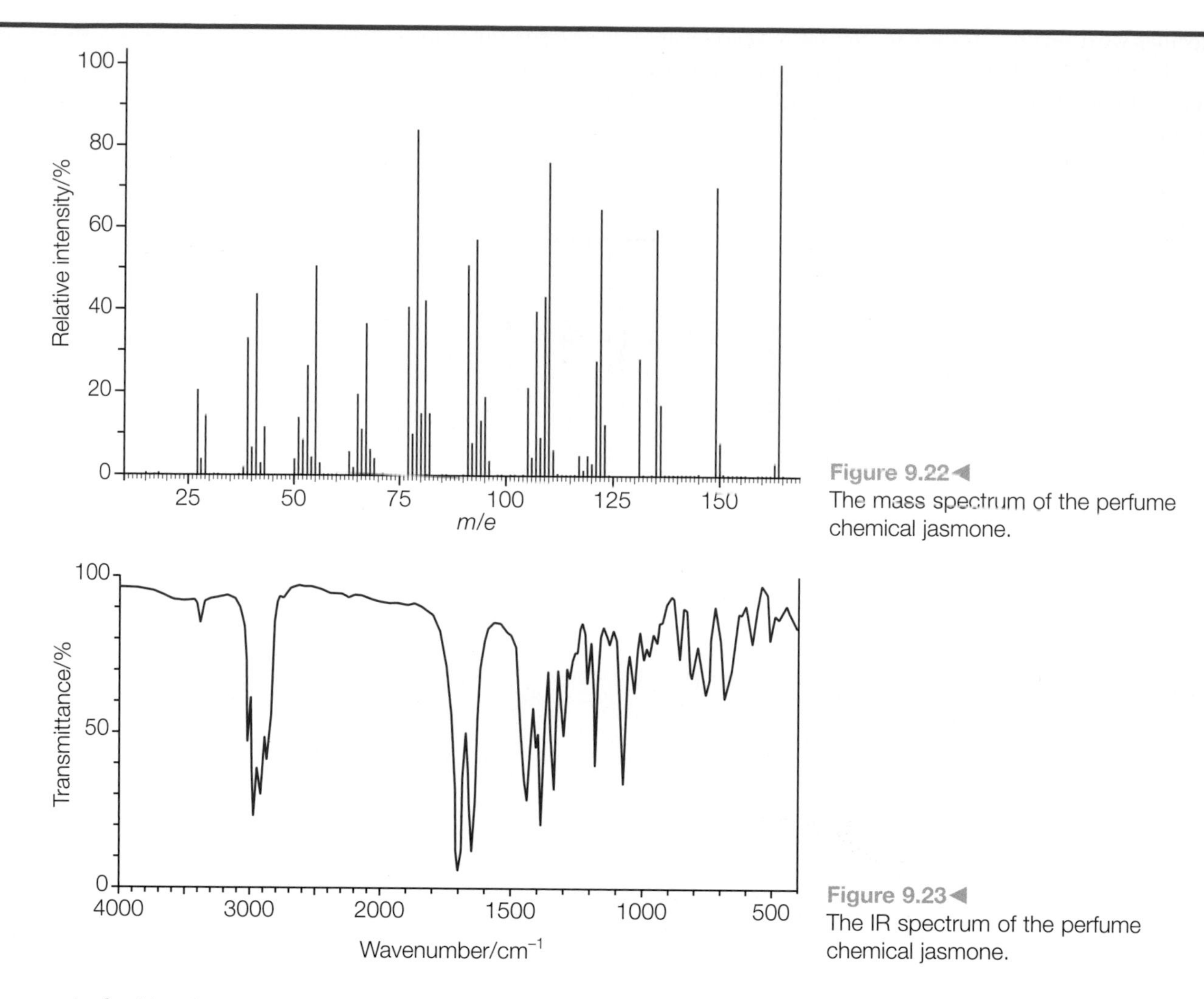

Figure 9.22◄ The mass spectrum of the perfume chemical jasmone.

Figure 9.23◄ The IR spectrum of the perfume chemical jasmone.

1 a) Use the mass spectrum of jasmone to determine its relative molecular mass.

b) What is the molecular formula of jasmone?

c) How many double bonds and/or rings are there in the molecule?

2 Refer to the IR spectrum in Figure 9.23 and the IR correlation tables in the Edexcel Data booklet. What functional groups are present in the molecule?

3 Suggest a possible structure for jasmone that is consistent both with the information from the spectra and with these facts:

- a type of nmr based on carbon-13 atoms shows that there are 11 different chemical environments for carbon atoms in the molecule
- other evidence suggests the presence of a five-membered ring of carbon atoms
- the full name of the compound is *cis*-jasmone.

4 Describe what you would expect to observe if you tested a sample of *cis*-jasmone with:

a) a solution of bromine in an organic solvent

b) Tollens' reagent

c) 2,4-dinitrophenylhydrazine.

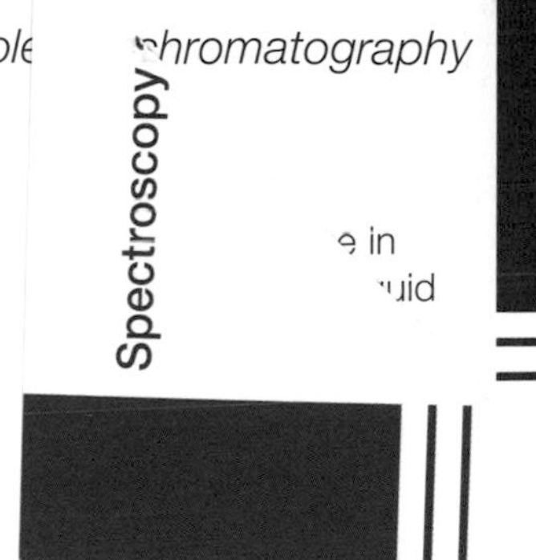

9.4 Principles of chromatography

In 1903, the Russian botanist Michel Tswett developed the technique of column chromatography to study plant pigments. 'Chroma' means colour. The name chromatography was chosen because the technique was first used to separate coloured chemicals from mixtures. Tswett extracted the coloured chemicals from leaves and added a sample to the top of a column of powdered chalk. Then he allowed a hydrocarbon solvent to flow through the column to separate the colours in the mixture.

There is now a range of chromatography techniques which can be used to:

- separate and identify the components of a mixture of chemicals
- check the purity of a chemical
- identify the impurities in a chemical preparation
- purify a chemical product.

Every type of chromatography has a stationary phase and a mobile phase that flows through it. Chemicals in a mixture separate because they differ in the extent to which they mix with the mobile phase or stick to the stationary phase.

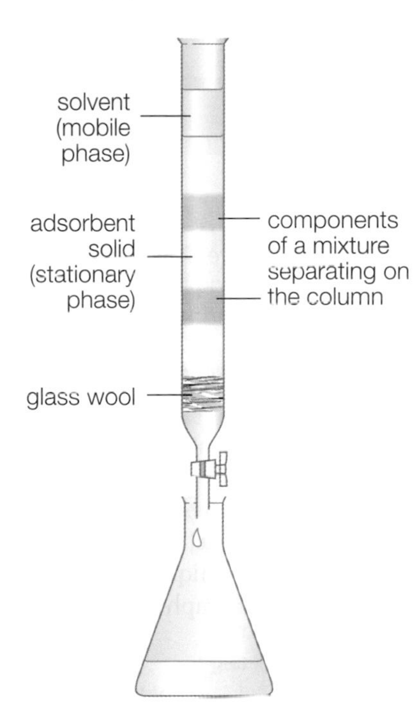

Figure 9.24▲
Column chromatography. A solution of the mixture to be analysed is added to the top of the column. Then a solvent is run slowly through the column. The substances in the mixture separate and emerge at different times from the column.

Figure 9.25▲
Coloured chemicals from a plant leaf separating on a chromatography column.

Definitions

The **stationary phase** in chromatography may be a solid or a liquid held by a solid support. The **mobile phase** moves through the stationary phase and may be a liquid or a gas.

In column chromatography the liquid flowing through the column is the **eluent**. It washes the components of the mixture through the column. This is the process of **elution**.

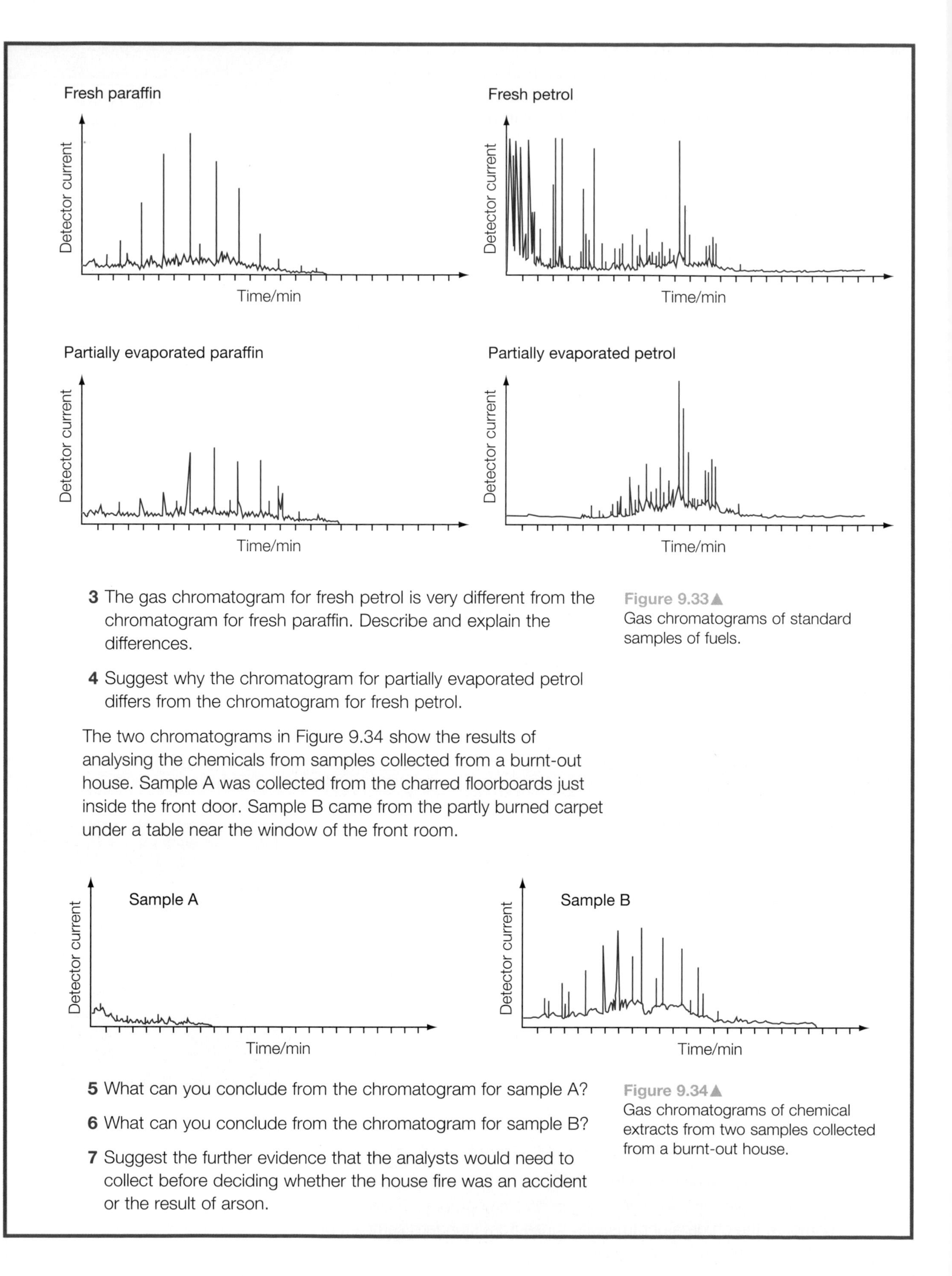

Figure 9.33 ▲ Gas chromatograms of standard samples of fuels.

3 The gas chromatogram for fresh petrol is very different from the chromatogram for fresh paraffin. Describe and explain the differences.

4 Suggest why the chromatogram for partially evaporated petrol differs from the chromatogram for fresh petrol.

The two chromatograms in Figure 9.34 show the results of analysing the chemicals from samples collected from a burnt-out house. Sample A was collected from the charred floorboards just inside the front door. Sample B came from the partly burned carpet under a table near the window of the front room.

Figure 9.34 ▲ Gas chromatograms of chemical extracts from two samples collected from a burnt-out house.

5 What can you conclude from the chromatogram for sample A?

6 What can you conclude from the chromatogram for sample B?

7 Suggest the further evidence that the analysts would need to collect before deciding whether the house fire was an accident or the result of arson.

Chromatography combined with mass spectrometry

Coupling gas chromatography (GC) with mass spectrometry (MS) gives a very powerful system for separating, identifying and measuring complex mixtures of chemicals. The combined technique (GC–MS) is widely used in drug detection, investigation of fires and environmental monitoring, as well as in airport security. Some space probes carry tiny GC–MS systems for analysing samples collected in space or on the surface of planets such as Mars.

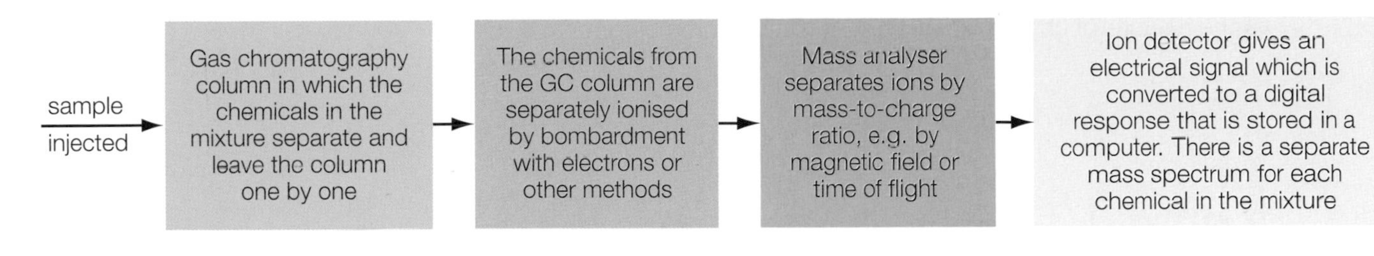

Figure 9.35▲
A schematic diagram showing the key features of a GC–MS system.

GC–MS overcomes some of the limitations of gas chromatography. Similar compounds often have similar retention times in GC, which means that they cannot be identified by chromatography alone even if the conditions are carefully standardised. Also, GC alone cannot identify any new chemicals because there are no standards that can be used to determine retention times under given conditions.

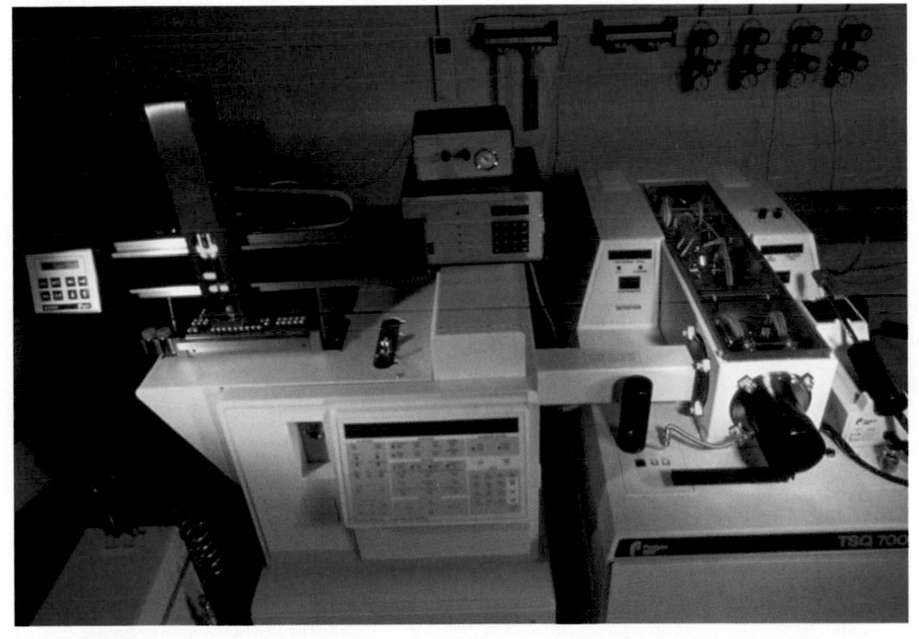

Figure 9.36▲
A gas chromatography machine (left) connected to a mass spectrometer (right) in a forensic laboratory. This equipment is sensitive enough to detect minute quantities of illegal drugs in the hair of a suspect – weeks after any drugs were taken.

GC–MS produces a mass spectrum for each of the chemicals separated on the GC column. These spectra can be used like fingerprints to identify the compounds because every chemical has a unique mass spectrum.

A computer connected to a GC–MS system stores all the data. The computer can be linked to a library of spectra of known compounds. The computer compares the mass spectrum of each chemical in a mixture to mass spectra in the library. It automatically reports a list of likely identifications along with the probability that the matches are correct.

Activity

Solving a pollution problem with GC–MS

With complex mixtures, computers can help an analyst to look for a 'needle in a haystack'. Figure 9.37 shows the chromatogram from an investigation of the air in a home where the family was feeling very sick. During the analysis by GC–MS the computer stored 700 mass spectra as the mixture of chemicals emerged from the chromatography column.

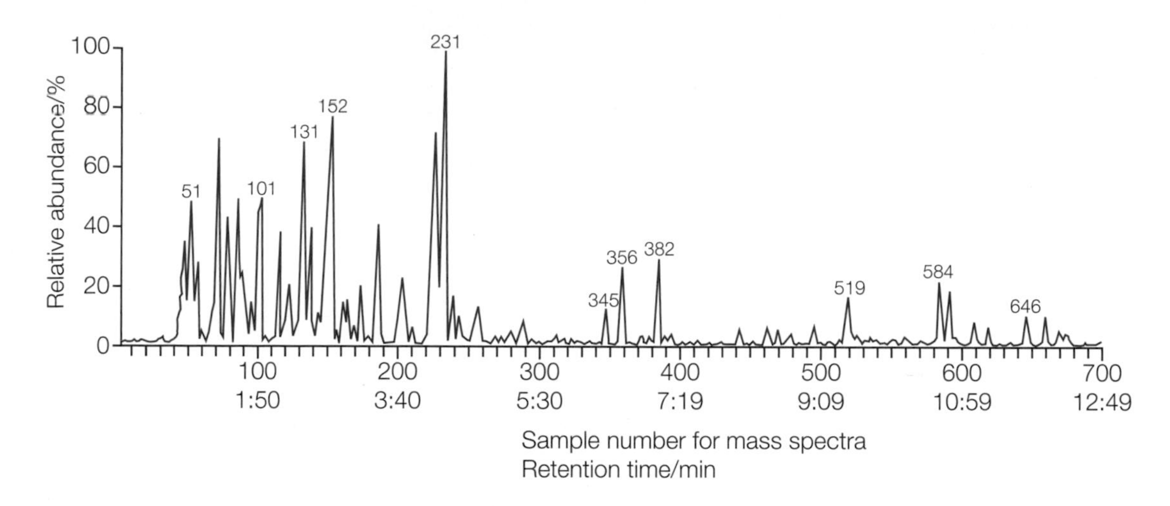

Figure 9.37▲
The gas chromatogram of chemicals sampled from the air in a house. The numbers 1–700 on the time axis indicate the points at which mass spectra were recorded and stored. Underneath these numbers are the retention times.

The analysts suspected that the chemicals causing the family's sickness might have come from petrol so they asked the computer to plot a chromatogram showing only those chemicals producing a peak with mass-to-charge ratio of 91 in their spectra. The result is shown in Figure 9.38, which indicates that methylbenzene, ethylbenzene and three dimethylbenzenes were present in the mixture. These are all chemicals that are distinctive for the mixture of hydrocarbons found in petrol. With this evidence the investigators carried out further searches and tracked down the source of the petrol vapour.

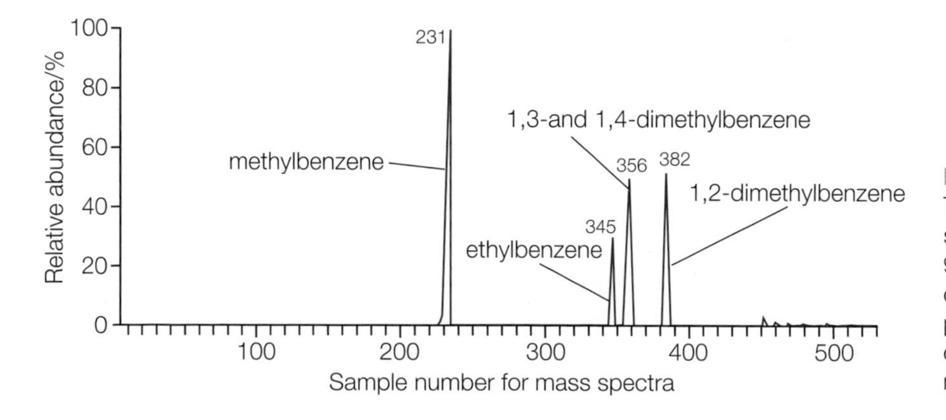

Figure 9.38◀
The GC–MS printout for the same sample as in Figure 9.37 but showing only the chemicals with a prominent peak with a mass-to-charge ratio of 91 in their mass spectra.

1 Why, in a mass spectrometer, does each chemical:

a) have to be ionised

b) pass through a region with electric and/or magnetic fields

c) produce a spectrum with several peaks?

2 Suggest the identity of the ion fragment with a mass-to-charge ratio of 91 in the mass spectra of methyl benzene, $C_6H_5–CH_3$, and related compounds.

3 How does the computer identify a chemical with a mass spectrum recorded at a particular retention time?

4 Suggest two reasons why forensic scientists find GC–MS particularly valuable.

5 HPLC can also be combined with MS. Give an example of a sample that could be analysed by HPLC–MS but not by GC–MS and explain your choice.

REVIEW QUESTIONS

www
Extension questions

1 Draw up a table to summarise and compare two methods of chromatography using an enlarged version of Table 9.2. (8)

Type of chromatography	Method of separation: adsorption or relative solubility	Example of an application	Advantages	Limitations
HPLC				
GC				

Table 9.2▲

2 Figure 9.39 shows the proton nmr spectrum of a compound of carbon, oxygen and hydrogen.

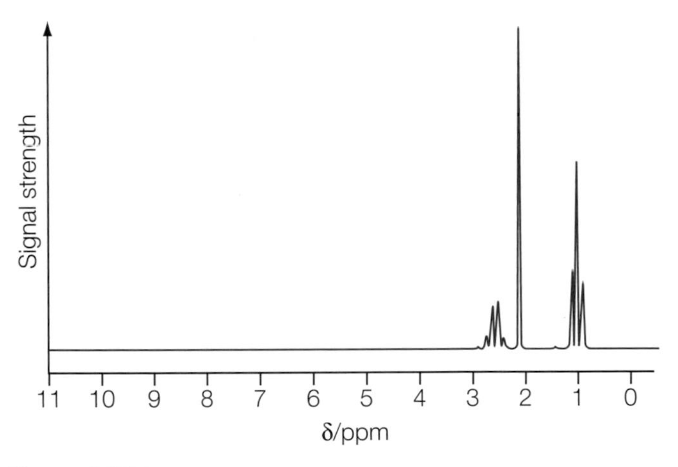

Figure 9.39▲

An integration trace gives the ratios for the peaks as shown in Table 9.3.

Chemical shift /ppm	Relative values from the integration trace
1.0	0.9
2.1	0.9
2.5	0.6

Table 9.3▲

a) i) How many chemical environments for hydrogen atoms are there in the molecule? (1)

ii) What is the ratio of the numbers of each type of hydrogen atom? (1)

b) Suggest likely chemical environments of the protons in the molecule with the help of the chart of chemical shift values in the Edexcel Data booklet. (2)

c) What can you deduce from the splitting patterns of the peaks at chemical shifts 1.0 and 2.5? (2)

d) The compound gives an orange precipitate with 2,4-dinitrophenylhydrazine but does not react with Fehling's solution or phosphorus(V) chloride. What does this tell you about the functional groups in the molecule? (3)

e) Suggest a structure for the compound. (1)

3 Two isomeric ketones with the formula $C_5H_{10}O$ have the mass spectra shown in Figures 9.40a and 9.40b.

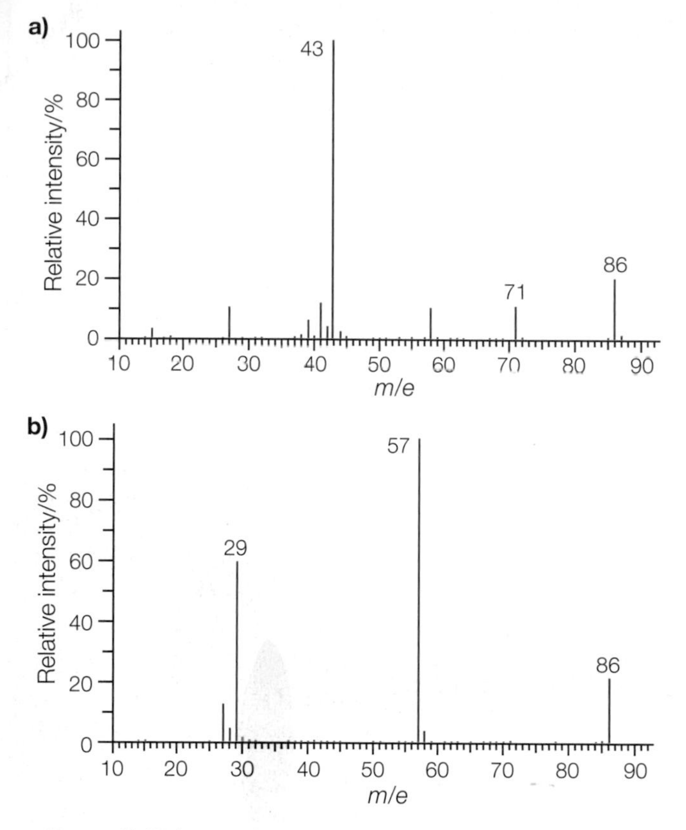

Figure 9.40▲

a) Draw the skeletal formula of the two ketones and name them. (2)

b) Why do both spectra have peaks at *m/e* values of 86? (2)

c) i) Suggest possible identities for the four fragments in the two spectra with *m/e* values of 29, 43, 57 and 71. (4)

ii) Hence show which spectrum belongs to which compound. (2)

d) Use the Edexcel chart of chemical shifts to predict the appearance of the proton nmr spectrum of pentan-3-one. (2)

4 Gas chromatography can be used to analyse the chemicals formed during the production of beer. Figure 9.41 shows a GC chromatogram for a sample taken during beer making. The instrument was fitted with a capillary column with solid stationary phase.

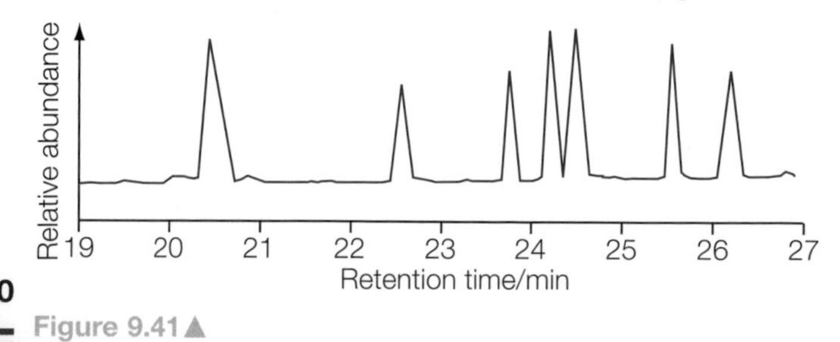

Figure 9.41▲

Compound	Retention time/min
Methanol	19.5
Ethanol	20.5
Propan-1-ol	20.9
Propan-2-ol	22.5
2-methylpropan-1-ol	22.7
Butan-2-ol	24.6
Ethanal	20.2
Propanal	24.5
Butanal	25.5
Propanone	23.8
Ethanoic acid	24.2
Butanoic acid	26.2

Table 9.4▲
Retention times for GC under the conditions used to analyse the beer sample in Figure 9.41.

a) Suggest a reason for the trend in the retention times of the four primary alcohols. (2)

b) Suggest a reason why 2-methylpropan-1-ol has a shorter retention time than butan-2-ol. (2)

c) i) Use Figure 9.41 and Table 9.4 to identify the chemicals in the beer sample. (3)

ii) How certain can you be of the answers you have given in **i)**? Which peaks were hard to identify and why? (2)

d) The mixture was analysed by GC–MS. Figure 9.42 shows the mass spectrum of the fifth peak to emerge from the GC column. Use this mass spectrum to decide whether this peak is butan-2-ol or propanal and explain how you decide on your answer. (4)

e) How does the chemical with the shortest retention time form during brewing? (2)

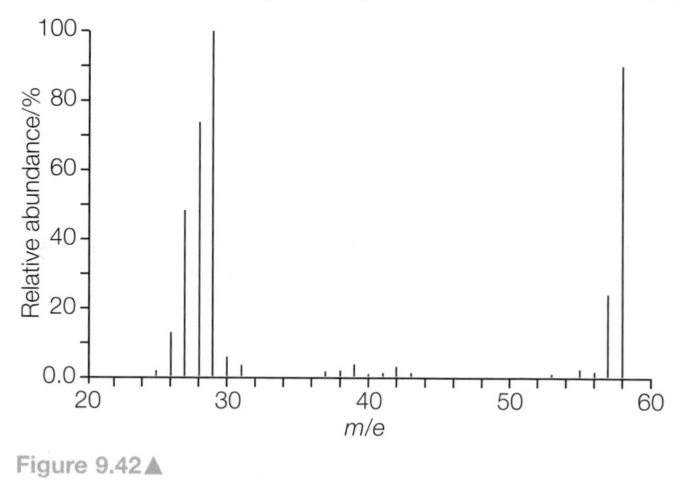

Figure 9.42▲

Unit 5

General Principles of Chemistry II – Transition Metals and Organic Nitrogen Chemistry

10 Redox equilibria

Redox reactions are very important in the natural environment, in living things and in modern technology. It should be no surprise that the Earth, with its oxygen atmosphere, has an extensive range of redox chemistry.

Redox reactions, like all reactions, tend towards a state of dynamic equilibrium. Redox reactions involve electron transfer and chemists have developed electrochemical cells based on redox changes. Some of these cells have great practical and technological importance while others, particularly fuel cells, are becoming a serious alternative to oil-based fuels for vehicles. Measurements of the voltages of cells help to assess the feasibility and likelihood of redox reactions.

10.1 Redox reactions

About one thousand billion (10^{12}) moles of oxygen are removed from the atmosphere every year in oxidising ions such as Fe^{2+} in weathered rocks and in oxidising molecules such as hydrogen sulfide, carbon monoxide and methane in volcanic gases.

Figure 10.1▶ Volcanoes release millions of tonnes of reducing gases into the atmosphere where they react with oxygen. This photo shows Mount St Helens in British Columbia erupting in 1980.

Redox reactions are also involved in the metabolic pathways of respiration. These pathways produce the molecule adenosine triphosphate (ATP). ATP transfers the energy released by oxidising food into movement, growth and all the other activities in living things that need a source of energy.

In addition, the voltages of chemical cells are obtained from the energy of redox reactions, and redox is also involved in manufacturing processes that use electrolysis to make products such as chlorine and aluminium.

Definitions

Redox stands for **Red**uction + **Ox**idation.

Oxidation involves the loss of electrons or an increase in oxidation number.

Reduction involves the gain of electrons or a decrease in oxidation number.

Remember the mnemonic **OILRIG** – **O**xidation **I**s **L**oss; **R**eduction **I**s **G**ain of electrons.

An **oxidation number** is a number assigned to an atom or ion to describe its relative state of oxidation or reduction.

Definitions of redox

Descriptions and theories of oxidation and reduction have developed over the years and, although there are now several definitions of redox, oxidation and reduction always go together.

From AS level, you should recall that oxidation originally meant addition of oxygen or removal of hydrogen, but the term now covers all reactions in which atoms or ions lose electrons. Chemists have further extended the definition of oxidation to include molecules by defining oxidation as a change that makes the oxidation number of an element more positive, or less negative.

Similarly, reduction originally meant removal of oxygen or addition of hydrogen, but the term now covers all reactions in which atoms, molecules or ions gain electrons. Defining reduction as a change in which the oxidation number of an element decreases further extends the concept of reduction.

Oxidation states and oxidation numbers

Most elements in the p block and d block of the periodic table form compounds in which their atoms have different oxidation states. Displaying the compounds of an element on an oxidation state diagram provides a 'map' of its chemistry and shows the different oxidation numbers that it can have (Figure 10.2).

There are strict rules for assigning oxidation numbers and these are shown in Table 10.1.

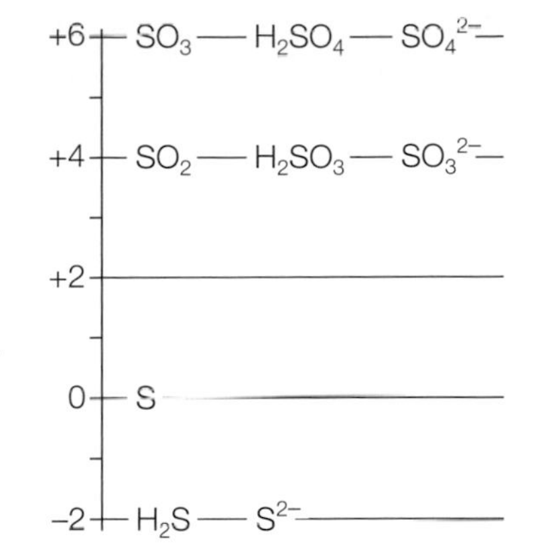

Figure 10.2▲
The oxidation numbers of sulfur in its different oxidation states.

Oxidation number rules

1. The oxidation number of the atoms in uncombined elements is zero.
2. In simple ions, the oxidation number of the element is the charge on the ion.
3. In neutral molecules, the sum of the oxidation numbers of the constituent elements is zero.
4. In ions containing two or more elements, the charge on the ion is the sum of the oxidation numbers.
5. In any compound, the more electronegative element has a negative oxidation number and the less electronegative element has a positive oxidation number.
6. The oxidation number of hydrogen in all its compounds is +1, except in metal hydrides in which it is −1.
7. The oxidation number of oxygen in all its compounds is −2, except in peroxides in which it is −1 and OF_2 in which it is +2.
8. The oxidation number of chlorine in all its compounds is −1, except in compounds with oxygen and fluorine in which it is positive.

Table 10.1◄
Rules for assigning oxidation numbers.

Test yourself

1 Describe, in terms of gain or loss of electrons, the redox reactions in these examples:
  a) the changes at the electrodes during the manufacture of aluminium from molten (liquid) aluminium oxide
  b) the reaction of iron with chlorine to form iron(III) chloride.

2 Explain why the oxidation number of oxygen is:
  a) +2 in OF_2
  b) −1 in peroxides such as Na_2O_2.

Half-equations

Half-equations are ionic equations used to describe either the gain or the loss of electrons during a redox process. Half-equations help to show what is happening during a redox reaction. Two half-equations can be combined to give the full equation for a redox reaction.

For example, zinc metal can reduce Cu^{2+} ions in copper(II) sulfate solution, forming copper metal and Zn^{2+} ions in zinc sulfate solution. This can be shown as two half-equations:

- electron loss (oxidation) $Zn(s) \rightarrow Zn^{2+}(aq) + 2e^-$
- electron gain (reduction) $Cu^{2+}(aq) + 2e^- \rightarrow Cu(s)$

Balancing the number of electrons lost by Zn with the number gained by $Cu^{2+}(aq)$, we can add the two half-equations to get the full equation:

$$Zn(s) + Cu^{2+}(aq) \rightarrow Zn^{2+}(aq) + Cu(s)$$

Test yourself

3 State the changes in oxidation number when concentrated sulfuric acid reacts with potassium bromide.
$4KBr(s) + 2H_2SO_4(aq) \rightarrow K_2SO_4(aq) + K_2SO_3(aq) + Br_2(aq) + 2HBr(aq) + H_2O(l)$

4 What is the oxidation number of each element in:
  a) KIO_3 b) N_2O_5 c) H_2O_2 d) SF_6 e) NaH?

5 Which sulfur compounds in Figure 10.2 can:
  a) act as an oxidising agent or as a reducing agent depending on the conditions
  b) only act as an oxidising agent
  c) only act as a reducing agent?

6 Draw similar charts to show the main oxidation states of
  a) nitrogen b) chlorine.

7 Are the named elements below oxidised or reduced in the following conversions:
  a) magnesium to magnesium sulfate
  b) iodine to aluminium iodide
  c) hydrogen to lithium hydride
  d) iodine to iodine monochloride, ICl?

Activity

Redox reactions in the Space Shuttle

Unlike most vehicles on Earth, spacecraft must carry oxidising agents as well as fuels. These mixtures of fuels plus oxidising agents are called propellants.

In order to launch the Space Shuttle from ground level into the Earth's upper atmosphere, a solid propellant mixture of powdered aluminium and ammonium chlorate(VII), NH_4ClO_4, is used. The reaction is

$$3Al(s) + 3NH_4ClO_4(s) \rightarrow Al_2O_3(s) + AlCl_3(s) + 3NO(g) + 6H_2O(g)$$

1 What is **a)** the fuel **b)** the oxidising agent in the solid propellant?

2 What are the oxidation numbers of each element in NH_4ClO_4?

3 a) Which elements are oxidised in the reaction?

b) State the change in oxidation numbers of these elements.

4 Which element is reduced in the reaction and what is the change in its oxidation number?

5 What actually propels the Space Shuttle?

Figure 10.3▲
The Space Shuttle *Endeavour* takes off on its journey to the International Space Station.

Propulsion of the Shuttle from the upper atmosphere into orbit is achieved using a mixture of liquid hydrogen and liquid oxygen. Once the hydrogen and oxygen are ignited, they continue to vaporise and burn continuously, producing a clean water-vapour exhaust.

6 Write an equation for the combustion of liquid hydrogen and oxygen during the second stage.

Once in orbit, the shuttle needs a propulsion system in which the fuel and oxidising agent ignite spontaneously on mixing, and which can be readily started and stopped for manoeuvrability.

In this stage, the fuel is liquid methylhydrazine (CH_3NHNH_2) with liquid dinitrogen tetroxide (N_2O_4) as the oxidiser. The reaction produces water vapour, carbon dioxide and nitrogen.

7 Write a balanced equation for the reaction between CH_3NHNH_2 and N_2O_4 in this stage.

8 Explain why CH_3NHNH_2 acts as the reducing agent and why N_2O_4 acts as the oxidising agent during the reaction.

10.2 Oxidation numbers and stoichiometry

Using oxidation numbers it is possible to balance the amounts of substances involved in a chemical reaction and work out an equation. This is done by setting the decrease in oxidation number of the element reduced equal to the increase in oxidation number of the element oxidised. The amounts of different substances involved in a chemical reaction is sometimes described as the reaction stoichiometry.

We can illustrate how oxidation numbers can help with the stoichiometry of redox reactions using the oxidation of iron(II) ions by manganate(VII) ions in acid solution.

Step 1 Write the formulae of the atoms, ions and molecules involved in the reaction.

$$MnO_4^- + H^+ + Fe^{2+} \rightarrow Mn^{2+} + H_2O + Fe^{3+}$$

(Notice that the oxygen in oxoanions such as MnO_4^- and $Cr_2O_7^{2-}$ usually gets converted to water by H^+ ions in the acid solution, while Mn and Cr in the oxoanions are converted to stable simple ions.)

Step 2 Identify the elements which change in oxidation number and the extent of change.

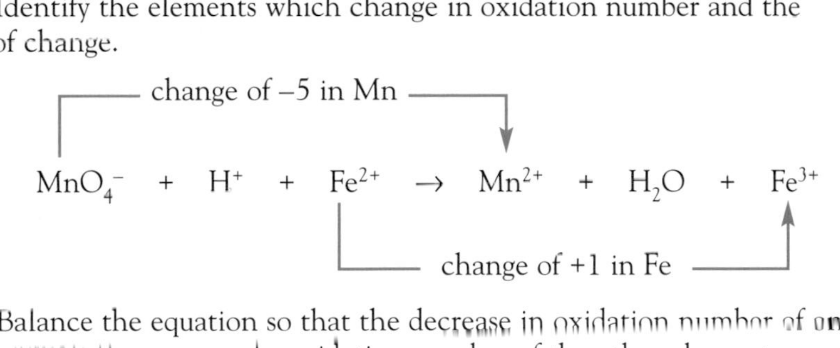

Step 3 Balance the equation so that the decrease in oxidation number of one element equals the increase in oxidation number of the other element

In this example, the decrease of –5 in the oxidation number of manganese is balanced by five Fe^{2+} ions each increasing their oxidation number by +1.

$$MnO_4^- + H^+ + 5Fe^{2+} \rightarrow Mn^{2+} + H_2O + 5Fe^{3+}$$

Step 4 Balance for oxygen and hydrogen.
In this example, the four oxygen atoms of the MnO_4^- ion join with eight hydrogen ions to form four water molecules.

$$MnO_4^- + 8H^+ + 5Fe^{2+} \rightarrow Mn^{2+} + 4H_2O + 5Fe^{3+}$$

Step 5 Finally, check that the overall charges on each side of the equation balance and then add state symbols.

The net charge on the left is 17+ which is the same as that on the right. So, the equation for the reaction is:

$$MnO_4^-(aq) + 8H^+(aq) + 5Fe^{2+}(aq) \rightarrow Mn^{2+}(aq) + 4H_2O(l) + 5Fe^{3+}(aq)$$

Half-equations and stoichiometry

If you know the half-equations for the substances concerned in a redox reaction, these provide an even easier way of writing the full equation for the reaction. In this case, you simply balance the number of electrons given up in one half-equation with the number taken in the other

For example, when hydrogen peroxide in acid solution reacts with iron(II) ions, the half-equations are:

$$H_2O_2(aq) + 2H^+(aq) + 2e^- \rightarrow 2H_2O(l)$$

$$Fe^{2+}(aq) \rightarrow Fe^{3+}(aq) + e^-$$

Multiplying the second equation by 2 balances the number of electrons given up with those taken and the overall equation can be obtained by adding the two half-equations together.

$$H_2O_2(aq) + 2H^+(aq) + 2e^- \rightarrow 2H_2O(l)$$
$$2Fe^{2+}(aq) \rightarrow 2Fe^{3+}(aq) + 2e^-$$

Overall: $H_2O_2(aq) + 2H^+(aq) + 2Fe^{2+}(aq) \rightarrow 2H_2O(l) + 2Fe^{3+}(aq)$

Oxidising agents and reducing agents

Oxidising agents (oxidants) are chemical reagents which can oxidise other substances. They do this either by taking electrons away from these substances or by increasing their oxidation number. Common oxidising agents include oxygen, chlorine, bromine, hydrogen peroxide, the manganate(VII) ion in potassium manganate(VII) and the dichromate(VI) ion in potassium or sodium dichromate(VI) (Figure 10.4).

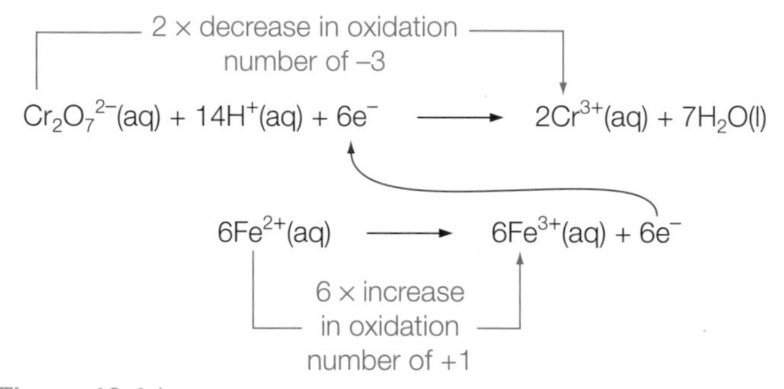

Figure 10.4▲
Dichromate(VI) ions act as oxidising agents by taking electrons from iron(II) ions in acid solution. An oxidising agent is itself reduced when it reacts.

Note

Iodine is only very slightly soluble in water, but it dissolves in a solution containing iodide ions to form the tri-iodide ion, $I_3^-(aq)$.

$$I_2(s) + I^-(aq) \rightarrow I_3^-(aq)$$

A reagent labelled 'iodine solution' is normally $I_2(s)$ in $KI(aq)$ which forms $KI_3(aq)$. The $I_3^-(aq)$ ion is yellow-brown which explains the colour change when iodine is produced from iodide ions.

Some reagents change colour when they are oxidised which makes them useful for detecting oxidising agents. In particular, a colourless solution of iodide ions is oxidised to iodine which turns the solution to a yellow-brown colour (see margin note).

$2I^-(aq) \rightarrow I_2(aq) + 2e^- \longrightarrow$ electrons taken by the oxidising agent

This can be a very sensitive test if starch is also present because starch forms an intense blue-black colour with iodine. Moistened starch-iodide paper is a version of this test which can detect oxidising gases such as chlorine and bromine vapour.

Definitions

Oxidising agents (oxidants) take electrons and are reduced when they react.

Reducing agents (reductants) give up electrons and are oxidised when they react.

Reducing agents (reductants) are chemical reagents that can reduce other substances. They do this either by giving electrons to these substances or by decreasing their oxidation number. Common reducing agents include metals such as zinc and iron, often with acid, sulfite ions (SO_3^{2-}), iron(II) ions and iodide ions.

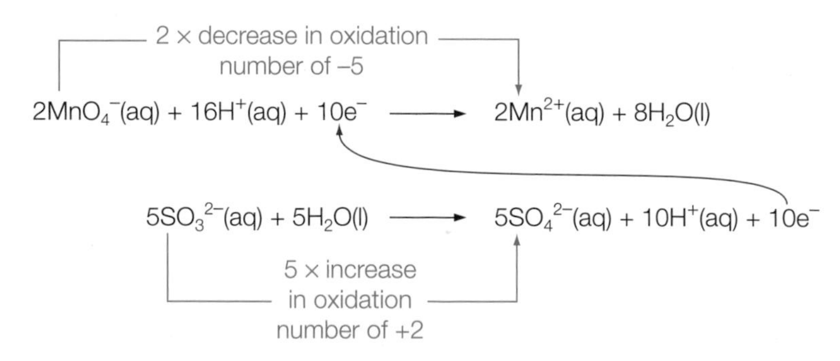

Figure 10.5▲
Sulfite ions act as reducing agents by giving electrons to manganate(VII) ions. A reducing agent is itself oxidised when it reacts.

Some reagents change colour when they are reduced, which makes them useful for detecting reducing agents (Figure 10.6 and Figure 10.7).

Figure 10.6▲
A test for reducing agents.
The test: add a solution of purple potassium manganate(VII) acidified with dilute sulfuric acid to the reducing agent.
The result: the purple solution turns colourless as purple MnO_4^- ions are reduced to very pale pink Mn^{2+} ions.

Figure 10.7▲
Another test for reducing agents.
The test: add orange dichromate(VI) solution acidified with dilute sulfuric acid to the reducing agent.
The result: the orange solution turns green as orange $Cr_2O_7^{2-}$ ions are reduced to green Cr^{3+} ions.

Test yourself

8 Write half-equations to show what happens when the following act as oxidising agents:
  a) Fe^{3+}(aq) b) Br_2(aq)
  c) H_2O_2(aq) in acid solution.
9 Write half-equations to show what happens when the following act as reducing agents:
  a) Zn(s) b) I^-(aq) c) Fe^{2+}(aq)
10 Explain why moist starch-iodide paper can be used as a very sensitive test for chlorine.
11 a) Write the half-equations involved in the reaction between Fe^{3+} ions in a solution of iron(III) chloride and iodide ions in potassium iodide solution.
  b) Write a full redox equation for the reaction involved.
  c) Describe the change you would see in the solution when the reaction occurs.
  d) How many moles of Fe^{3+} ions react with 1 mole of I^- ions?

Test yourself

12 a) Write the half-equations involved when dichromate(VI) ions in acid solution react with sulfite ions, SO_3^{2-}, to form chromium(III) ions and sulfate ions.
  b) Write a full, balanced redox equation for the reaction.
  c) Describe the change you would see in the solution when the reaction occurs.
  d) How many moles of sulfite react with one mole of dichromate(VI)?
13 Write balanced equations for each of the following redox reactions:
  a) manganese(IV) oxide with hydrochloric acid to form manganese(II) ions and chlorine
  b) copper metal with nitrate ions in nitric acid to form copper(II) ions and nitrogen dioxide gas.

Practical guidance

10.3 Redox titrations

In redox titrations, oxidising agents react with reducing agents. During the titration, you will usually measure the volume of a standard solution of an oxidising agent or a reducing agent that reacts exactly with a measured volume of the other reagent.

Definitions

A **standard solution** is a solution with an accurately known concentration. The method of preparing a standard solution is to dissolve a weighed sample of a primary standard in water and then make the solution up to a definite volume in a graduated flask.

A **primary standard** is a chemical which can be weighed out accurately to make up a standard solution. A primary standard must:

- be very pure
- not gain or lose mass when exposed to the air
- have a relatively high molar mass so weighing errors are minimised
- react exactly and rapidly as described by the chemical equation.

Primary standards for redox titrations include sodium thiosulfate.

Measuring reducing agents – potassium manganate(VII) titrations

Potassium manganate(VII) is often chosen to measure reducing agents because it can be obtained as a pure, stable solid which reacts in acid solution exactly as in the following equation:

$$MnO_4^-(aq) + 8H^+(aq) + 5e^- \rightarrow Mn^{2+}(aq) + 4H_2O(l)$$

Because of these properties, potassium manganate(VII) is used to make up standard solutions.

No indicator is required in potassium manganate(VII) titrations. On adding the potassium manganate(VII) solution from a burette, the purple MnO_4^- ions change rapidly to pale pink Mn^{2+} ions which look colourless in dilute solution. At the end-point, one drop of excess MnO_4^- is sufficient to produce a permanent red-purple colour.

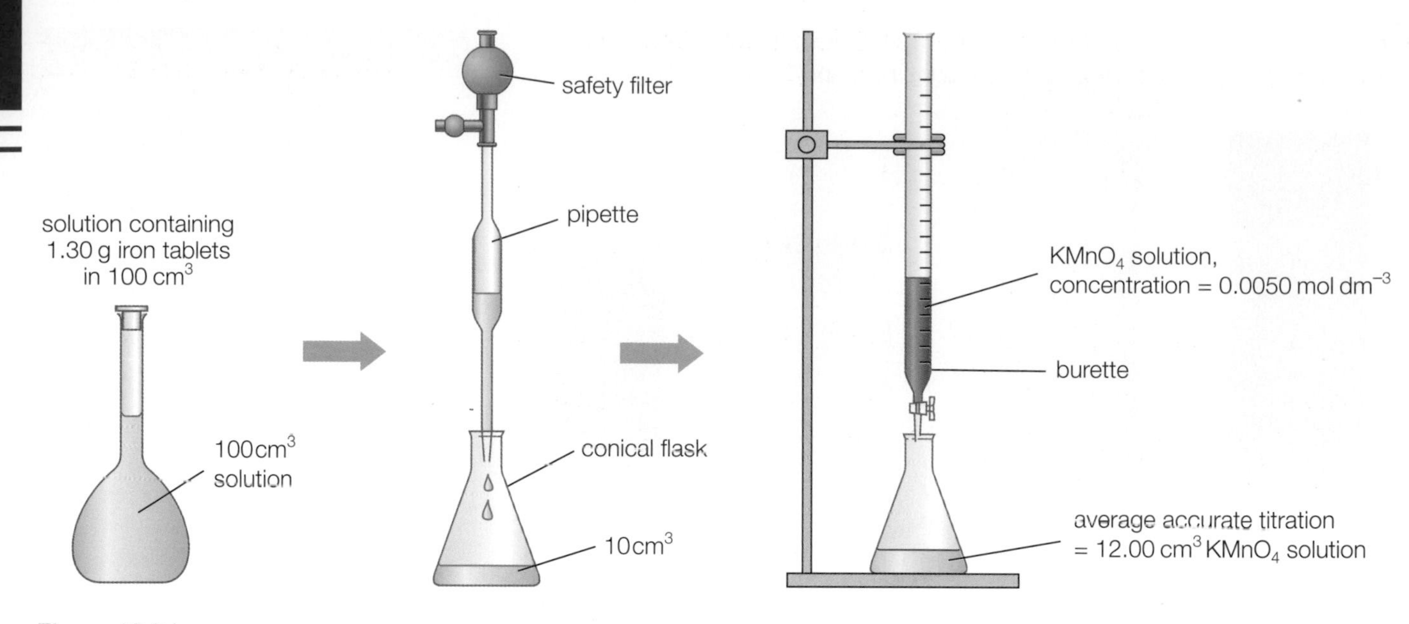

Figure 10.8▲
Finding the percentage of iron in iron tablets.

Worked example

Two iron tablets (mass 1.30 g) containing iron(II) sulfate were dissolved in dilute sulfuric acid and made up to 100 cm^3 (Figure 10.8). 10 cm^3 of this solution required 12.00 cm^3 of a standard solution of 0.0050 mol dm^{-3} $KMnO_4$ to produce a faint red colour. What is the percentage of iron in the iron tablets? (Fe = 55.8)

Notes on the method

1. Write the half-equations and work out the amounts in moles of Fe^{2+} and MnO_4^- which react.
2. Calculate the amount of MnO_4^- which reacts in the titration, and hence the amount of Fe^{2+} which reacts.
3. Work out the amount of Fe^{2+} in the whole solution, and hence in the tablets dissolved.
4. Calculate the percentage of iron in the tablets.

Answer

1. The half-equations for the reaction are:

$$MnO_4^- + 8H^+ + 5e^- \rightarrow Mn^{2+} + 4H_2O \quad \text{and}$$
$$(Fe^{2+} \rightarrow Fe^{3+} + e^-) \times 5$$

$\therefore$ 5 mol Fe^{2+} react with 1 mol MnO_4^-

2. Amount of MnO_4^- reacting in the titration $= \frac{12}{1000} \text{dm}^3 \times 0.0050 \text{ mol dm}^{-3}$

$\therefore$ Amount of Fe^{2+} reacting in the titration $= \frac{12}{1000} \times 0.0050 \times 5 \text{ mol}$

3. Amount of Fe^{2+} in 100 cm^3 of solution (2 tablets)

$$= \frac{12}{1000} \times 0.0050 \times 5 \times 10 \text{ mol}$$

$$= \frac{3}{1000} \text{ mol}$$

4. Mass of Fe^{2+} in 2 tablets $= \frac{3}{1000} \text{ mol} \times 55.8 \text{ g mol}^{-1}$

$= 0.1674 \text{ g}$

$\therefore$ % of iron in the tablets $= \frac{0.1674}{1.30} \times 100 = 12.9\%$

Measuring oxidising agents – iodine/thiosulfate titrations

The measurement of oxidising agents usually involves a combination of potassium iodide and sodium thiosulfate.

The oxidising agent to be estimated (e.g. iron(III) ions, copper(II) ions, chlorine, manganate(VII) ions in acid) is first added to excess potassium iodide. This produces iodine.

$$2Fe^{3+}(aq) + 2e^- \rightarrow 2Fe^{2+}$$
$$2I^-(aq) \rightarrow I_2(aq) + 2e^-$$

The iodine stays in solution in excess potassium iodide forming a yellow-brown solution.

The amount of iodine produced is then determined by titration with a standard solution of sodium thiosulfate, $Na_2S_2O_3(aq)$, which reduces the iodine back to colourless iodide ions.

$$I_2(aq) + 2e^- \rightarrow 2I^-(aq)$$
$$2S_2O_3^{2-}(aq) \rightarrow S_4O_6^{2-}(aq) + 2e^-$$

thiosulfate tetrathionate

The greater the amount of oxidising agent added, the more iodine is formed and the more thiosulfate is needed from the burette to react with it. On adding thiosulfate from the burette, the colour of the iodine solution becomes paler. Near the end-point the solution is a very pale yellow. Adding a little soluble starch solution as an indicator at this point gives a sharp colour change from dark blue to colourless at the end-point.

The following activity illustrates the use of iodine/thiosulfate titrations to determine the concentration of supermarket bleaches and then considers the accuracy, reliability and validity of the final results.

Activity

Supermarket bleaches

The active reagent in household bleaches is sodium chlorate(I), NaClO. To increase the cleaning power of these bleaches manufacturers usually add detergents, and to improve their smell they add perfumes. Sodium chlorate(I) is a strong oxidising agent which bleaches by oxidising coloured materials to colourless or white substances.

Figure 10.9▲ Supermarket bleaches.

The half-equation when sodium chlorate(I) acts as an oxidising agent is

$$ClO^-(aq) + 2H^+(aq) + 2e^- \rightarrow Cl^-(aq) + H_2O$$

A student was asked to determine the concentration of sodium chlorate(I) in a supermarket bleach.

Using a measuring cylinder, 100 cm³ of the bleach was added to a graduated flask and made up to a volume of 1000 cm³. 10.0 cm³ of the diluted solution was then pipetted into a conical flask followed by the addition of excess potassium iodide.

The iodine produced was finally titrated with 0.10 mol dm⁻³ sodium thiosulfate solution, giving an average accurate titration of 26.60 cm³.

1 Write a half-equation for the oxidation of iodide ions to iodine by chlorate(I) ions.

2 Write a balanced equation for the reaction of chlorate(I) ions with iodide ions in acid solution to form iodine, chloride ions and water.

3 Write a balanced equation for the reaction of iodine with thiosulfate ions during the titration.

4 Using the equations in **2** and **3**, calculate the number of moles of thiosulfate which react with the iodine produced by one mole of chlorate(I) ions.

5 Calculate the number of moles of thiosulfate in the average accurate titration, and hence the number of moles of sodium chlorate(I) in 10 cm^3 of the diluted bleach.

6 Calculate the mass of sodium chlorate(I) in 100 cm^3 of undiluted bleach. (Na = 23.0, Cl = 35.5, O = 16.0)

7 What precautions should the student take to ensure that the result is accurate?

8 What could the student do to improve the reliability of the result?

9 What must the student do to ensure that the result is valid?

10 Use the practical guidance entitled 'Errors and uncertainty' on the Dynamic Learning Student website to calculate:

Practical guidance

a) the uncertainty and percentage uncertainty in

i) the volume of undiluted bleach taken

ii) the volume of diluted bleach pipetted

iii) the volume of thiosulfate titrated

iv) the concentration of the thiosulfate solution

b) the total percentage uncertainty in the mass of sodium chlorate(I) in 100 cm^3 of undiluted bleach.

11 Finally, write your result for the mass of sodium chlorate(I) in undiluted bleach in the form $x \pm y$ g per 100 cm^3.

Test yourself

14 Write half-equations for the reactions between acidified potassium manganate(VII) and:

a) iron(II) sulfate solution

b) hydrogen peroxide solution.

15 Use your answers to **14** to calculate the volume of 0.02 mol dm^{-3} potassium manganate(VII) solution required to oxidise 20 cm^3 of:

a) 0.10 mol dm^{-3} iron(II) sulfate solution

b) 0.200 mol dm^{-3} hydrogen peroxide solution.

16 All the iron in 1.34 g of some iron ore was dissolved in acid and reduced to iron(II) ions. The solution was then titrated with 0.020 mol dm^{-3} potassium manganate(VII) solution. The titre was 26.75 cm^3. Calculate the percentage by mass of iron in the ore. (Fe = 55.8)

17 0.275 g of an alloy containing copper was dissolved in nitric acid and then diluted with water, producing a solution of copper(II) nitrate. An excess of potassium iodide was then added. The copper(II) ions reacted with the iodide ions to form a precipitate of copper(I) iodide and iodine. In a titration, the iodine reacted with 22.50 cm^3 of 0.140 mol dm^{-3} sodium thiosulfate solution. (Cu = 63.5)

a) Write an equation for:

i) the reaction of copper(II) ions with iodide ions to form copper(I) iodide and iodine.

ii) iodine with sodium thiosulfate during the titration.

b) Calculate the percentage by mass of copper in the alloy.

10.4 Electrode potentials

Redox reactions involve the transfer of electrons from a reducing agent to an oxidising agent. The electron transfer can be shown by writing half-equations. So, for example, when zinc is added to copper(II) sulfate solution, Zn atoms give up electrons forming Zn^{2+} ions. At the same time, the electrons are transferred to Cu^{2+} ions which form Cu atoms.

The two half-equations for the reaction are:

$$Zn(s) \rightarrow Zn^{2+}(aq) + 2e^-$$

and $Cu^{2+}(aq) + 2e^- \rightarrow Cu(s)$

The overall balanced equation is

$$Zn(s) + Cu^{2+}(aq) \rightarrow Zn^{2+}(aq) + Cu(s)$$

Test yourself

18 Write two ionic half-equations and the overall balanced equation for each of the following redox reactions. In each example, state which atom, ion or molecule is oxidised and which is reduced:

a) magnesium metal with copper(II) sulfate solution

b) aqueous chlorine with a solution of potassium bromide

c) a solution of silver nitrate with copper metal.

Electrochemical cells

Instead of mixing two reagents, it is possible to carry out a redox reaction in an electrochemical cell so that electron transfer takes place along a wire connecting the two electrodes. This harnesses the energy from the redox reaction to produce an electrical potential difference (voltage).

One of the first useable cells was based on the reaction of zinc metal with aqueous copper(II) ions (Figure 10.10). In this cell, zinc is oxidised to zinc(II) ions as copper(II) ions are reduced to copper metal.

$$Zn(s) \rightarrow Zn^{2+}(aq) + 2e^-$$
$$Cu^{2+}(aq) + 2e^- \rightarrow Cu(s)$$

In electrochemical cells, the two half-reactions happen in separate half-cells. The electrons flow from one cell to the other through a wire connecting the electrodes. A salt bridge connecting the two solutions completes the electrical circuit.

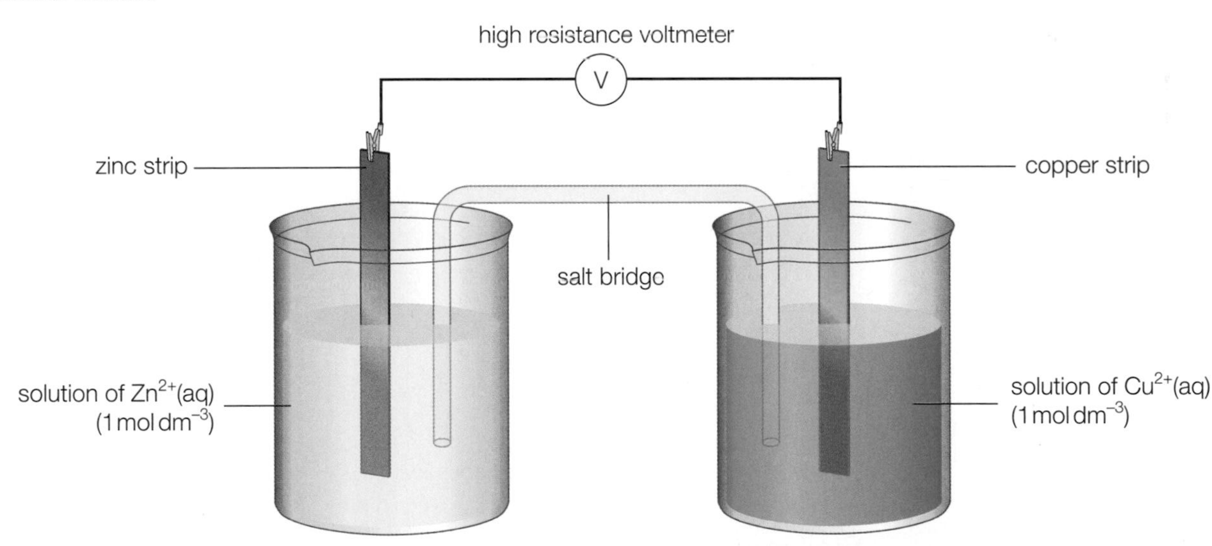

Figure 10.10▲ An electrochemical cell based on the reaction of zinc metal with aqueous copper(II) ions. In this cell, electrons tend to flow from the negative zinc electrode to the positive copper electrode through the external circuit.

The salt bridge makes an electrical connection between the two halves of the cell by allowing ions to flow while preventing the two solutions from mixing. At its simplest, a salt bridge consists of a strip of filter paper soaked in saturated potassium nitrate solution and folded over each of the two beakers. All potassium salts and all nitrates are soluble so the salt bridge does not react to produce precipitates with any of the ions in the half-cells. In more permanent cells, a salt bridge may consist of a porous solid such as sintered glass.

Definition

The **e.m.f. (electromotive force)** of a cell measures the maximum 'voltage' produced by an electrochemical cell. The symbol for e.m.f. is E and its SI unit is the volt (V). The e.m.f. is the energy transferred in joules per coulomb of charge flowing through the circuit connected to a cell. Cell e.m.f.s are at a maximum when no current flows because under these conditions no energy is lost due to the internal resistance of the cell as the current flows.

Chemists measure the tendency for the current to flow in the external circuit by using a high-resistance voltmeter to measure the maximum cell e.m.f. when no current is flowing.

In Figure 10.10, electrons tend to flow out of the zinc electrode (negative) through the external circuit to the copper electrode (positive). The maximum voltage of the cell, usually called its electromotive force (e.m.f.), or cell potential, is 1.10 V under standard conditions.

Standard conditions

In order to compare the voltages (e.m.f.s) developed by different electrochemical cells, scientists carry out the measurements under standard conditions. These standard conditions for electrochemical measurements are the same as those for thermochemical measurements which we met during the AS course. They are:

- temperature 298 K (25 °C)
- gases at a pressure of 1 atmosphere (1.013×10^5 Pa $\approx$ 100 kPa)
- solutions at a concentration of 1.0 mol dm^{-3}.

Chemists have also developed a convenient shorthand called a cell diagram for describing cells. The cell diagram for the cell in Figure 10.10 is shown in Figure 10.11 with an explanation below each entry. Under standard conditions, the symbol for the e.m.f. of the cell is $E^{\ominus}_{cell}$ and this is called the standard e.m.f. of the cell.

Zn(s) |	Zn^{2+}(aq)	⋮	Cu^{2+}(aq) |	Cu(s)	$E^{\ominus}_{cell}$ = +1.10 V
metal electrode on the left	materials in contact with left electrode	salt bridge	materials in contact with right electrode	metal electrode on the right	The cell e.m.f. (potential) The sign of E is the charge on the right-hand electrode

Figure 10.11▲
A cell diagram for the cell composed of the Zn(s) | Zn^{2+}(aq) and Cu(s) | Cu^{2+}(aq) half-cells.

If the cell e.m.f. is positive, the reaction in the cell tends to go according to the cell diagram reading from left to right. As a current flows in the external circuit connecting the two electrodes in Figure 10.11, zinc atoms turn into zinc ions and go into solution, while copper ions turn into copper atoms and deposit on the copper electrode (Figure 10.12).

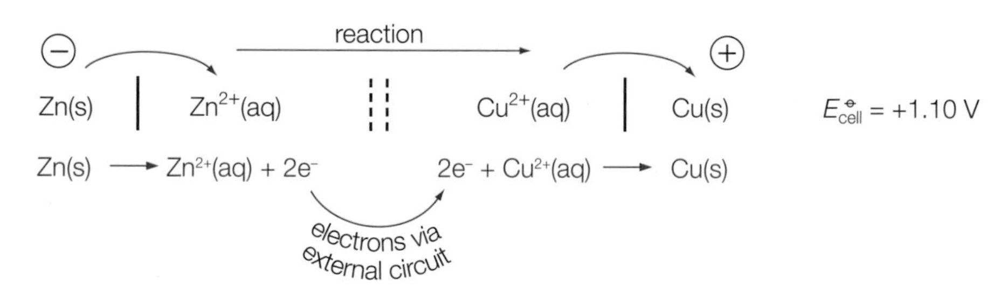

Figure 10.12▲
The direction of change in an electrochemical cell.

Test yourself

19 Consider a cell based on the redox reaction below which tends to go in the direction shown.

$$Cu(s) + 2Ag^+(aq) \rightarrow Cu^{2+}(aq) + 2Ag(s)$$

The potential difference between the electrodes is 0.46 V.

a) Write the half-equations for the electrode processes when the cell supplies a current.

b) Write the conventional cell diagram for the cell, including the value for $E^{\ominus}_{cell}$.

Standard electrode potentials

The study of many cells has shown that a half-electrode, such as the Cu^{2+}(aq) | Cu(s) electrode, makes the same contribution to the cell e.m.f. in *any* cell, so long as the measurements are made under the same conditions. But there is no way of measuring the e.m.f. of an isolated, single electrode because it has only one terminal.

Chemists have solved this problem by selecting a standard electrode system as a reference electrode against which they can compare all other electrode systems. The chosen reference electrode is the standard hydrogen electrode.

By convention, the electrode potential of the standard hydrogen electrode is zero. This is represented as

$Pt[H_2(g)]|2H^+(aq) ¦ E^\ominus = 0.00\,V$

The standard electrode potential for any half-cell is measured relative to a standard hydrogen electrode under standard conditions (Figure 10.13). A standard hydrogen electrode sets up an equilibrium between hydrogen ions in solution (1 mol dm^{-3}) and hydrogen gas (1 atm pressure) at 298 K on the surface of a platinum electrode coated with platinum black.

Note

It is only in a standard hydrogen electrode that the platinum metal is covered with finely divided platinum black. This helps to maintain an equilibrium between hydrogen gas and hydrogen ions and ensure a reversible reaction between them.

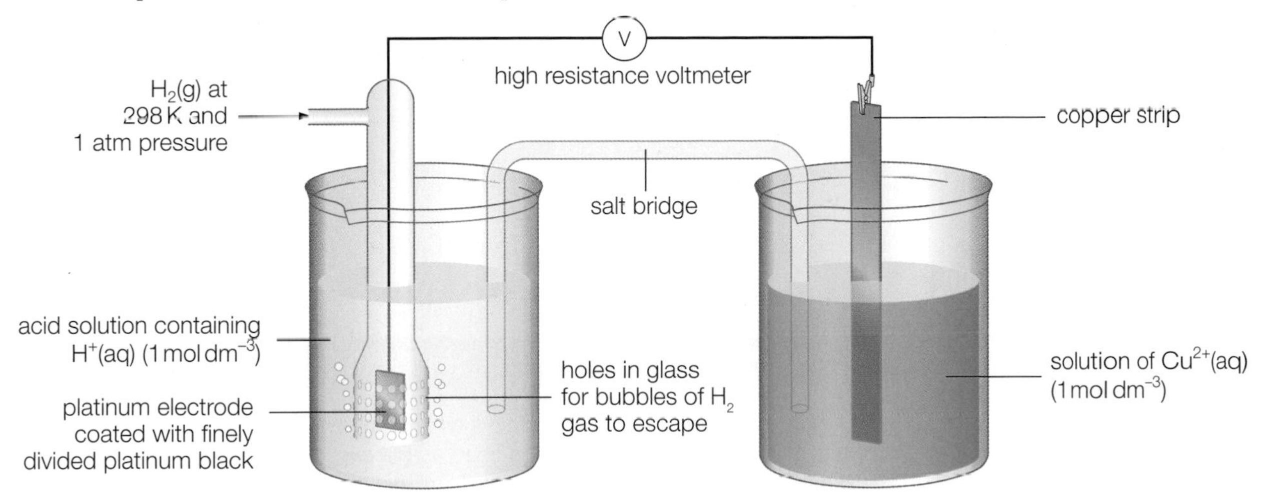

Figure 10.13▲
The e.m.f. of this cell under standard conditions is, by definition, the standard electrode potential of the $Cu^{2+}(aq)|Cu(s)$ electrode.

By convention, when a standard hydrogen electrode is the left-hand electrode in an electrochemical cell, the cell e.m.f. is the electrode potential of the right-hand electrode.

So, the conventional cell diagram for the cell which defines the standard electrode potential of the $Cu^{2+}(aq)|Cu(s)$ electrode is:

$Pt[H_2(g)]|2H^+(aq) ¦¦ Cu^{2+}(aq)|Cu(s) \qquad E^\ominus = +0.34\,V$

The electrode and its standard electrode potential are often represented more simply as:

$$\underset{\text{oxidised form}}{Cu^{2+}(aq)} + 2e^- \rightleftharpoons \underset{\text{reduced form}}{Cu(s)} \qquad E^\ominus = +0.34\,V$$

This also serves to emphasise that standard electrode potentials represent reduction processes.

A hydrogen electrode is difficult to set up and maintain, so it is much easier to use a secondary standard such as a silver/silver chloride electrode or a calomel electrode as a reference electrode. These electrodes are available commercially and are reliable to use. They have been calibrated against a standard hydrogen electrode. Calomel is an old-fashioned name for mercury(I) chloride. The cell reaction and electrode potential relative to a hydrogen electrode for a calomel electrode are:

$$Hg_2Cl_2(s) + 2e^- \rightleftharpoons 2Hg(l) + 2Cl^-(aq) \qquad E^\ominus = +0.27\,V$$

For the reverse reaction, $E^\ominus$ has the opposite sign so,

$$2Hg(l) + 2Cl^-(aq) \rightleftharpoons Hg_2Cl_2(s) + 2e^- \qquad E^\ominus = -0.27\,V$$

Figure 10.13 shows how to measure the standard electrode potential of metals in contact with their ions in aqueous solution.

It is also possible to measure the standard electrode potentials of electrode systems in which both the oxidised and reduced forms are ions in solution, such as ions of the same element in different oxidation states. In these cases, the electrode in the system is platinum (Figure 10.14).

Definitions

A **standard hydrogen electrode** is a half-cell in which a 1 mol dm^{-3} solution of hydrogen ions is in equilibrium with hydrogen gas at 1 atm pressure on the surface of a platinum electrode coated with platinum black at 298 K.

The **standard electrode (reduction) potential**, $E^\ominus$, of a standard half-cell is the e.m.f. of that half-cell relative to a standard hydrogen electrode under standard conditions. Standard electrode potentials are sometimes called standard redox potentials.

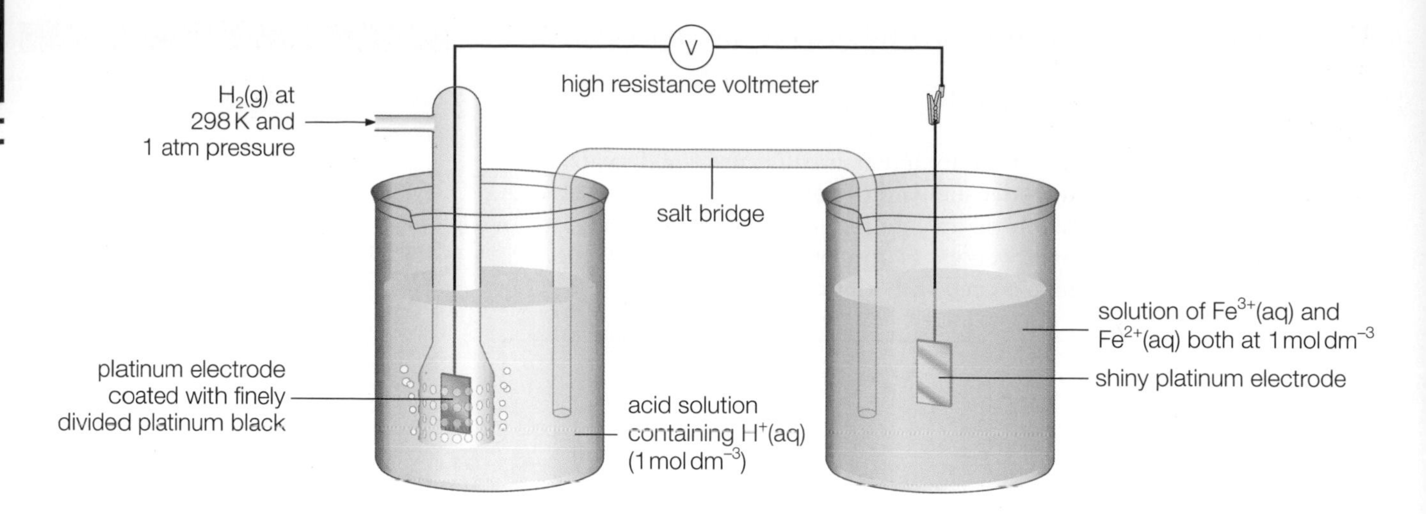

Figure 10.14▲
The apparatus in a cell for measuring the standard electrode potential of the redox reaction $Fe^{3+}(aq) + e^- \rightleftharpoons Fe^{2+}(aq)$.

The cell diagram for the cell in Figure 10.14 is:

$$Pt[H_2(g)] \mid 2H^+(aq) \mathrel{\vdots\vdots} Fe^{3+}(aq),\ Fe^{2+}(aq) \mid Pt \qquad E^\ominus = +0.77\,V$$

Test yourself

20 Suggest why a hydrogen electrode is difficult to set up and maintain.

21 Why do you think that platinum is used as the electrode for systems in which both the oxidised and reduced forms are ions in solution, such as Fe^{3+}(aq) and Fe^{2+}(aq)?

22 What are the half-equations and standard electrode potentials of the right-hand electrode in each of the following cells?

a) $Pt[H_2(g)] \mid 2H^+(aq) \mathrel{\vdots\vdots} Sn^{2+}(aq) \mid Sn(s) \qquad E^\ominus = -0.14\,V$

b) $Pt[H_2(g)] \mid 2H^+(aq) \mathrel{\vdots\vdots} Br_2(aq),\ 2Br^-(aq) \mid Pt(s) \qquad E^\ominus = +1.07\,V$

c) $Pt \mid [2Hg(l) + 2Cl^-(aq)],Hg_2Cl_2(s) \mathrel{\vdots\vdots} Cr^{3+}(aq) \mid Cr(s) \qquad E^\ominus = -1.01\,V$

23 The standard electrode potential for the $Cu^{2+}(aq) \mid Cu(s)$ electrode is +0.34 V. For the cell, $Cu(s) \mid Cu^{2+}(aq) \mathrel{\vdots\vdots} Pb^{2+}(aq) \mid Pb(s)$, the standard cell e.m.f., $E^\ominus_{cell}$, equals −0.47 V.
What is the standard electrode potential for the $Pb^{2+}(aq) \mid Pb(s)$ electrode?

24 a) What is the e.m.f. when a standard calomel electrode is connected to a standard $Cu^{2+}(aq) \mid Cu(s)$ electrode?

b) Write half-equations for the reactions at the electrodes.

10.5 Cell e.m.f.s and the direction of change

Chemists use standard electrode potentials:

- to calculate the e.m.f.s (standard potentials) of electrochemical cells
- to predict the direction (feasibility) of redox reactions.

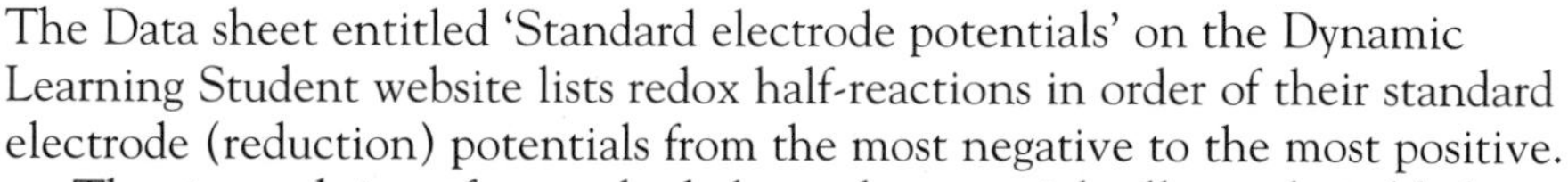

The Data sheet entitled 'Standard electrode potentials' on the Dynamic Learning Student website lists redox half-reactions in order of their standard electrode (reduction) potentials from the most negative to the most positive.

The size and sign of a standard electrode potential tells you how likely it is that a half-reaction will occur. The more positive the standard electrode potential, the more likely it is that the half-reaction will occur.

So, the half-reaction $H_2O_2(aq) + 2H^+(aq) + 2e^- \rightarrow 2H_2O(l)$ with a standard electrode potential of +1.77 volts near the bottom of the table is much more likely than the half-reaction $Li^+(aq) + e^- \rightarrow Li(s)$ with a standard electrode potential of −3.03 volts at the top of the list. Indeed, this last half-reaction will be much more likely to occur in the opposite direction when it would contribute a voltage of +3.03 volts to any electrochemical cell (Figure 10.15).

Using a table of standard electrode potentials, it is possible to calculate cell e.m.f.s ($E^{\ominus}_{cell}$ values) by combining the standard electrode potentials of the two half-cells which make up the full cell. This provides a prediction of the expected direction of chemical change for the redox reaction in the cell.

Look again at Figure 10.11 and the value of the cell e.m.f., $E^{\ominus}_{cell}$. The value of +1.10 volts arises from the sum of the $E^{\ominus}$ values for two half-reactions:

$Cu^{2+}(aq) + 2e^- \rightarrow Cu(s) \quad E^{\ominus} = +0.34\,V$

and the reverse of:

$Zn^{2+}(aq) + 2e^- \rightarrow Zn(s) \quad E^{\ominus} = -0.76\,V$

So, we have:

$$Cu^{2+}(aq) + 2e^- \rightarrow Cu(s) \qquad E^{\ominus} = +0.34\,V$$
$$Zn(s) \rightarrow Zn^{2+}(aq) + 2e^- \qquad E^{\ominus} = +0.76\,V$$

$$\text{Overall: } Zn(s) + Cu^{2+}(aq) \rightarrow Zn^{2+}(aq) + Cu(s) \qquad E^{\ominus}_{cell} = +1.10\,V$$

From this discussion, you will see that

$$E^{\ominus}_{cell} = -E^{\ominus}_{(\text{left-hand electrode})} + E^{\ominus}_{(\text{right-hand electrode})}$$
$$= E^{\ominus}_{(\text{right-hand electrode})} - E^{\ominus}_{(\text{left-hand electrode})}$$

Positive values of $E^{\ominus}_{cell}$ indicate that a reaction is likely to go in the direction of the overall equation. The more positive the value of $E^{\ominus}_{cell}$, the more likely the reaction. On the other hand, negative values for $E^{\ominus}_{cell}$ indicate that reactions will not happen.

Worked example

Draw the cell diagram for a cell based on the two half-equations below. Work out the e.m.f. of the cell and write the overall equation for the reaction which is likely to occur (the spontaneous reaction).

WWW Data

$Fe^{3+}(aq) + e^- \rightleftharpoons Fe^{2+}(aq) \qquad E^{\ominus} = +0.77\,V$
$Cu^{2+}(aq) + 2e^- \rightleftharpoons Cu(s) \qquad E^{\ominus} = +0.34\,V$

Notes on the method
Write the cell diagram with the more positive electrode on the right. Then use the equation:

$$E^{\ominus}_{cell} = E^{\ominus}_{(\text{right-hand electrode})} - E^{\ominus}_{(\text{left-hand electrode})}$$

to calculate the cell e.m.f.

Answer
The $Fe^{3+}(aq)$, $Fe^{2+}(aq)$ electrode is the more positive so it should be on the right-hand side of the cell diagram. Both the oxidised and reduced forms are in solution so a shiny platinum electrode is needed.

$Fe^{3+}(aq), Fe^{2+}(aq)\,|\,Pt(s)$

The left-hand electrode is $Cu^{2+}(aq)\,|\,Cu(s)$ so the copper metal can also be the conducting electrode. The cell diagram is therefore:

$Cu(s)\,|\,Cu^{2+}(aq)\,\vdots\vdots\,Fe^{3+}(aq), Fe^{2+}(aq)\,|\,Pt(s)$

and $E^{\ominus}_{cell} = (+0.77\,V) - (+0.34\,V) = +0.43\,V$

Balancing the two half-equations in the direction of the cell diagram gives the overall equation:

$Cu(s) + 2Fe^{3+}(aq) \rightarrow Cu^{2+}(aq) + 2Fe^{2+}(aq)$

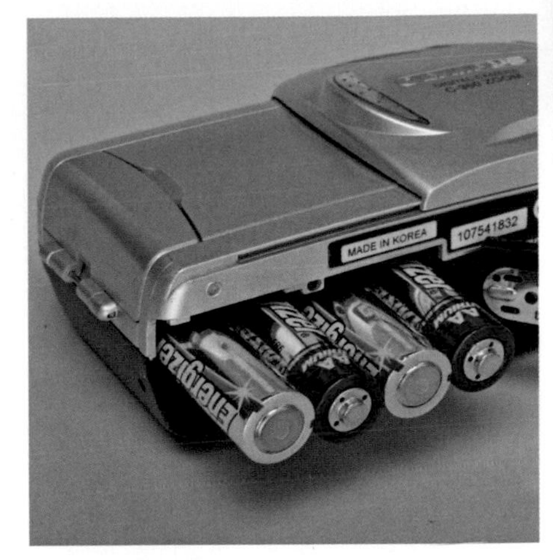

Figure 10.15▲
The four lithium batteries in this camera have total e.m.f. (potential) of about 6 volts.

Test yourself

25 Draw the cell diagram for a cell based on each of the following pairs of half-equations. For each example look up the standard electrode potentials, work out the e.m.f. of the cell and write the overall equation for the reaction which tends to happen (the spontaneous reaction):
- **a)** $V^{3+}(aq) + e^- \rightleftharpoons V^{2+}(aq)$
  $Zn^{2+}(aq) + 2e^- \rightleftharpoons Zn(s)$
- **b)** $Br_2(aq) + 2e^- \rightleftharpoons 2Br^-(aq)$
  $I_2(aq) + 2e^- \rightleftharpoons 2I^-(aq)$
- **c)** $Cl_2(aq) + 2e^- \rightleftharpoons 2Cl^-(aq)$
  $PbO_2(s) + 4H^+(aq) + 2e^- \rightleftharpoons Pb^{2+}(aq) + 2H_2O(l)$

26 Using the standard electrode potential data on the Dynamic Learning Student website, arrange the following sets of metals in order of decreasing strength as reducing agents:
- **a)** Ca, K, Li, Mg, Na
- **b)** Cu, Fe, Pb, Sn, Zn.

27 Using the standard electrode potential data on the Dynamic Learning Student website, arrange the following sets of molecules or ions in order of decreasing strength as oxidising agents in acid solution:
- **a)** $Cr_2O_7^{2-}$, Fe^{3+}, H_2O_2, MnO_4^-
- **b)** Br_2, Cl_2, ClO^-, H_2O_2, O_2.

The electrochemical series

A list of electrode systems set out in order of their electrode potentials (as on the Data sheet: 'Standard electrode potentials' on the Dynamic Learning Student website) is a useful guide to the behaviour of oxidising and reducing agents. It is an electrochemical series.

The metal ion|metal electrodes with highly negative electrode potentials involve half-reactions for group 1 metal ions and metals (Table 10.2). Lithium is the most reactive of these metals when it reacts as a reducing agent forming metal ions. Consequently, the reverse reaction of Li^+ ions forming Li metal is least likely and this results in the most negative standard electrode potential.

The metal ion|metal electrodes with positive electrode potentials involve half-reactions for d-block metal ions and metals low in the reactivity series, such as copper and silver. These metals are relatively unreactive as reducing agents and they do not react with dilute acids to form hydrogen gas. However, their ions are readily reduced to the metal and this results in positive standard electrode potentials.

So, as you might expect, the order of metal ion/metal systems in Table 10.2 closely corresponds to the reactivity series for metals and the reactions shown by metal/metal ion displacement reactions (Figure 10.16).

The electrode potentials of the half-equations involving halogen molecules and halide ions are positive. The $F_2(aq)|2F^-(aq)$ system is the most positive showing that fluorine is the most reactive of the halogens as an oxidising agent. The next most reactive halogen is chlorine, then bromine and finally iodine is the least reactive. This corresponds to the order of reactivity of the halogens and the results of their displacement reactions.

Metal ion/metal electrode	Standard electrode potential, $E^\ominus$/V
$Li^+(aq)\,|\,Li(s)$	−3.03
$K^+(aq)\,|\,K(s)$	−2.92
$Na^+(aq)\,|\,Na(s)$	−2.71
$Al^{3+}(aq)\,|\,Al(s)$	−1.66
$Zn^{2+}(aq)\,|\,Zn(s)$	−0.76
$Fe^{2+}(aq)\,|\,Fe(s)$	−0.44
$Pb^{2+}(aq)\,|\,Pb(s)$	−0.13
$Cu^{2+}(aq)\,|\,Cu(s)$	+0.34
$Ag^+(aq)\,|\,Ag(s)$	+0.80

Table 10.2▲
The standard electrode potentials of some common metals.

Note

In whatever order electrode (reduction) potentials are tabulated, it is always true that:

- the half-cell with the most positive electrode potential has the greatest tendency to gain electrons so it is the most powerfully oxidising
- the half-cell with the most negative electrode potential has the greatest tendency to give up electrons so it is the most powerfully reducing.

The limitations of predictions from $E^\ominus$ data

In some cases, the predictions from standard electrode potentials may not be borne out in practice.

Although an overall positive value of $E^\ominus$ for a redox reaction suggests that the reaction should take place, in practice the reaction may be too slow. The important point to appreciate is that $E^\ominus$ values relate to the relative stabilities of reactants and products.

Therefore, a positive value for $E^\ominus$ indicates that the products are more stable than the reactants and the reaction is energetically feasible. But $E^\ominus$ values do not give any indication about the rates of reactions or their kinetic feasibility.

For example, $E^\ominus$ values predict that $Cu^{2+}(aq)$ should oxidise $H_2(g)$ to H^+ ions:

$$Cu^{2+}(aq) + H_2(g) \rightarrow Cu(s) + 2H^+(aq) \qquad E^\ominus = +0.34 \text{ volts}$$

However, nothing happens when hydrogen is bubbled into copper(II) sulfate solution because the activation energy is so high and the reaction rate is effectively zero.

A second important point about $E^\ominus$ values is that they relate only to standard conditions. Changes in concentration, temperature and pressure affect electrode potentials. In particular, all electrode (reduction) potentials become more positive if the concentration of reactant ions is increased and less positive if their concentration is reduced. This means that some reactions which are not possible under standard conditions occur under non-standard conditions and vice versa.

Figure 10.16▲
Crystals of copper metal deposited on a strip of zinc in copper(II) sulfate solution. By setting up a cell as in Figure 10.10, it is possible to separate the two halves of a redox displacement reaction involving zinc and copper(II) ions. The same reaction will also occur, as in this photo, if the two are just mixed together.

For example, under standard conditions, MnO_2 will not oxidise 1.0 mol dm^{-3} HCl(aq) to Cl_2.

$$MnO_2(s) + 4H^+(aq) + 2e^- \rightarrow Mn^{2+}(aq) + 2H_2O(l) \qquad E^\ominus = +1.23\,V$$
$$2Cl^-(aq) \rightarrow Cl_2(g) + 2e^- \qquad E^\ominus = -1.36\,V$$

Overall:

$$MnO_2(s) + 4H^+(aq) + 2Cl^-(aq) \rightarrow Mn^{2+}(aq) + 2H_2O(l) + Cl_2(g) \quad E^\ominus = -0.13\,V$$

But if MnO_2 is heated with *concentrated* HCl, the electrode potentials of both half-equations become more positive, the overall $E^\ominus$ becomes positive and chlorine is produced.

So, our predictions from cell e.m.f.s about the feasibility (probability) of redox reactions may not occur in practice due to kinetic effects (slow reaction rates) or non-standard conditions of concentration and temperature.

Test yourself

28 Why is it not possible to measure the electrode potential for the $Na^+(aq)|Na(s)$ system using the method illustrated in Figure 10.13?

29 Predict which of the following pairs of reagents will react. Write the overall balanced equation for the reactions that you predict will tend to go:

a) $Zn(s) + Ag^+(aq)$ **b)** $Fe(s) + Ca^{2+}(aq)$ **c)** $Cr(s) + H^+(aq)$
d) $Ca(s) + H^+(aq)$ **e)** $Ag(s) + H^+(aq)$ **f)** $I_2(aq) + Cl^-(aq)$
g) $Cl_2(aq) + I^-(aq)$

30 a) Use the data provided on the Dynamic Learning Student website to find the standard electrode potentials for the two reactions below.
$Cu^{2+}(aq) + e^- \rightarrow Cu^+(aq)$ $Cu^+(aq) + e^- \rightarrow Cu(s)$
b) Using the $E^\ominus$ values show that $Cu^+(aq)$ will disproportionate (be oxidised and reduced simultaneously) under standard conditions.

10.6 How far and in which direction?

From Topic 3, we know that the direction and extent of a chemical reaction can be described by its equilibrium constant, K_c. The larger the value of the equilibrium constant, the greater is the proportion of products to reactants at equilibrium.

Entropy change and equilibrium constants

From Section 3.4, we also know that there is a connection between the total entropy change of a reaction, ΔS_{total}, and its equilibrium constant, K_c. This connection is summarised in the equation

$$\Delta S_{total} = R \ln K_c$$

where R is the gas constant = 8.31 J K^{-1} mol^{-1} and $\ln K_c$ is the logarithm to base e of the equilibrium constant.

Entropy change and electrode potentials

For redox reactions, electrode potentials offer an alternative way of deciding the direction and extent of a reaction (Section 10.5). The more positive the value of the e.m.f. of a cell based on the reaction, E_{cell}, the more likely the reaction is to go.

In addition to this, scientists have found that the total entropy change of a redox reaction and E_{cell} are closely related by the equation

$$\Delta S_{total} = \frac{zF}{T} \times E_{cell}$$

In this equation, z is the number of moles of electrons transferred in the cell reaction, F is the charge on one mole of electrons = 96 500 coulombs mol^{-1} and T is the temperature in kelvin.

Notice from these two equations that:

$$\Delta S_{total} = R \ln K_c = \frac{zF}{T} \times E_{cell}$$

This shows that ΔS_{total}, K_c and E_{cell} are directly related to each other. They can all answer two key questions for any reaction:

- Will the reaction go? and
- How far will it go?

Table 10.3 shows how the values of these three predictors are related to the extent of a reaction.

Table 10.3 Predicting the direction and extent of chemical reactions from the values of ΔS_{total}, E_{cell} and K_c.

ΔS_{total} /J mol^{-1} K^{-1}	E_{cell}/V	K_c (units depend on the reaction)	Extent of reaction
More positive than +200	More positive than +0.6	$>10^{10}$	Goes to completion
≈ +40	≈ +0.1	≈ 10^{2}	Equilibrium with more products than reactants
≈ 0	≈ 0	≈ 1	Roughly equal amounts of reactants and products
≈ −40	≈ −0.1	≈ 10^{-2}	Equilibrium with more reactants than products
More negative than −200	More negative than −0.6	$> 10^{-10}$	No reaction

The fact that ΔS_{total}, K_c and E_{cell} are all directly related to each other has had a profound impact on scientific thinking and on chemists in particular. It has brought together concepts of entropy, equilibrium and electrochemistry, showing that ideas developed in different areas and different contexts of chemistry are all related to the over-riding concept of the thermodynamic feasibility of chemical reactions.

10.7 Modern storage cells

Definition

A **battery** is two or more electrochemical cells connected in series.

From alarm clocks to wrist watches, from radios to calculators, most of us rely on the electrical power of cells and batteries.

Electrochemical cells are a useful and economic way of storing and, when convenient, using the energy from chemical reactions. Because of this, they are sometimes called storage cells. However, the impractical cells discussed in the previous two sections have been replaced by more convenient, portable cells for everyday use.

Activity

The dry cell

Probably the commonest, cheapest and most convenient cell in use today is the modern dry cell (Figure 10.17) which is an adaptation of a cell developed by the French chemist Georges Leclanché in the nineteenth century. The dry cell is used in a wide range of small electrical appliances such as radios, torches and alarm clocks.

The cell diagram for a dry cell is

$Zn(s)\,|\,Zn^{2+}(aq)\,\vdots\vdots\,NH_4^{+}(aq),[2NH_3(g) + H_2(g)]\,|\,C_{(graphite)}$ $\quad E_{cell} = 1.5\,V$

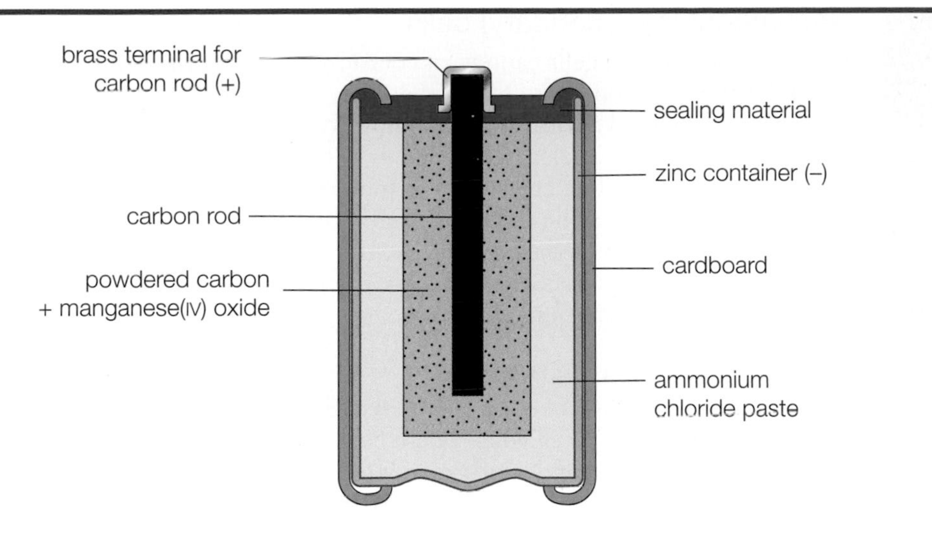

Figure 10.17 ◄
A modern dry cell.

The negative terminal in the cell is zinc. The positive terminal is a carbon (graphite) rod where ammonium ions are converted to ammonia and hydrogen when the cell is delivering a current.

A single dry cell can produce a voltage (e.m.f.) of 1.5 volts, although batteries of these cells giving 100 volts or more have been used. In recent years, a modified form of the dry cell, known as the alkaline cell, has become more common. You can study the alkaline cell in Review question 2 at the end of this topic.

1 Why is the dry cell more convenient than the cells described in the previous two sections?

2 Why is ammonium chloride used as a paste rather than as a dry solid?

3 Write a half-equation for the reaction at:

a) the zinc terminal **b)** the carbon (graphite) terminal.

4 Assuming that the electrode potential for the $Zn^{2+}(aq)|Zn(s)$ half-cell is –0.76 volts, calculate the electrode potential of the other half-cell in a dry cell.

5 The surface area of the positive terminal is increased by surrounding the carbon rod with a mixture of powdered graphite and manganese(IV) oxide. The purpose of the manganese(IV) oxide is to oxidise the hydrogen produced at the electrode to water. This prevents bubbles of hydrogen from coating the carbon terminal and reducing its efficiency.

Figure 10.18 ▲
A selection of the different sizes and types of dry cells and alkaline cells used in radios, clocks, torches and toys.

a) Why do you think it is important to increase the surface area of the carbon terminal?

b) Write an equation for the reaction of manganese(IV) oxide with hydrogen to produce water.

c) Why does ammonia not cause the same problems as hydrogen in reducing the efficiency of the carbon terminal?

6 Use the equations and information in Section 10.6 to calculate ΔS_{total} for the reactions at the electrodes in a dry cell at 27 °C.

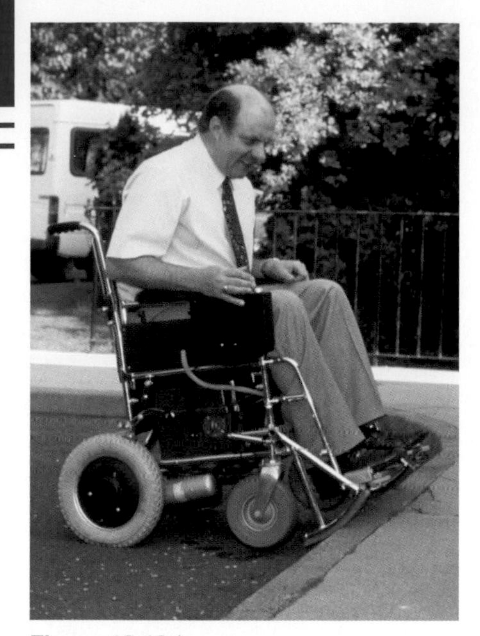

Figure 10.19▲
Most battery-operated wheelchairs are powered by lead–acid cells.

Rechargeable (secondary) cells

Dry cells and alkaline cells cannot, of course, provide continuous supplies of electrical energy indefinitely. Nor is there any way of recharging these cells so that they can be used again. There are, however, cells that can be recharged and used again. These are called secondary cells in contrast to primary cells like dry cells and alkaline cells, which cannot be recharged.

When a secondary cell is recharged, an electric current passes through it in the opposite direction to the current that the cell produces. Recharging is an example of electrolysis as chemical reactions occur to reform the chemicals that make up the electrodes.

Lead–acid cells (car batteries)

Almost all car batteries are composed of six lead–acid cells in series giving a total battery potential of 12 volts. These batteries are only used to provide electricity for the lights and windscreen wipers and a spark to ignite the fuel when a vehicle is started. In contrast, milk floats and battery-operated wheelchairs are actually powered by lead–acid cells.

In recent years, the increasing cost of petrol and diesel has made the electrically powered car a more viable alternative. Unfortunately, attempts to develop and market a vehicle of this type using lead–acid cells have not been successful. The main reasons for this are:

- the high cost of lead–acid batteries
- the low power/weight ratio of lead–acid batteries
- the limited mileage before recharging is necessary.

The negative terminal in a lead–acid cell is lead. This gives up electrons, forming lead(II) ions when the cell is working normally (discharging).

$$Pb(s) \rightarrow Pb^{2+}(aq) + 2e^-$$

The positive terminal is lead coated with lead(IV) oxide. During discharge, the lead(IV) oxide reacts with H^+ ions in the sulfuric acid electrolyte and takes electrons.

$$PbO_2(s) + 4H^+(aq) + 2e^- \rightarrow Pb^{2+}(aq) + 2H_2O(l)$$

Notice that lead(II) ions, $Pb^{2+}(aq)$, are formed at both terminals during discharge. These react with sulfate ions, $SO_4^{2-}(aq)$, in the electrolyte forming insoluble lead(II) sulfate on each terminal.

$$Pb^{2+}(aq) + SO_4^{2-}(aq) \rightarrow PbSO_4(s)$$

Test yourself

31 Why have attempts to develop family cars powered by lead–acid batteries been unsuccessful?

32 The milk float in Figure 10.20 is powered by lead–acid batteries. Why is this both possible and convenient?

Figure 10.20▶

The formation of insoluble lead(II) sulfate creates another potential problem for lead–acid cells. If the cells are discharged for long periods, the precipitate of lead(II) sulfate becomes coarser and thicker and the process cannot be reversed when the cells are recharged.

When a lead–acid cell is recharged, the current is reversed and the reactions at each terminal are reversed. This turns Pb^{2+} ions back to lead metal at one terminal and back to PbO_2 at the other, with sulfate ions going back into the electrolyte.

Test yourself

33 a) Use the Data sheet headed 'Standard electrode potentials' on the Dynamic Learning Student website to write down the value of $E^\ominus$ for a $Pb^{2+}(aq)\,|\,Pb(s)$ half-cell.

b) What is the approximate cell potential for one lead–acid cell?

c) What is the approximate electrode potential of the $[PbO_2(s) + 4H^+(aq)]$, $[Pb^{2+}(aq) + 2H_2O(l)]\,|\,Pb(s)$ half-cell in a lead–acid cell?

d) Write a balanced equation for the overall reaction in a lead–acid cell when it is supplying current.

e) Write the half-equations for the processes at the two terminals when the cell is being recharged.

Lithium cells

Modern mobile phones and laptop computers use lithium batteries. The advantages of electrodes based on lithium are that the metal has a low density, therefore cells based on lithium electrodes can be relatively light. Also, lithium is very reactive which means that the electrode potential of a lithium half-cell is relatively high and each cell has a large e.m.f.

The difficulty to overcome is that lithium is so reactive that it readily combines with oxygen in the air forming a layer of non conducting oxide on the surface of the metal. The metal also reacts rapidly with water. Research workers have solved these technical problems by developing electrodes with lithium atoms and ions inserted into the crystal lattices of other materials. In addition, the electrolyte is a polymeric material rather than an aqueous solution.

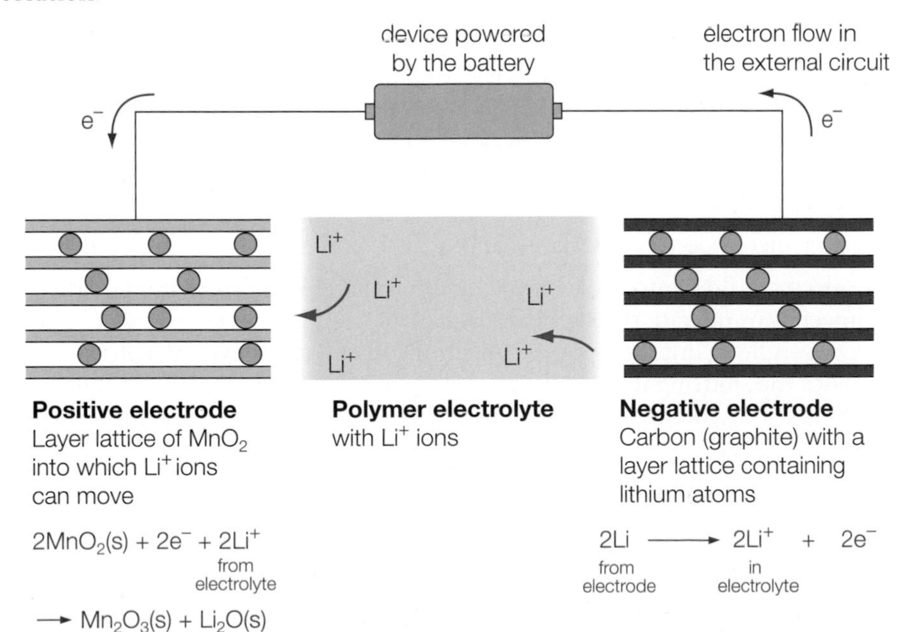

Figure 10.21▲
A schematic diagram of a lithium battery discharging. The electrode processes are reversible, so the battery can be recharged.

10.8 Fuel cells

Fuel cells are electrochemical cells in which the chemical energy of a fuel is converted directly into electrical energy. Fuel cells differ from typical electrochemical cells, such as dry cells and lead–acid cells, in having a continuous supply of reactants from which to produce a steady electric current. Fuel cells use a variety of fuels including hydrogen, hydrocarbons (such as methane) and alcohols. Inside a fuel cell, energy from the redox reaction between a fuel and oxygen is used to create a potential difference (voltage). Hydrogen and alcohol fuel cells have been used in the development of electric cars and in space exploration.

The hydrogen–oxygen fuel cell

One of the most important fuel cells is the hydrogen–oxygen fuel cell (Figure 10.22).

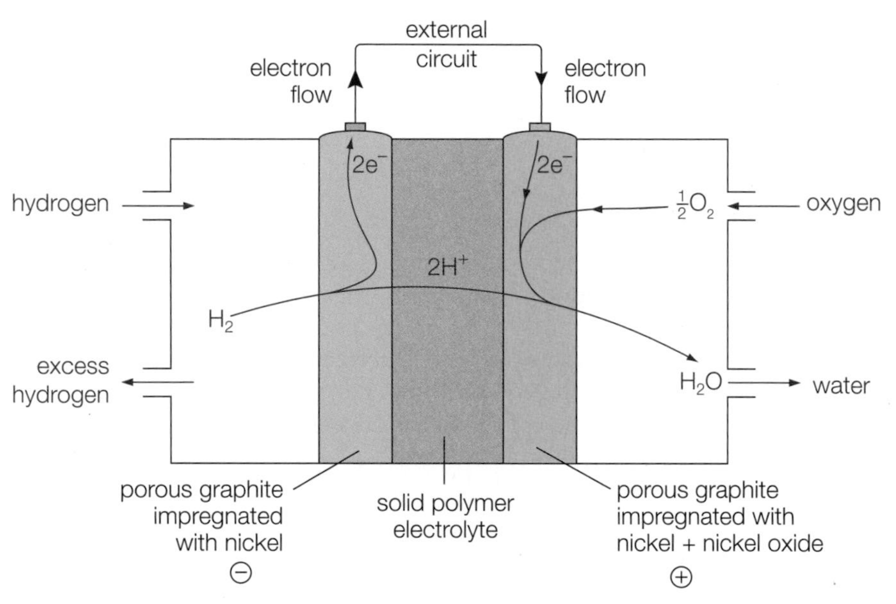

Figure 10.22 A hydrogen–oxygen fuel cell.

In the hydrogen–oxygen fuel cell, the negative electrode is porous graphite impregnated with nickel, and the positive electrode is porous graphite impregnated with nickel and nickel(II) oxide. The nickel and nickel oxide catalyse the breakdown of hydrogen and oxygen molecules into single atoms.

Hydrogen gas flows onto the negative electrode where the H_2 molecules split into single H atoms on the nickel catalyst. The H atoms then lose electrons and form H^+ ions.

Negative electrode $H_2 \rightarrow 2H \rightarrow 2H^+ + 2e^-$ Equation 1

The electrons flow into the external circuit as an electric current while the hydrogen ions migrate through the electrolyte.

Oxygen flows onto the positive electrode where the nickel/nickel(II) oxide catalyses the splitting into single oxygen atoms. These oxygen atoms then combine with hydrogen ions from the electrolyte and electrons to form water.

Positive terminal $O + 2H^+ + 2e^- \rightarrow H_2O$ Equation 2

The overall reaction in the hydrogen–oxygen fuel cell (obtained by adding Equations 1 and 2) is

$$H_2 + O \rightarrow H_2O$$

When alcohol (ethanol) is used as the energy source in place of hydrogen, the following half-reactions occur at the terminals.

Negative terminal $C_2H_5OH(l) + H_2O(l) \rightarrow CH_3COOH(aq) + 4H^+(aq) + 4e^-$

Positive terminal $[2H^+(aq) + O_2(g) + 2e^- \rightarrow H_2O(l)] \times 2$

The overall reaction in this alcohol–oxygen fuel cell is therefore:

$$C_2H_5OH(l) + O_2(g) \rightarrow CH_3COOH(aq) + H_2O(l)$$

These fuel cells are no different in principle from more familiar electrochemical cells. The innovation is that new reactants (such as H_2 or C_2H_5OH and O_2) are constantly fed into the cell and the products (H_2O and CH_3COOH) are drawn off. This continuous flow of materials allows the cell potential to remain constant and the power output is uninterrupted.

The great advantage of all fuel cells is that they convert chemical energy directly into electricity and in doing so achieve a remarkable efficiency of about 70%. In comparison, modern power plants and petrol engines using fossil fuels have a conversion efficiency from chemical energy to electrical energy or kinetic energy of only about 40%.

10.9 Fuel cell vehicles (FCVs)

As calls for a cleaner environment become louder and reserves of fossil fuels become scarcer, the search for alternative fuels is on. One possible alternative to petrol and diesel is to run vehicles on hydrogen or alcohol using fuel cells. These are far more efficient than conventional engines.

A further advantage of hydrogen fuel cells in electric vehicles is that water vapour is the only exhaust gas – a very attractive 'green' proposition.

The idea of an electric car powered by hydrogen or alcohol is not new, but until recently it has not been feasible due to technological problems. These include:

- the weight of fuel cells
- the high cost of fuel cells
- the finite life of fuel cells, which require regular replacement because of deterioration of the catalyst caused by impurities in the reactants
- the cost of producing hydrogen or alcohol. (See Section 19.4 in *Edexcel Chemistry for AS*.)

An additional problem with hydrogen-powered vehicles is the difficulty of carrying and transporting sufficient hydrogen.

Figure 10.23▲
A technician assembles a fuel cell for testing. The fuel cell is constructed from lightweight polymeric materials.

In recent years, compact lightweight fuel cells have been developed along with the technology to store large quantities of hydrogen much more conveniently and safely as metal hydrides.

The efficiency of fuel cells is so much better than petrol and diesel engines that fuel cell cars of the future might well use hydrogen-rich fuels such as ethanol, natural gas and petrol in place of hydrogen. Although fuel cells have been constructed that use methane, methanol and ethanol directly, scientists are now working on the development of compact, catalytic 'reformers' which will produce hydrogen from these fuels as on-board components of FCVs. Although there are clear benefits in the use of fuel cells and fuel cell vehicles, there are also logistical problems in their development and use.

Activity

A greener future for transport!

Figure 10.24 ◀ Technicians assembling the components of an electric fuel cell car based on a modified Honda.

Read Section 10.9, 'Fuel cell vehicles (FCVs)', again and then answer the following questions.

1 Why are compounds such as ethanol, natural gas and petrol described as hydrogen-rich fuels?

2 What are the major problems in developing FCVs powered by hydrogen and hydrogen-rich fuels?

3 What are the major advantages of electric fuel cell vehicles over conventional petrol-engine and diesel-engine vehicles?

4 What is the main advantage of fuel cell vehicles powered by hydrogen rather than hydrogen-rich fuels?

5 Write equations for the oxidation of hydrogen and petrol in fuel cells, assuming that they are completely oxidised and that petrol is pure octane.

6 a) Use the Data sheet headed 'Mean bond enthalpies and bond lengths' on the Dynamic Learning Student website to find the mean bond enthalpies for the different bonds in hydrogen, oxygen, octane, carbon dioxide and water.

b) Make a table showing the bonds broken and the bonds formed during the complete combustion of hydrogen and then calculate a value for the enthalpy change of combustion of hydrogen.

c) Calculate the energy produced during the complete combustion of one gram of hydrogen.

7 Repeat the calculations in questions **6b)** and **6c)** for petrol (octane) in place of hydrogen.

8 Bearing in mind that fuel cells are approximately 70% efficient and petrol engines are about 40% efficient, how many times more efficient is a fuel cell vehicle powered by hydrogen compared to a vehicle with a petrol engine, per gram of fuel?

Figure 10.25▲
The limited edition Honda FCX Clarity, powered by fuel cells, was first leased in 2008 in Southern California.

10.10 Ethanol fuel cells in modern breathalysers

Although alcoholic drinks can make you more relaxed and sociable, they can lead to serious road accidents when someone drinks and drives. Because of the worrying increase in accidents due to drink-driving in the 1960s, roadside breathalyser testing by the police was introduced in the UK in 1967.

Soon after taking an alcoholic drink, ethanol, C_2H_5OH, passes through the stomach wall and into the bloodstream. The ethanol concentration in the blood equilibrates fairly rapidly with that in the air in the lungs and, at body temperature, results show that

$$\frac{[C_2H_5OH(\text{blood})]}{[C_2H_5OH\ (\text{air in lungs})]} = \frac{2300}{1}$$

So, by measuring the concentration of ethanol in the breath, it is possible to calculate the concentration of ethanol in the blood.

When breathalyser tests started in 1967, drivers had to take a deep breath and blow though a previously sealed tube until they had inflated a 1 dm^3 bag.

The tube was packed with orange crystals of potassium dichromate(VI), $K_2Cr_2O_7$, plus sulfuric acid on solid, unreactive silicon dioxide. If there was ethanol in the driver's breath, this was oxidised to ethanal and then ethanoic acid.

$$\underset{\text{ethanol}}{CH_3CH_2OH} \rightarrow \underset{\text{ethanal}}{CH_3CHO} + 2H^+ + 2e^-$$

$$\underset{\text{ethanal}}{CH_3CHO} + H_2O \rightarrow \underset{\text{ethanoic acid}}{CH_3COOH} + 2H^+ + 2e^-$$

At the same time, the orange dichromate(VI) ions, $Cr_2O_7^{2-}$, were reduced to green chromium(III) ions, Cr^{3+}.

$$\underset{\text{orange}}{Cr_2O_7^{2-}} + 14H^+ + 6e^- \rightarrow \underset{\text{green}}{2Cr^{3+}} + 7H_2O$$

By measuring the length of crystals in the tube which had changed in colour from orange to green, it was possible to estimate the concentration of ethanol in the driver's breath.

The breathalyser tubes, containing orange dichromate(VI), were quick and convenient to use, but chromium(VI) compounds are toxic and it was difficult to obtain accurate measurements of the concentrations of ethanol from the colour changes. Because of this a safer, more reliable method was introduced in 1980 using an ethanol/oxygen fuel cell.

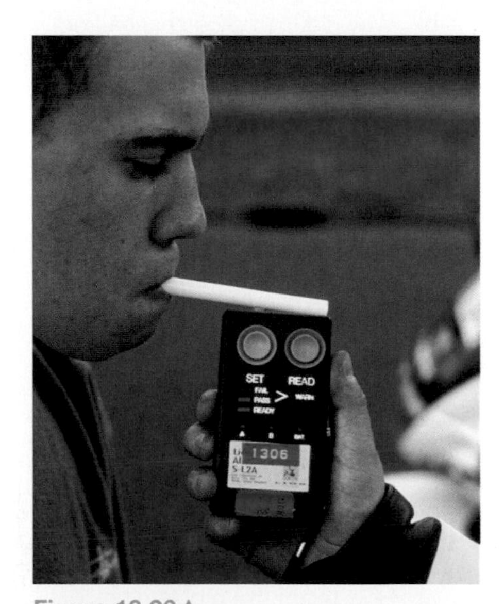

Figure 10.26▲
Roadside 'breathalyser' tests involve the use of an electronic instrument containing an ethanol/oxygen fuel cell.

In this cell, the electrodes are made of a precious metal such as platinum or silver. Between the electrodes, there is a permeable membrane of porous glass containing the electrolyte, sodium hydroxide.

When a driver breathes into the instrument, ethanol is catalytically oxidised to ethanoic acid at the negative terminal.

Negative terminal $CH_3CH_2OH(g) + H_2O(l) \rightarrow CH_3COOH(aq) + 4H^+(aq) + 4e^-$

At the positive terminal, oxygen is reduced to water.

Positive terminal $O_2(g) + 4H^+(aq) + 4e^- \rightarrow 2H_2O(l)$

The instrument is first calibrated by passing air with known ethanol concentrations onto the negative electrode and measuring the resulting voltages. As the ethanol concentration increases, the cell voltage also increases, so the roadside breathalyser can be modified to show the ethanol concentration rather than the cell voltage.

Although the results of the ethanol fuel cell breathalyser are accurate, they are not usually presented as evidence in court because the instrument does not provide a printout when the test is carried out.

When the fuel cell breathalyser shows an ethanol concentration above the legal limit, the driver must take a second breath test at a police station. Here, the ethanol concentration is measured accurately by absorption of infrared radiation. Figure 10.27 shows the infrared absorption spectrum of ethanol. The absorption at 2950 cm^{-1}, which relates to the stretching vibrations of C–H bonds, is used to measure the concentration of ethanol in a driver's breath.

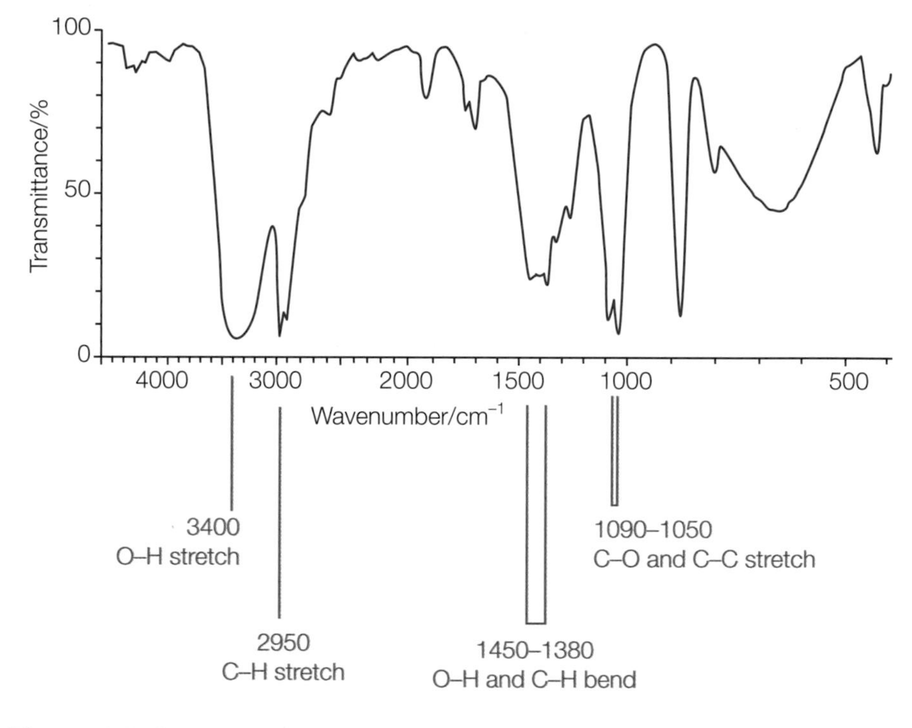

Figure 10.27▲
The infrared absorption spectrum of ethanol.

Infrared radiation is passed through a sample of the driver's breath and the intensity of the radiation transmitted at 2950 cm^{-1} is compared with the same radiation transmitted by a sample of ethanol-free air.

The amount of radiation absorbed by the driver's breath is proportional to the concentration of ethanol in it, and the IR spectroscope can be calibrated to give a direct printout of the result which can be used as evidence in court.

Test yourself

34 Hydrogen is said to have 'the highest density of energy per unit mass of any fuel'. What does this expression mean?

35 Why is liquid hydrogen unsuitable for use as a fuel in motor vehicles?

36 a) Why were roadside breath tests for ethanol introduced in 1967?

b) Why did the police change from using breathalysers based on chromium compounds to those based on an ethanol fuel cell?

c) Why are IR absorption tests used in addition to ethanol fuel cell tests when a driver is over the legal limit?

37 The infrared absorption spectrum of ethanol shows a prominent O–H bond absorption at 3340 cm^{-1}. Why is this absorption not used in measuring the concentration of ethanol in a driver's breath?

38 Propanone, CH_3COCH_3, is often present in the breath of diabetics.

a) What problems will this create in breathlysing someone who is a diabetic?

b) How might this problem be solved using IR spectroscopy?

REVIEW QUESTIONS

Extension questions

1 Use the standard electrode potentials in the table below to answer the questions that follow.

I	$Fe^{3+}(aq) + e^- \rightarrow Fe^{2+}(aq)$	$E^\ominus = +0.77\,V$
II	$Cu^{2+}(aq) + 2e^- \rightarrow Cu(s)$	$E^\ominus = +0.34\,V$
III	$2H^+(aq) + 2e^- \rightarrow H_2(g)$	$E^\ominus = 0.00\,V$
IV	$O_2(g) + 4H^+(aq) + 4e^- \rightarrow 2H_2O(l)$	$E^\ominus = +0.40\,V$

a) An electrochemical cell was arranged using systems I and II.

i) Write half-equations for the reactions that occur in each half-cell when a current flows. Say which half-equation involves oxidation and which involves reduction. (2)

ii) Calculate the change in oxidation number of the oxidised and reduced elements in each half-cell. (2)

iii) Determine the e.m.f. of the cell. (1)

b) Fuel cells using systems III and IV are increasingly being used to generate electricity.

i) Construct an overall equation for the cell reaction and show your working. (2)

ii) From which half-cell do electrons flow into the external circuit? (1)

iii) State two advantages and two disadvantages of using fuel cells based on systems III and IV to generate energy rather than using fossil fuels. (4)

2 A more efficient and more expensive form of the dry cell is the alkaline cell.
The reaction at the negative terminal is again the oxidation of zinc, but in contact with OH^- ions to form zinc oxide.
The positive terminal is manganese(IV) oxide which is reduced to manganese(III) oxide.

$$2MnO_2(s) + H_2O(l) + 2e^- \rightarrow Mn_2O_3(s) + 2OH^-(aq)$$

a) An alkaline cell is an example of a primary cell. What is meant by a primary cell? (1)

b) How does a secondary cell differ from a primary cell? (2)

c) Write a half-equation for the reaction at the negative terminal when an alkaline cell is used. (2)

d) State the changes in oxidation number at each terminal. (2)

e) Why is an alkaline cell in continuous use more efficient than a dry cell? (2)

3 A student set up the electrochemical cell shown in Figure 10.28.

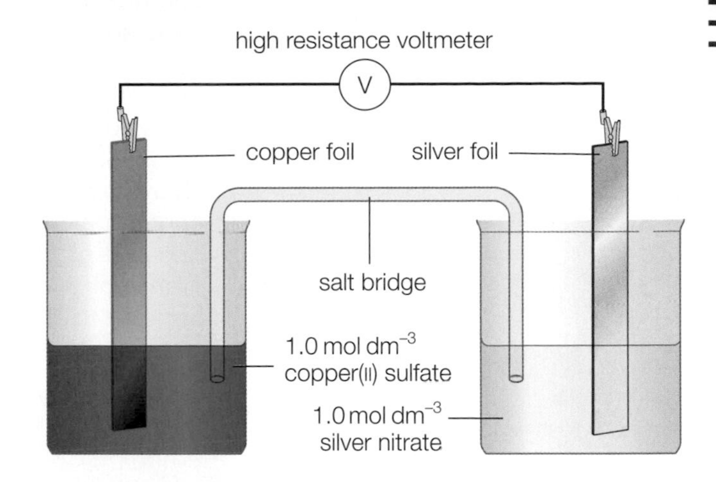

Figure 10.28▲

You are provided with the following standard electrode potentials:

$Cu^{2+}|Cu \quad E^\ominus = +0.34\,V$
$Ag^+|Ag \quad E^\ominus = +0.80\,V$

a) How could the student have made a salt bridge? (1)

b) Write half-equations to show the reactions that occurred in:

i) the $Cu|Cu^{2+}$ half-cell (1)

ii) the $Ag|Ag^+$ half-cell. (1)

c) Write an equation for the overall cell reaction. (1)

d) Calculate the e.m.f. (potential difference) for this cell. (2)

e) At which electrode does reduction occur? Explain your answer. (2)

f) The student found that the cell e.m.f. was less than the calculated value. Suggest two reasons for this. (2)

4 a) What are the principal differences between a hydrogen–oxygen fuel cell and a conventional electrochemical cell like the 'dry cell'? (3)

b) Outline and explain two advantages that would be gained by generating electricity using fuel cells rather than in thermal power stations. (4)

c) What major advantage would hydrogen-powered fuel cells have over other fuel cells? (2)

d) Suggest **three** obstacles to the present development and production of hydrogen-powered fuel cell vehicles. (3)

11 Transition metals

The transition metals are vital to life and bring colour to our lives. They are also metals of great engineering and industrial importance. Chemically, these elements, which occupy the d block of the periodic table, are more alike than might be expected. Across the ten transition metals from scandium to zinc in period 4, the similarities are as striking as the differences. Chemists explain the characteristics of transition metals in terms of the electronic configurations of their atoms. Transition metal chemistry is colourful because of the range of oxidation states and complex ions. Transition metals matter because their properties are fundamental, not only to life but also to modern technology.

Figure 11.1▶
A stained glass window in Ely Cathedral. The commonest colorants in stained glass are the oxides of transition metals.

11.1 Electronic configurations

As the shells of electrons around the nuclei of atoms get further from the nucleus, they become closer in energy. Therefore, the difference in energy between the second and third shells is less than that between the first and second. When the fourth shell is reached there is, in fact, an overlap between the orbitals of highest energy in the third shell (the 3d orbitals) and that of lowest energy in the fourth shell (the 4s orbital) (Figure 11.2).

The 3d sub-shell is on average nearer the nucleus than the 4s sub-shell, but at a higher energy level. So, once the 3s and 3p sub-shells are filled, the next electrons go into the 4s sub-shell because it occupies a lower energy level than the 3d sub-shell.

This means that potassium and calcium have the electron structure $[Ar]4s^1$ and $[Ar]4s^2$ respectively (Table 11.1).

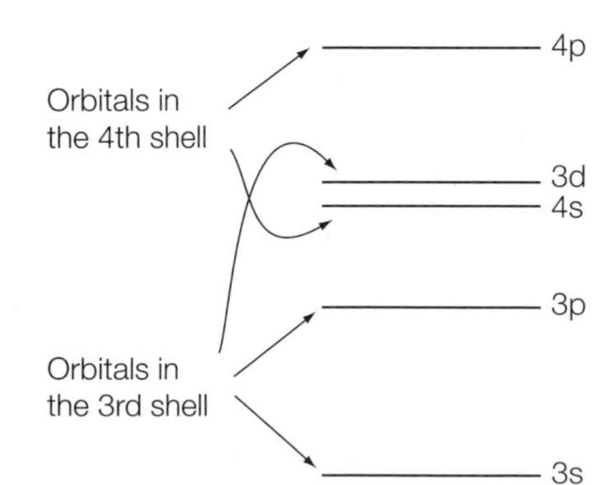

Figure 11.2▲
The relative energy levels of orbitals in the third and fourth shells.

Element	Symbol	Electronic structure	
		spdf notation	Electrons-in-boxes notation (3d, 4s)
Potassium	K	$[Ar]4s^1$	[Ar] [ ][ ][ ][ ][ ] [↑]
Calcium	Ca	$[Ar]4s^2$	[Ar] [ ][ ][ ][ ][ ] [↑↓]
Scandium	Sc	$[Ar]3d^14s^2$	[Ar] [↑][ ][ ][ ][ ] [↑↓]
Titanium	Ti	$[Ar]3d^24s^2$	[Ar] [↑][↑][ ][ ][ ] [↑↓]
Vanadium	V	$[Ar]3d^34s^2$	[Ar] [↑][↑][↑][ ][ ] [↑↓]
Chromium	Cr	$[Ar]3d^54s^1$	[Ar] [↑][↑][↑][↑][↑] [↑]
Manganese	Mn	$[Ar]3d^54s^2$	[Ar] [↑][↑][↑][↑][↑] [↑↓]
Iron	Fe	$[Ar]3d^64s^2$	[Ar] [↑↓][↑][↑][↑][↑] [↑↓]
Cobalt	Co	$[Ar]3d^74s^2$	[Ar] [↑↓][↑↓][↑][↑][↑] [↑↓]
Nickel	Ni	$[Ar]3d^84s^2$	[Ar] [↑↓][↑↓][↑↓][↑][↑] [↑↓]
Copper	Cu	$[Ar]3d^{10}4s^1$	[Ar] [↑↓][↑↓][↑↓][↑↓][↑↓] [↑]
Zinc	Zn	$[Ar]3d^{10}4s^2$	[Ar] [↑↓][↑↓][↑↓][↑↓][↑↓] [↑↓]

Table 11.1 ◀
Electron configurations from potassium to zinc in period 4 of the periodic table. ([Ar] represents the electronic configuration of argon.)

Look carefully at Table 11.1. In period 4, the d-block elements run from scandium ($1s^2,2s^22p^6,3s^23p^63d^1,4s^2$) to zinc ($1s^2, 2s^22p^6,3s^23p^63d^{10},4s^2$). But, notice that the electronic configurations of chromium and copper do not fit the general pattern. The explanation of these irregularities lies in the stability associated with half-filled and filled sub-shells. So, the electronic structure of chromium, $[Ar]3d^54s^1$, with half-filled sub-shells and an equal distribution of charge around the nucleus, is more stable than the electronic structure $[Ar]3d^44s^2$.

Similarly, the electronic structure of copper, $[Ar]3d^{10}4s^1$, with a filled 3d sub-shell and a half-filled 4s sub-shell is more stable than $[Ar]3d^94s^2$.

Along the series of d-block elements from scandium to zinc, the number of protons in the nucleus increases by one from one element to the next. However, the added electrons go into an inner d sub-shell, but the outer electrons are always in the 4s sub-shell. This means that there are clear similarities amongst the transition elements. Changes in their chemical properties across the series are much less marked than the big changes across a series of p-block elements such as aluminium to argon.

Notice how the energy-level model for electronic structure can help to account for the similarities in properties of transition metals. Later, in Section 11.5, we will see the limitations of the energy-level model and the need for more sophisticated explanations.

Figure 11.3 ▼
Specimens of some d-block elements. The chemistry of an element is determined to a large extent by its outer shell electrons because they are the first to get involved in reactions. All the d-block elements have their outer electrons in the 4s sub-shell.

Test yourself

1 Write the full s,p,d,f electronic configuration of
  a) a scandium atom
  b) a scandium(III) ion
  c) a manganese atom
  d) a manganese(II) ion.
2 Look at the electronic structures of iron and copper in Table 11.1.
  a) Write the electronic structure of an iron(II) ion.
  b) Write the electronic structure of an iron(III) ion.
  c) Which ion, Fe^{2+} or Fe^{3+}, would you expect to be the more stable? Explain your choice.
  d) Write the formula for the ion of copper that you would expect to be the most stable. Explain your choice.

Ions of the transition metals

When transition metals form their ions, electrons are lost initially from the 4s sub-shell and not the 3d sub-shell. This may seem somewhat illogical because, prior to holding any electrons, the 4s level is more stable than the 3d level. But once the 3d sub-shell is occupied by electrons, these 3d electrons, being closer to the nucleus, repel the 4s electrons to a higher energy level. The 4s electrons are, in fact, repelled to an energy level higher than those occupying the 3d sub-shell. So, when transition metals form ions, they lose electrons from the 4s before the 3d level. This further emphasises the fact that transition metals have similar chemical properties dictated by the behaviour of the 4s electrons in their outer shells.

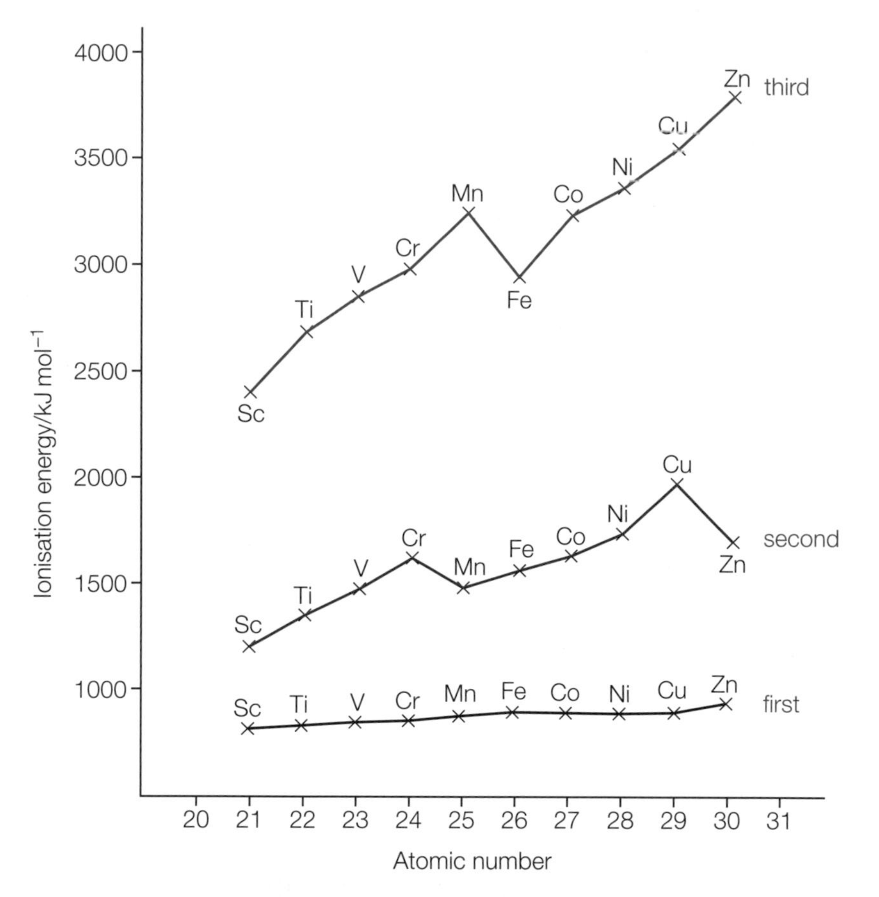

Figure 11.4▶ Graphs of the first, second and third ionisation energies of the elements from scandium to zinc in the periodic table.

Activity

Studying the ionisation energies of transition metals

Earlier in this section, we used the energy-level model to predict the electronic structures of d-block elements in period 4 and their ions. Experimental evidence for these electronic configurations can be obtained from the ionisation energies of the elements concerned.

Look carefully at Figure 11.4, which shows graphs of the first, second and third ionisation energies of the elements from scandium to zinc.

1 Write the electronic structures of the following atoms and ions using [Ar] for the electronic structure of argon.

a) Zn **b)** Cu^+ **c)** Zn^+ **d)** Cr^+ **e)** Mn^+.

2 Write an equation for:

a) the second ionisation energy of chromium

b) the third ionisation energy of iron.

3 Explain the general trend in ionisation energies as atomic number increases.

4 a) How does the first ionisation energy of zinc compare with those of the other d-block elements in period 4?

b) What does this tell you about zinc relative to the other elements?

c) How does this relate to the electronic configuration of zinc atoms?

5 a) What does the high second ionisation energy of copper, relative to its neighbours in the periodic table, tell you about copper?

b) How does this relate to the electronic configuration of copper atoms and ions?

6 a) What does the high second ionisation energy of chromium, relative to its neighbours in the periodic table, tell you about chromium?

b) How does this relate to the electronic configuration of chromium atoms and ions?

7 a) Which elements have relatively high third ionisation energies compared with their neighbours in the d-block elements of period 4?

b) How do these relatively high third ionisation energies provide further evidence for the proposed electronic configurations of the elements concerned?

Tutorial

11.2 Defining the transition metals

The simplest and neatest way to define the transition metals would be to say that they are the elements in the d block of the periodic table. But this simple definition leads to the inclusion of scandium and zinc as transition metals and ignores the fact that these two metals have some clear differences from the metals between them in the periodic table from titanium to copper. For instance:

- Scandium and zinc have only one oxidation state in their compounds (scandium +3, zinc +2), whereas the elements from titanium to copper have two or more.
- The compounds of scandium to zinc are usually white, unlike those of transition metals which are generally coloured.
- Scandium, zinc and their compounds show little catalytic activity.

As scandium and zinc do not show the typical properties of transition metals, chemists looked for a more satisfactory definition. This definition should exclude scandium and zinc, but include all the elements from titanium to copper. In order to achieve this, chemists describe transition metals as those elements that form one or more stable ions with incompletely filled d orbitals.

Definition

A **transition metal** is an element that has one or more stable ions with incompletely filled d orbitals.

Characteristics of the transition metals

In general, transition elements share a number of common properties. (See the Data sheet: 'Properties of selected elements – d-block metals' on the Dynamic Learning Student website.)

- They are hard metals with useful mechanical properties, high melting and high boiling temperatures.
- They show variable oxidation numbers in their compounds.
- They form coloured ions in solution.
- They can act as catalysts both as the elements and as their compounds.
- They form complex ions involving monodentate, bidentate and polydentate ligands (Section 11.7).

In the next five sections, we look in turn at each of these five characteristics of transition metals.

11.3 The transition elements as metals

Most of the transition elements have a close-packed structure in which each atom has twelve nearest neighbours. In addition, transition elements have relatively low atomic radii because an increasingly large nuclear charge is attracting electrons that are being added to an inner sub-shell. The dual effect of close packing and small atomic radii results in strong metallic bonding. So, transition metals have higher melting temperatures, higher boiling temperatures, higher densities and higher tensile strength than s-block metals such as calcium and p-block metals such as aluminium and lead. A plot of physical properties against atomic number often has two peaks or two troughs associated with a half-filled and then a filled d sub-shell (Figure 11.5).

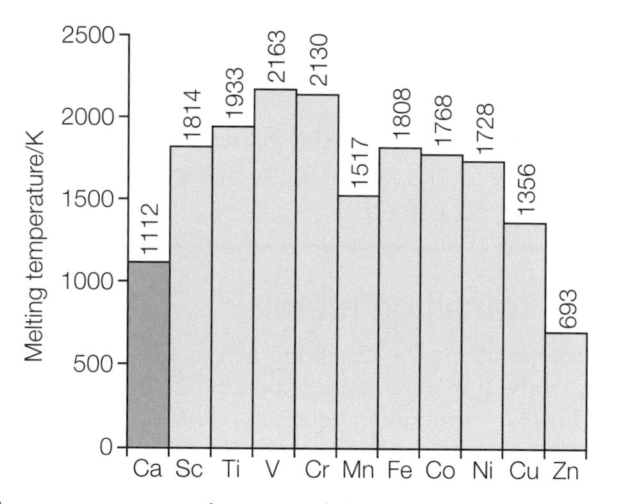

Figure 11.5▶
A plot of melting temperature against atomic number for the elements calcium to zinc in the periodic table.

Element	Standard electrode potential, $E^{\ominus}$ $M^{2+}(aq)\mid M(s)/V$
Vanadium	−1.20
Chromium	−0.91
Manganese	−1.19
Iron	−0.44
Cobalt	−0.28
Nickel	−0.25
Copper	+0.34

Table 11.2▲
Standard electrode potentials of the transition metals from V to Cu.

The transition metals are much less reactive than the s-block metals. However, their electrode potentials (Table 11.2) suggest that all of them, except copper, should react with dilute strong acids such as 1 mol dm^{-3} hydrochloric acid and sulfuric acid. In practice, many of the metals react very slowly with dilute acids because the metal is protected by a thin, unreactive layer of oxide. Chromium provides a very good example of this. Despite the predictions from its standard electrode potential, it is used as a protective, non-rusting metal owing to the presence of an unreactive, non-porous layer of chromium(III) oxide, Cr_2O_3.

Copper is the least reactive of the transition metals in period 4. It does not react with dilute non-oxidising acids, such as dilute HCl and dilute H_2SO_4, and it oxidises only very slowly in moist air. Copper is also a good conductor of electricity, which leads to its use in electricity cables, and for domestic water pipes. Copper's mechanical properties are enhanced by making alloys such as brass and bronze.

Test yourself

3 Why can scandium and zinc be described as d-block elements, but not as transition metals?

4 Suggest a reason why zinc only forms compounds in the +2 oxidation state.

5 a) What is the general trend in standard electrode potentials of the $M^{2+}(aq) | M(s)$ systems for the transition metals in Table 11.2?

b) What does this suggest about the reactivity of transition metals across period 4 in the periodic table?

6 Explain why the atomic radius falls from 0.15 nm in titanium to 0.14 nm in vanadium and then 0.13 nm in chromium.

Figure 11.6▲
This young woman is playing a saxophone made of brass. Brass is an alloy of 60–80% copper and 20–40% zinc. It is easily worked, has an attractive gold colour and does not corrode.

11.4 Variable oxidation numbers

Most of the d-block elements in period 4 form a range of compounds in which they are present in different oxidation states. The main reason for this is that the transition metals from titanium to copper have electrons of similar energy in both the 3d and 4s levels. This means that each of these elements can form ions of roughly the same stability in aqueous solution or in crystalline solids by losing different numbers of electrons.

The formulae of the common oxides of the elements from scandium to zinc are shown in Figure 11.7 along with the stable oxidation states of each element in its compounds. The main oxidation states of the elements are shown in larger and bold blue print.

	Sc	Ti	V	Cr	Mn	Fe	Co	Ni	Cu	Zn
Common oxides	Sc_2O_3	Ti_2O_3	V_2O_3	Cr_2O_3	MnO	FeO	CoO	NiO	Cu_2O	ZnO
		TiO_2	V_2O_5	CrO_3	MnO_2	Fe_2O_3	Co_2O_3		CuO	
					Mn_2O_7					
					+7					
				+6	+6					
			+5							
		+4	+4		**+4**					
	+3	**+3**	**+3**	**+3**	+3	**+3**	+3			
		+2	+2	+2	**+2**	**+2**	**+2**	**+2**	**+2**	**+2**
									+1	

Figure 11.7▲
Oxidation states and common oxides of the elements scandium to zinc with the main oxidation states in larger and bold blue print.

The elements at each end of the series in Figure 11.7 give rise to only one oxidation state. The elements near the middle of the series have the greatest range of oxidation states. Most of the elements form compounds in the +2 state corresponding to the use of both 4s electrons in bonding.

The +2 state is a main oxidation state for all elements in the second half of the series, whereas +3 is a main oxidation state for all elements in the first part. Across the series, the +2 state becomes more stable relative to the +3 state.

From scandium to manganese, the highest oxidation state corresponds to the total number of electrons in the 3d and 4s energy levels. However, these higher oxidation states never exist as simple ions. Typically, they occur in compounds in which the metal is covalently bonded to an electronegative atom, usually oxygen, as in the dichromate(VI) ion, $Cr_2O_7^{2-}$, and the manganate(VII) ion, MnO_4^-.

One of the most attractive and effective demonstrations of the range of oxidation states in a transition element can be shown by shaking a solution of ammonium vanadate(V), NH_4VO_3, in dilute sulfuric acid with zinc. Before adding zinc, H^+ ions in the sulfuric acid react with VO_3^- ions to form dioxovanadium(V) ions and the solution is yellow.

$$VO_3^-(aq) + 2H^+(aq) \rightarrow VO_2^+(aq) + H_2O(l)$$

When the yellow solution, containing dioxovanadium(V) ions, is shaken with zinc, it is reduced first to blue oxovanadium(IV) ions, $VO^{2+}(aq)$, then to green vanadium(III) ions, $V^{3+}(aq)$, and finally to violet vanadium(II) ions, $V^{2+}(aq)$ (Figure 11.8).

Figure 11.8▲
The oxidation states of vanadium showing the colours of its ions in the +5, +4, +3 and +2 oxidation states.

Chromium forms compounds in three oxidation states, +2, +3 and +6. In the +3 state, chromium exists as Cr^{3+} ions which can be both oxidised and reduced.

Under alkaline conditions, hydrogen peroxide oxidises green chromium(III) ions, $Cr^{3+}(aq)$, to yellow chromium(VI) in chromate ions, $CrO_4^{2-}(aq)$.

$$H_2O_2(aq) + 2e^- \rightarrow 2OH^-(aq)$$
$$Cr^{3+}(aq) + 8OH^-(aq) \rightarrow CrO_4^{2-}(aq) + 4H_2O(l) + 3e^-$$

In contrast to this, zinc reduces green $Cr^{3+}(aq)$ to blue-violet $Cr^{2+}(aq)$ ions.

$$Zn(s) \rightarrow Zn^{2+}(aq) + 2e^-$$
$$Cr^{3+}(aq) + e^- \rightarrow Cr^{2+}(aq)$$

Chromium(II) ions are powerful reducing agents which are rapidly converted to chromium(III) by oxygen in the air.

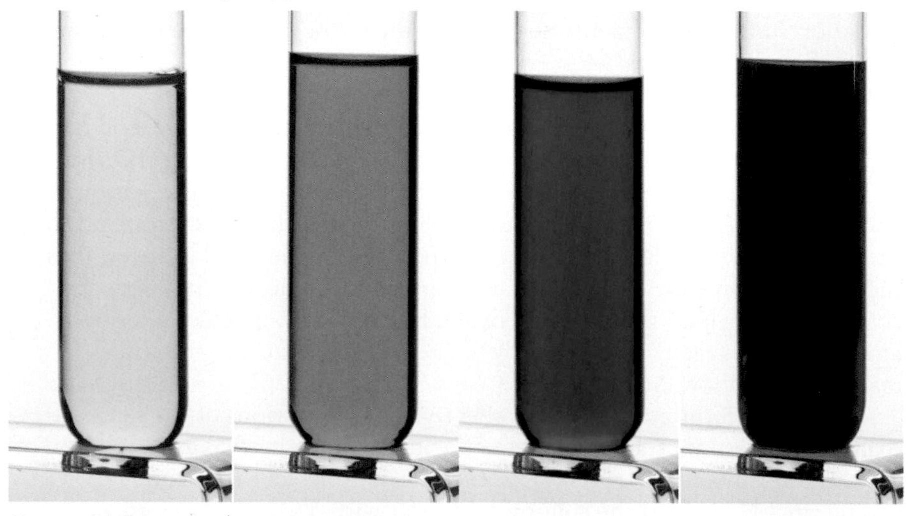

Figure 11.9▲
Solutions containing ions in the three oxidation states of chromium. From left to right, these test tubes contain CrO_4^{2-}, $Cr_2O_7^{2-}$, Cr^{3+} and Cr^{2+} ions in aqueous solution.

WWW Data

Test yourself

7 Write down four generalisations about the oxidation states of transition metals based on Figure 11.7 and the text in Section 11.4.

8 Give examples of compounds other than oxides of:
- **a)** chromium in the +3 and +6 states
- **b)** manganese in the +2 and +7 states
- **c)** iron in the +2 and +3 states
- **d)** copper in the +1 and +2 states.

9 Explain why the oxidation state of vanadium in VO_2^+ ions is +5.

10 **a)** Write half-equations for
- **i)** the reduction of dioxovanadium(V) ions, VO_2^+, to oxovanadium(IV) ions in acid solution
- **ii)** the oxidation of zinc to zinc(II) ions.

b) Use the Data sheet: 'Standard electrode potentials' to show that zinc will reduce VO_2^+ ions to VO^{2+} ions in acid solution.

11 **a)** Write a half-equation involving electrons for:
- **i)** the oxidation of $Cu^+(aq)$ to $Cu^{2+}(aq)$
- **ii)** the reduction of $Cu^+(aq)$ to $Cu(s)$.

b) Use the Data sheet: 'Standard electrode potentials' on the Dynamic Learning Student website to find the standard electrode potentials for the two half-equations in part **a)**.

c) Using your data in part **b)**, explain why $Cu^+(aq)$ ions will disproportionate in aqueous solution.

Photochromic sunglasses (Figure 11.10), in which the glass darkens on exposure to strong light and becomes clearer when the light intensity falls, depend on a redox reaction involving Ag^+ and Cu^+ ions. The glass contains small particles of silver chloride, AgCl, and copper(I) chloride, CuCl. When the light is strong, Ag^+ ions oxidise Cu^+ to Cu^{2+}.

$$Ag^+ + Cu^+ \rightarrow Ag + Cu^{2+}$$

The silver produced reflects light from the glass and cuts out transmission (glare) to the eyes. In conditions of low light intensity, Cu^{2+} ions oxidise Ag back to Ag^+ ions and the glass becomes transparent again.

Figure 11.10▲
Photochromic sunglasses become darker in strong light.

11.5 Coloured ions

Most coloured compounds get their colour by absorbing some of the radiation in the visible region of the electromagnetic spectrum with wavelengths between 400 nm and 700 nm. When light hits a substance, part is absorbed, part is transmitted (if the substance is transparent) and part is usually reflected. If all the light is absorbed, the substance looks black. If all the light is reflected, the substance looks white. If very little light is absorbed and all the radiations in the visible region of the electromagnetic spectrum are transmitted equally, the substance will be colourless like water.

However, many compounds, and particularly those of transition metals, absorb radiations in only certain areas of the visible spectrum. This means that the substances take on the colour of the light that they transmit or reflect. For example, if a material absorbs all radiations in the green-blue-violet region of the spectrum, it will appear red-orange in white light (Figure 11.11).

Colour of compound	Wavelength absorbed/nm	Colour of light absorbed
greenish yellow	400–430	violet
yellow to orange	430–490	blue
red	490–510	blue-green
purple	510–530	green
violet	530–560	yellow-green
blue	560–590	yellow
greenish blue	590–610	orange
blue-green to green	610–700	red

Figure 11.11◄
A chart showing complementary colours in the left and right-hand columns. The colour of a compound is the colour complementary to the light it absorbs.

It is the electrons in coloured compounds that absorb radiation and jump from their normal state to a higher excited state. According to the quantum theory, there is a fixed relationship between the size of the energy 'jump' and the wavelength of the radiation absorbed. In many compounds, the electron 'jumps' between one sub-level and the next are so large that the radiation absorbed is in the ultraviolet region of the spectrum. These compounds are therefore white or colourless because they are not absorbing any of the radiations in the visible region of the electromagnetic spectrum.

However, the colour of transition metal ions arises from the possibility of transitions between the orbitals within the d sub-shell.

In a free, gaseous atom or ion, the five 3d orbitals are all at the same energy level even though they do not all have the same shape. But when the ion of a d-block element is surrounded by other ions in a crystalline solid or by molecules, such as water in aqueous solutions, the differences in shape cause the five orbitals to split into two groups. Two of the 3d orbitals move to a slightly higher energy level than the other three. As a result, ions such as $Cu^{2+}(aq)$ appear coloured because light of a particular frequency can be

Note

The **quantum theory** states that radiation is emitted or absorbed in tiny, discrete amounts called energy quanta. Quanta have energy, $E = h\nu$ where h is Planck's constant and ν is the frequency of the radiation.

absorbed from visible light as electrons jump from a lower to a higher 3d orbital (Figure 11.12). If all the d orbitals are full, or empty, there is no possibility of electronic transitions between them.

five d orbitals with the same energy in a free gaseous ion

[↑↓] [↑↓] [↑↓] [↑↓] [↑]

$Cu^{2+}(g)$

the d orbitals split into two groups with different energies in an aqueous ion

[↑↓] [↑]

absorption of light at a particular frequency can promote an electron to a higher level

[↑↓] [↑↓] [↑↓]

$Cu^{2+}(aq)$

Figure 11.12▲
The energy between the separated d orbitals in an aqueous transition metal ion, like $Cu^{2+}(aq)$, allows electron transitions from a lower orbital to a higher orbital. The ion absorbs light with a particular frequency in the visible region of the electromagnetic spectrum.

In explaining the colour of transition metal ions, you should appreciate the limitations of our simple energy level model of electronic structure. The need for more sophisticated explanations is clear, bearing in mind the existence of sub-shells and the different shapes of orbitals within d sub-shells.

Figure 11.13▲
A member of the Chinese Opera wearing stage make-up. Coloured pigments used in their make-up include chromium(III) oxide (green), iron oxides (red and yellow) and manganese salts (violet).

Test yourself

12 a) Explain why Zn^{2+}, Cu^{+} and Sc^{3+} ions are usually colourless in solution and white in solids by writing out their electronic configurations.

b) What colours of light are absorbed most effectively by a Cu^{2+} ion?

11.6 Action as catalysts

Transition elements and their compounds play a crucial role as catalysts in industry. Table 11.3 lists some important examples of transition metals and their compounds as catalysts.

Table 11.3▶
Some important examples of transition metals and their compounds as catalysts.

Transition element/compound used as catalyst	Reaction catalysed
Vanadium(V) oxide, V_2O_5, or vanadate, VO_3^-	Contact process in the manufacture of sulfuric acid $2SO_2(g) + O_2(g) \rightleftharpoons 2SO_3(g)$
Iron or iron(III) oxide	Haber process to manufacture ammonia $N_2(g) + 3H_2(g) \rightleftharpoons 2NH_3(g)$
Nickel, platinum and palladium	Manufacture of low-fat spreads and margarine $RCH{=}CH_2(g) + H_2(g) \rightarrow RCH_2CH_3(g)$
Platinum or platinum–rhodium alloys	Conversion of NO and CO to CO_2 and N_2 in catalytic converters in vehicles $2CO(g) + 2NO(g) \rightarrow 2CO_2(g) + N_2(g)$
Platinum	Reforming straight-chain alkanes as cyclic alkanes and arenes $CH_3(CH_2)_5CH_3 \rightarrow CH_3{-}C_6H_5 + 4H_2$ heptane methylbenzene

Catalysts can be divided into two types – heterogeneous and homogeneous.

Heterogeneous catalysis

Heterogeneous catalysis involves a catalyst in a different state from the reactants it is catalysing. It is used in almost every large-scale manufacturing process such as the manufacture of ammonia in the Haber process (Table 11.3) in which nitrogen and hydrogen gas flow through a reactor containing lumps of iron or iron(III) oxide.

Platinum metal alloyed with other metals, such as rhodium, is used in catalytic converters. In a catalytic converter, a honeycombed ceramic structure is coated with a very thin layer of the expensive catalyst in order to increase its surface area in contact with exhaust gases (Figure 11.14).

Definition

A **heterogeneous catalyst** is a catalyst that is in a different state from the reactants. Generally, a heterogeneous catalyst is a solid, while the reactants are gases or in solution.

The advantage of heterogeneous catalysts is that they can be separated from the reaction products easily.

Figure 11.14 ◀
A catalytic converter provides an example of a heterogeneous catalyst. Pollutant molecules of NO and CO in the exhaust gases become attached to the surface of the platinum alloy where they react to form N_2 and CO_2. In this computer graphic, oxygen atoms are coloured red, nitrogen atoms are coloured blue and carbon atoms are coloured green.

Impurities in reactants can 'poison' heterogeneous catalysts, particularly metals, causing them to become less effective. Carbon monoxide 'poisons' the iron catalyst used in the Haber process and lead compounds 'poison' catalytic converters, so lead-free petrol must be used.

Heterogeneous catalysts work by adsorbing reactants at active sites on their surface. Nickel acts as a catalyst for the addition of hydrogen to unsaturated compounds with carbon–carbon double bonds (Table 11.3). Hydrogen molecules are adsorbed on the catalyst surface where they are thought to split into single atoms (free radicals). These highly reactive hydrogen atoms undergo addition with molecules of unsaturated compounds, like ethene, when the unsaturated compounds approach the catalyst surface (Figure 11.15).

Figure 11.15 ▼
A possible mechanism for the hydrogenation of an alkene using a nickel catalyst. The reaction takes place on the surface of the catalyst, which adsorbs hydrogen molecules and then splits them into atoms.

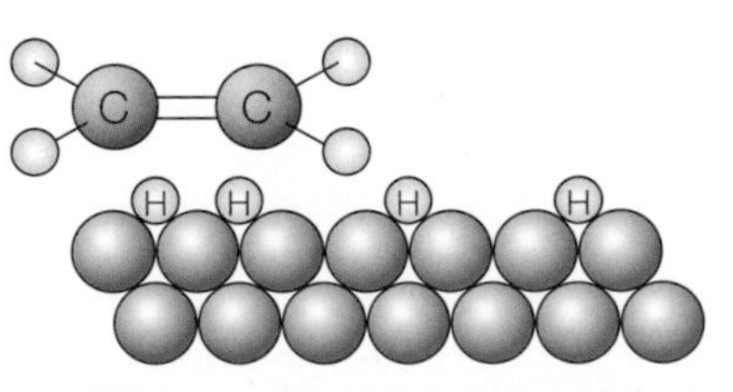

step 1

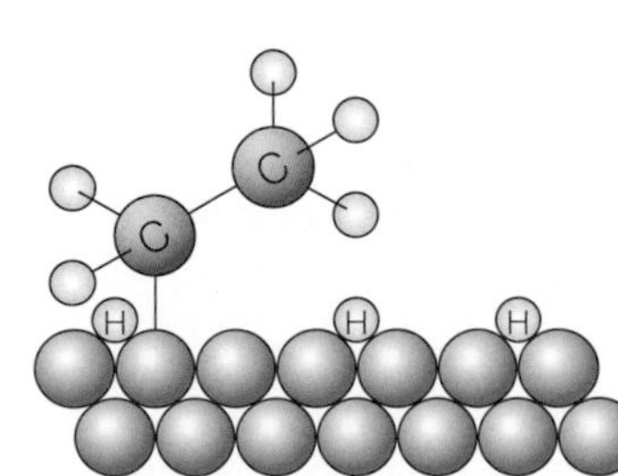

step 2

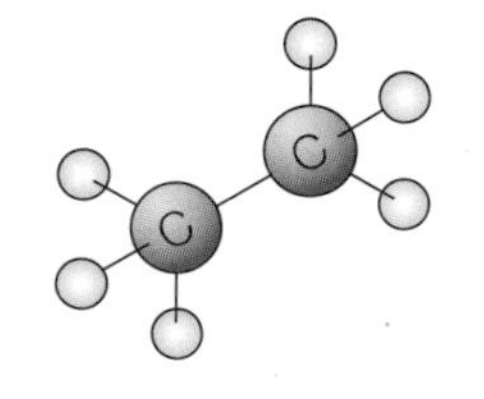

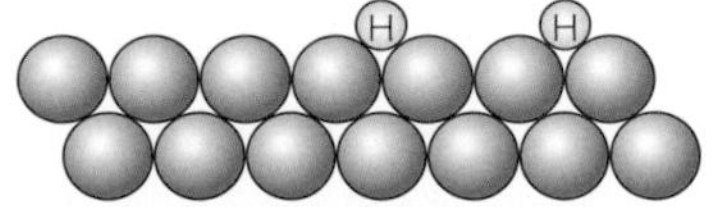

Ethene approaches the catalyst surface where hydrogen gas is adsorbed as single atoms

Ethene adds one hydrogen atom and the $CH_3CH_2•$ radical is attached to the surface

After adding a second hydrogen atom the hydrocarbon, now ethane, escapes from the surface

If a metal is to be a good catalyst for the addition of hydrogen, it must not adsorb the hydrogen so strongly that the hydrogen atoms become unreactive. This happens with tungsten. Equally, if adsorption is too weak there will be insufficient adsorbed atoms for the reaction to occur at a useful rate, and this is the case with silver. The strength of adsorption must have a suitable intermediate value, which is the case with nickel, platinum and palladium.

Definition

A **homogeneous catalyst** is a catalyst that is in the same state as the reactants. Usually, the reactant and the catalyst are dissolved in the same solution.

Homogeneous catalysis

Homogeneous catalysis involves a catalyst in the same state as the reactants it is catalysing. Homogeneous catalysis is most important in biological systems in which enzymes (proteins) act as catalysts in the metabolic processes of all organisms. Very often transition metal ions act as co-enzymes in these processes, enhancing the catalytic activity of the associated enzyme. Cytochrome oxidase is an important enzyme containing copper. This enzyme is involved when energy is released from the oxidation of food. In the absence of copper, cytochrome oxidase is totally ineffective and the animal or plant is unable to metabolise successfully.

Transition metal ions can also be effective as homogeneous catalysts themselves because they can gain and lose electrons, changing from one oxidation state to another. The oxidation of iodide ions by peroxodisulfate(VI) ions using iron(III) ions as a catalyst is a good example of this.

$$2I^-(aq) + S_2O_8^{2-}(aq) \rightarrow I_2(aq) + 2SO_4^{2-}(aq)$$

In the absence of Fe^{3+} ions the reaction is very slow, but with Fe^{3+} ions in the mixture the reaction is many times faster. A possible mechanism is that Fe^{3+} ions are reduced to Fe^{2+} as they oxidise iodide ions to iodine. Then the $S_2O_8^{2-}$ ions oxidise Fe^{2+} ions back to Fe^{3+} ready to oxidise more of the iodide ions, and so on.

Sometimes one of the products of a reaction can act as a catalyst for the process. This is called autocatalysis. An autocatalytic reaction starts slowly, but then speeds up as the catalytic product is formed. Mn^{2+} ions act as autocatalysts in the oxidation of ethanedioate ions, $C_2O_4^{2-}$, by manganate(VII) ions in acid solution. This is also an example of homogeneous catalysis.

$$2MnO_4^-(aq) + 16H^+(aq) + 5C_2O_4^{2-}(aq) \rightarrow 2Mn^{2+}(aq) + 8H_2O(l) + 10CO_2(g)$$

Test yourself

13 What is the advantage of using a solid heterogeneous catalyst in:
- a) a continuous industrial process
- b) an industrial batch process.

14 a) Write half-equations to explain the mechanism by which iron(III) ions catalyse the reaction between iodide ions and peroxodisulfate ions.
- b) Do you think Fe^{2+} ions will also catalyse this reaction? Explain your answer.

15 a) Suggest two methods of speeding up the reaction between $MnO_4^-(aq)$ and $C_2O_4^{2-}(aq)$ from the start of the reaction.
- b) What would you expect to see when a solution of potassium manganate(VII) is added to an acidified solution of potassium ethanedioate:
  - i) at the start of the reaction
  - ii) as the reaction gets underway?

Activity

The development of new catalysts

One of the real challenges and priority areas in chemical research today is the development of new and improved catalysts. Catalysts are particularly important in the drive for a 'greener world' and a 'greener future'. (Topic 18, *Edexcel Chemistry for AS*.)

- Catalysts speed up reactions and products are obtained faster.
- Catalysts allow processes to operate at lower temperatures with savings on energy and fuel.
- Catalysts make possible processes that have high atom economies and produce less waste.
- Catalysts can be highly selective so that only the desired product is formed without the obvious waste from side-reactions.

The manufacture of ethanoic acid (Section 18.3, *Edexcel Chemistry for AS*) illustrates the advantages of developing new catalysts very clearly.

Until the 1970s, the main method of manufacturing ethanoic acid was to oxidise hydrocarbons from crude oil in the presence of a cobalt(II) ethanoate catalyst. The process operated at 180–200 °C and 40–50 times atmospheric pressure. Only 35% by mass of the products was ethanoic acid.

Today, ethanoic acid is manufactured as the only product in the direct combination of methanol and carbon monoxide. The catalyst of iridium metal is mixed with ruthenium compounds which act as catalyst promoters and triple the rate of reaction (Figure 11.16).

In addition, the process operates at lower temperature and lower pressure, producing only ethanoic acid.

As scientists develop new techniques, like the improved catalytic process for ethanoic acid, and propose new ideas, it is important that their work is reported, checked and validated. This reporting and validating is carried out in three ways.

- *Through reports, journals and conferences* at which scientists discuss their work with others.
- *By peer review* in which others working in the same area look closely at the accuracy, reliability and validity of the experimental methods used, the results obtained and the conclusions drawn.
- *By replicating experiments*, which is perhaps the ultimate test of reliability and validity for any innovation. If other scientists repeat the work and achieve the same results, then its integrity is not questioned.

1 Why is the development of new catalysts important?

2 Assume that ethanoic acid can be manufactured by reacting ethane with oxygen.

a) Write an equation for the reaction involved.

b) Calculate the atom economy of the process.

c) Why is the atom economy of the process used to manufacture ethanoic acid before the 1970s much less than the figure you calculated in part **b)**?

d) Briefly explain how cobalt(II) ethanoate might act as a catalyst for the reaction.

3 a) Write an equation for the present-day manufacture of ethanoic acid with an iridium catalyst.

b) What type of catalyst is the iridium?

c) What is the atom economy of this process?

4 State three ways in which the present-day process to manufacture ethanoic acid is either 'greener' or more efficient than the pre-1970s process.

5 In what ways are the developments of new materials and methods reported by research chemists?

6 Why is the reporting and validating of new techniques and ideas in science so important?

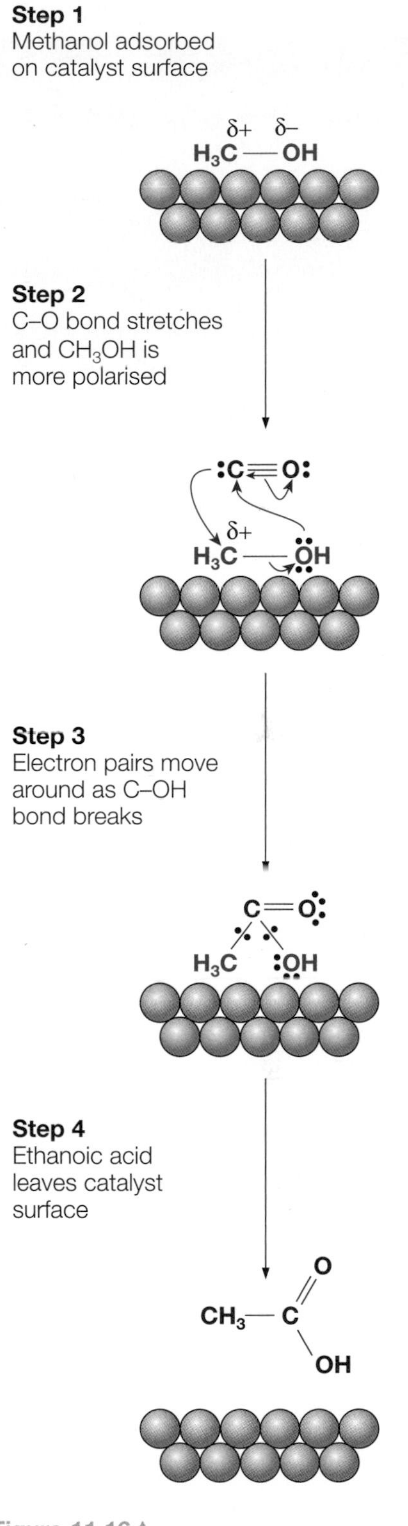

Figure 11.16▲
A possible mechanism for the action of iridium as a catalyst in the manufacture of ethanoic acid.

11.7 Formation of complex ions

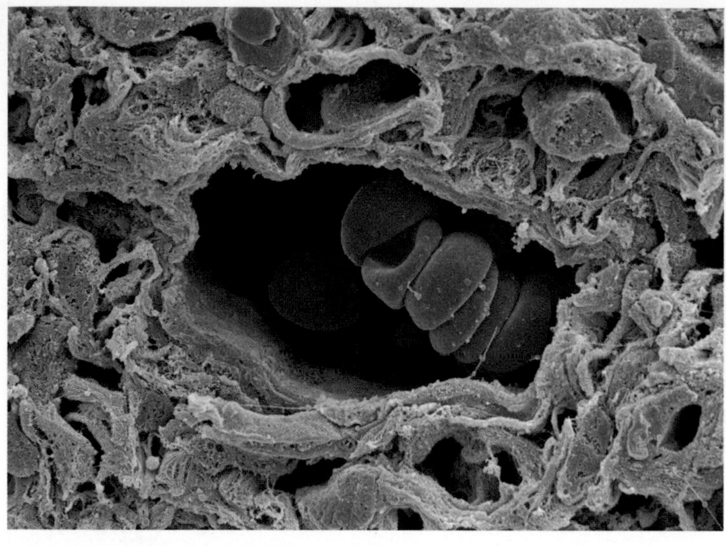

Figure 11.17▲
Red blood cells in a small artery. The blood cells are red because they contain haemoglobin combined with oxygen. Haemoglobin is composed of complex ions with haem groups, globin molecules and Fe^{2+} ions.

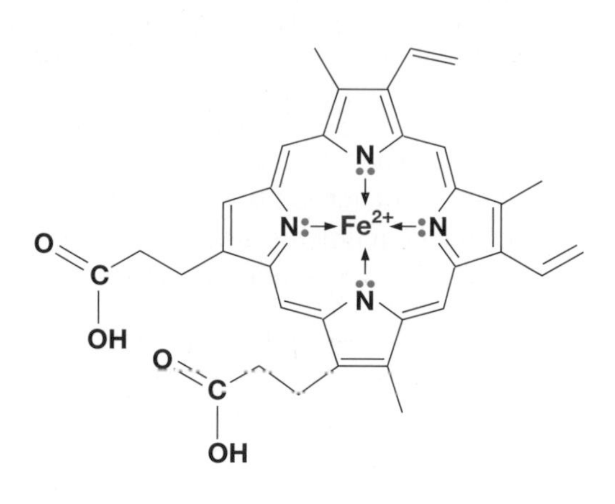

Figure 11.18▲
A haem group with its Fe^{2+} ion. Haem groups form four dative covalent bonds to the central Fe^{2+} ion. Molecules like haem which form dative covalent bonds to a central metal ion are examples of ligands.

During the AS course, we found that, in aqueous solution, H^+ ions were strongly attached to water molecules by dative covalent bonds forming oxonium ions, H_3O^+ (Figure 11.19).

In the same way as H^+, other cations can also exist in aqueous solution as hydrated ions. So Cr^{3+}(aq), Cu^{2+}(aq) and Ag^+(aq) can be represented more precisely as $[Cr(H_2O)_6]^{3+}$(aq), $[Cu(H_2O)_6]^{2+}$(aq) and $[Ag(H_2O)_2]^+$(aq) in aqueous solution. But notice that the larger size of these other cations relative to H^+ enables them to associate with up to six water molecules.

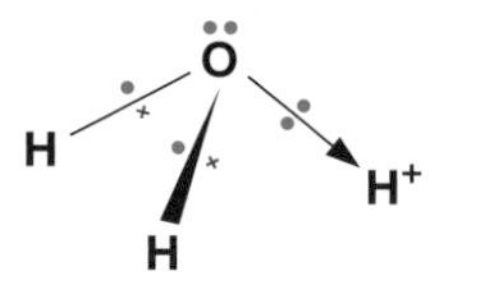

Figure 11.19▲
Dative covalent bonding in an aqueous H_3O^+ ion

Note

The terms 'hydronium ion' and 'hydroxonium ion' are sometimes used for the oxonium ion, H_3O^+.

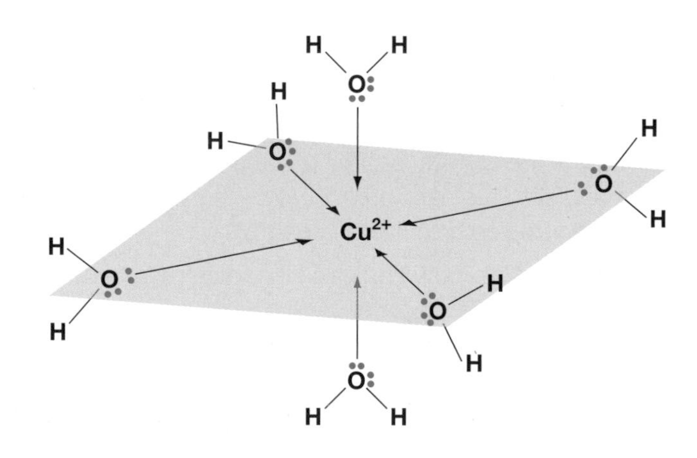

Figure 11.20▲
Dative covalent bonding in an aqueous Cu^{2+} ion.

Other polar molecules, besides water, can form dative covalent bonds with metal ions. For example, in excess ammonia solution, Cr^{3+} ions form $[Cr(NH_3)_6]^{3+}$, Cu^{2+} ions form $[Cu(NH_3)_4(H_2O)_2]^{2+}$ and Ag^+ ions form $[Ag(NH_3)_2]^+$. In addition to polar molecules, anions can also associate with cations using dative covalent bonds. For example, when anhydrous copper(II) sulfate is added to concentrated hydrochloric acid, the solution contains yellow $[CuCl_4]^{2-}$ ions.

Ions such as $[Cu(H_2O)_6]^{2+}$, $[Cu(NH_3)_4(H_2O)_2]^{2+}$ and $[CuCl_4]^{2-}$ in which a metal ion is associated with a number of molecules or anions are called complex ions, and the anions and molecules attached to the central metal ion are called ligands. Each ligand must have at least one lone pair of electrons which it uses to form a dative covalent bond with the metal ion. The number of ligands in a complex ion is typically two, four or six.

Chemists have an alternative name for dative covalent bonds which they often prefer when describing complex ions. The alternative name is 'co-ordinate bond' which also gives rise to the terms 'co-ordination compound' and 'co-ordination number'. A co-ordination compound is one which contains a complex ion, and the co-ordination number of a complex ion is the number of co-ordinate bonds from the ligands to the central metal ion.

Co-ordination compounds contain complexes which may be cations, anions or neutral molecules. Examples of co-ordination compounds include:

- $K_3[Fe(CN)_6]$ containing the negatively charged complex ion $[Fe(CN)_6]^{3-}$
- $CuSO_4.5H_2O$ containing the positively charged complex ion $[Cu(H_2O)_4]^{2+}$
- $Ni(CO)_4$ containing a neutral complex between nickel atoms and carbon monoxide molecules.

Definitions

A **complex ion** is an ion in which a number of molecules or anions are bound to a central metal cation by co-ordinate bonds.

A **ligand** is a molecule or anion bound to the central metal ion in a complex ion by co-ordinate bonding.

The **co-ordination number** of a metal ion in a complex is the number of co-ordinate bonds to the metal ion from the surrounding ligands.

Note

In aqueous solution, the copper(II) ion is surrounded by six water molecules forming the complex ion $[Cu(H_2O)_6]^{2+}$. In solid hydrated copper(II) sulfate ($CuSO_4.5H_2O$), however, there are only four water molecules co-ordinated with each copper(II) ion. The fifth water molecule in the solid copper(II) sulfate is associated with a sulfate ion, SO_4^{2-}.

Figure 11.21 ▲
Crystals of six co-ordination compounds. From left to right these are: $NiSO_4.7H_2O$, $FeSO_4.7H_2O$, $CoCl_2.6H_2O$, $CuSO_4.5H_2O$, $Cr_2(SO_4)_3.18H_2O$ and $K_3[Fe(CN)_6]$.

There are two common visible signs that a reaction has occurred during the formation of a new complex ion:

- a colour change
- an insoluble solid dissolving.

A familiar example of a colour change occurs when excess ammonia solution is added to copper(II) sulfate solution. Ammonia molecules displace water molecules from hydrated copper(II) ions forming $[Cu(NH_3)_4(H_2O)_2]^{2+}(aq)$ ions and the colour changes from pale blue to deep blue.

$$[Cu(H_2O)_6]^{2+}(aq) + 4NH_3(aq) \rightarrow [Cu(NH_3)_4(H_2O)_2]^{2+}(aq) + 4H_2O(l)$$

The test for chloride ions using aqueous silver nitrate followed by ammonia solution is an example of an insoluble solid dissolving as a complex ion forms. Adding silver nitrate to a solution of chloride ions produces a white precipitate of silver chloride, AgCl. This precipitate dissolves on adding ammonia solution as silver ions form the complex ion, $[Ag(NH_3)_2]^+(aq)$ with ammonia molecules.

$$AgCl(s) + 2NH_3(aq) \rightarrow [Ag(NH_3)_2]^+(aq) + Cl^-(aq)$$

11.8 Naming complex ions

There are four simple rules to follow when naming a complex ion.

1 Identify the number of ligands around the central cation using Greek prefixes: mono-, di-, tri-, tetra-, etc.
2 Name the ligand using names ending in -o for anions, e.g. chloro- for Cl^-, fluoro- for F^-, cyano- for CN^-, hydroxo- for OH^-. Use aqua for H_2O and ammine for NH_3.
3 Name the central metal ion using the normal name of the metal for positive and neutral complex ions and the old-fashioned Latinised name ending in –ate for negative complex ions – e.g. ferrate for iron, cuprate for copper, argentate for silver.
4 Finally, add the oxidation number of the central metal ion.

The examples in Table 11.4 illustrate how you should use the rules.

Formula of complex ion	1 Identify the number of ligands	2 Name the ligand	3 Name the central metal ion	4 Add the oxidation number of the central metal ion
$[Ag(NH_3)_2]^+$	di	ammine	silver	(I)
$[Cu(H_2O)_6]^{2+}$	hexa	aqua	copper	(II)
$[CuCl_4]^{2-}$	tetra	chloro	cuprate	(II)
$[Fe(CN)_6]^{3-}$	hexa	cyano	ferrate	(III)

Table 11.4▲
Writing the systematic names of complex ions.

Note

Notice that ammonia, NH_3, in complexes is described as 'ammine' whereas the $-NH_2$ group in organic compounds such as CH_3NH_2 is described as 'amine'.

Test yourself

16 What is the co-ordination number of:
a) Cu^{2+} ions in $[CuCl_4]^{2-}$
b) Cu^{2+} ions in $[Cu(H_2O)_6]^{2+}$
c) Fe^{3+} ions in $[Fe(CN)_6]^{3-}$?

17 Write the systematic names of the following complex ions:
a) $[Cu(NH_3)_4]^{2+}$ b) $[Zn(OH)_4]^{2-}$ c) $[AlH_4]^-$ d) $[Ni(H_2O)_6]^{2+}$

18 What is the oxidation state of the metal ion in the following complex ions?
a) $[NiCl_4]^{2-}$ b) $[Ag(NH_3)_2]^+$ c) $[Fe(H_2O)_6]^{3+}$ d) $[Fe(CN)_6]^{4-}$

19 The fixer used to remove unexposed and undeveloped silver bromide from photographic film contains thiosulfate ions, $S_2O_3^{2-}$. Each silver ion forms a complex ion with two thiosulfate ions as the silver bromide dissolves. Write an equation for this reaction.

11.9 The shapes of complex ions

The shapes of complex ions depend on the number of ligands around the central metal ion. There is no simple, definitive rule for predicting the shapes of complexes from their formulae, but:

- in complexes with a co-ordination number of six, the ligands usually occupy octahedral positions so that the six electron pairs around the central atom are repelled as far as possible (Figure 11.22)
- in complexes with a co-ordination number of four, the ligands usually occupy tetrahedral positions although there are a few complexes with fourfold co-ordination, such as $[Pt(NH_3)_2Cl_2]$, that have a square planar structure (Figure 11.22)
- in complexes with a co-ordination number of two, the ligands usually form a linear structure with the central metal ion (Figure 11.22).

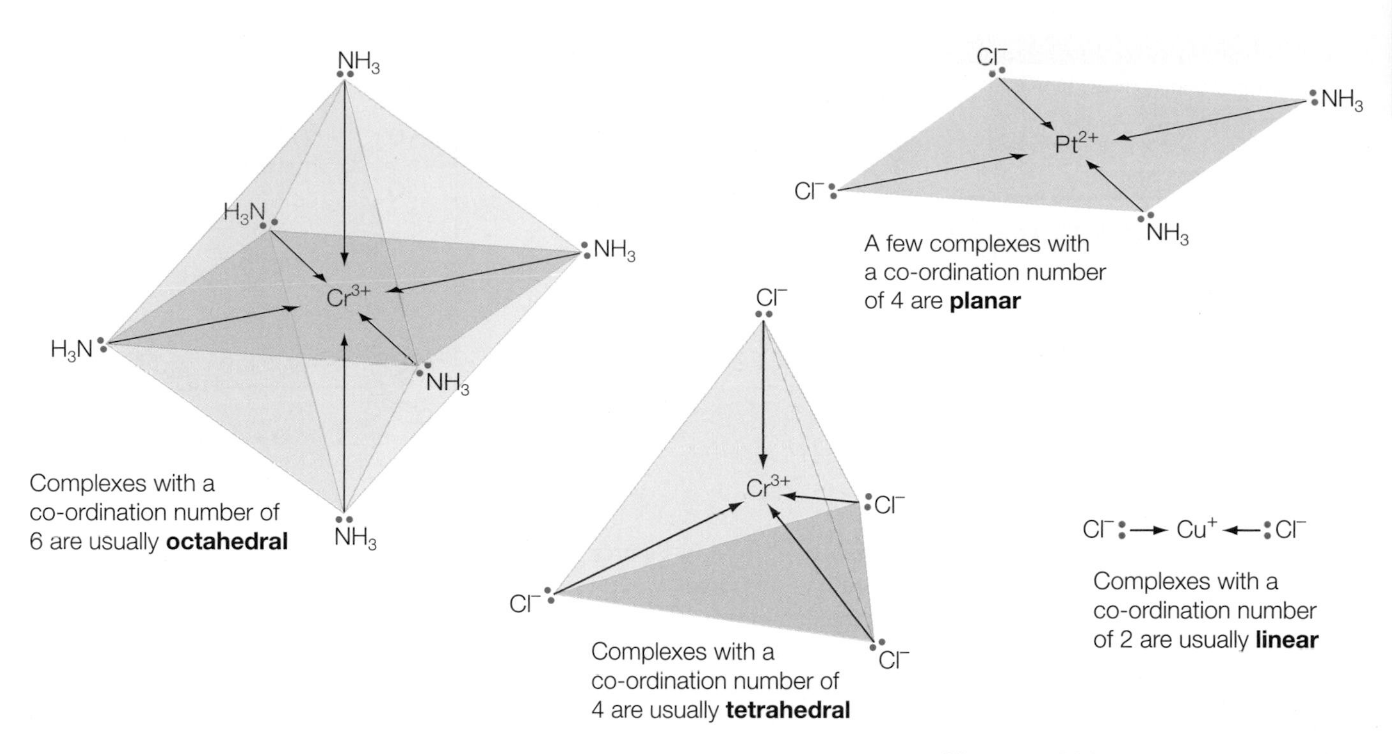

Figure 11.22▲
The shapes of complex ions.

Types of ligand

Most ligands use only one lone pair of electrons to form a co-ordinate bond with the central metal atom. These ligands are described as monodentate because they have only 'one tooth' to hold onto the central cation (*dens* is Latin for tooth). Examples of monodentate ligands include H_2O, NH_3, Cl^-, OH^- and CN^-.

Some ligands have more than one lone pair of electrons which can form co-ordinate bonds with the same metal ion. Bidentate ('two-toothed') ligands, for example, form two dative covalent bonds with metal ions in complexes. Bidentate ligands include 1,2-diaminoethane, $H_2NCH_2CH_2NH_2$, the ethanedioate ion, $C_2O_4^{2-}$, and amino acids.

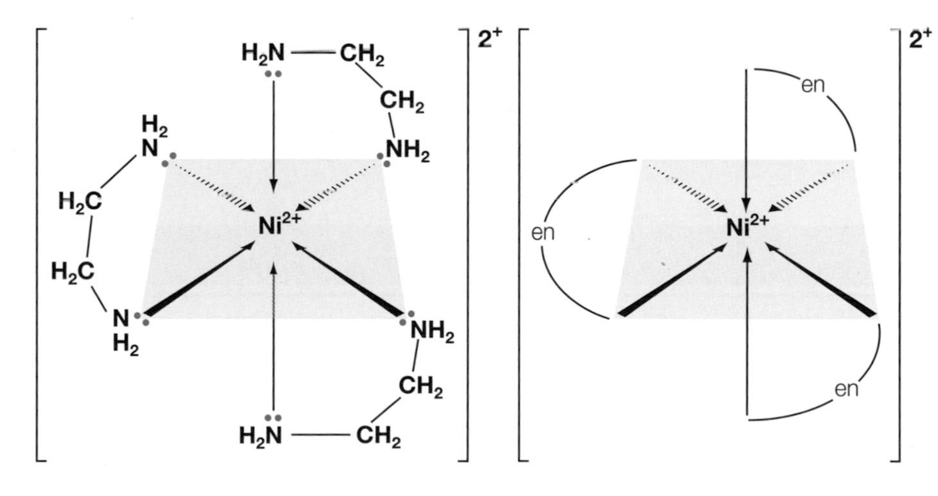

Figure 11.23◀
Representations of a complex formed by the bidentate ligand 1,2-diaminoethane with nickel(II) ions. Note the use of 'en' as an abbreviation for the ligand.

The hexadentate ligand edta is particularly impressive because it can form six co-ordinate bonds with the central metal ion in complexes. Edta is the common abbreviation for an ion which binds so firmly with metal ions that it holds them in solution and makes them chemically inactive. Edta is added to commercially produced salad dressings to extend their shelf life. The edta traps traces of metal ions which would otherwise catalyse the oxidation of vegetable oils. Edta is also an ingredient of bathroom cleaners to help remove scale by dissolving Ca^{2+} ions from the calcium carbonate left by hard water.

Note

Monodentate ligands may also be called unidentate ligands and polydentate ligands are sometimes called multidentate ligands.

Note

Edta crystals consist of the disodium salt of **e**thane**d**iamine**t**etraethanoic **a**cid. Chemists sometimes use the abbreviation Na_2H_2Y for the salt, where Y represents the 4– ion.

Definitions

Polydentate ligands are ligands which form more than one co-ordinate bond with the same metal ion.

Chelates are complex ions involving multidentate ligands.

Figure 11.25▲
The blue pigment in this painting by George Romney in 1763 is called Prussian blue. Its correct chemical name is iron(III) hexacyanoferrate(II).

Figure 11.24▶
The complex ion formed by edta with a Pb^{2+} ion. The edta can fold itself around metal ions, such as Pb^{2+}, so that four oxygen atoms and two nitrogen atoms form co-ordinate bonds to the metal ion. This is the ion formed when edta is used to treat lead poisoning. Edta forms such a stable complex with Pb^{2+} ions that they can be excreted through the kidneys.

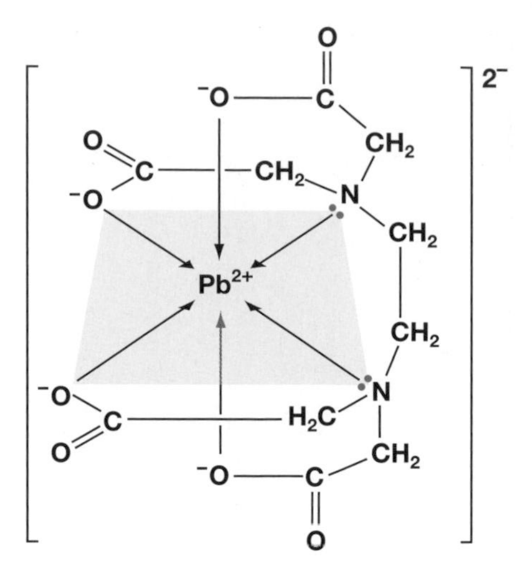

Ligands like those in Figures 11.23 and 11.24, which form more than one co-ordinate bond with metal ions, are sometimes called polydentate ligands and the complexes which these ligands form are called chelates (pronounced 'keelates'). The term chelate comes from a Greek word for a crab's claw, reflecting the claw-like way in which chelating ligands grip metal ions. Powerful chelating agents trap metal ions and effectively isolate them in solution.

Test yourself

20 Predict the likely shape of the following complex ions:
 a) $[Ag(CN)_2]^-$ **b)** $[Fe(CN)_6]^{3-}$ **c)** $[NiCl_4]^{2-}$ **d)** $[CrCl_2(H_2O)_4]^+$

21 Explain how the amino acid glycine (H_2NCH_2COOH) can act as a bidentate ligand.

22 a) Why is edta described as a hexadentate ligand?
 b) What is the overall shape of the edta complex in Figure 11.24?

23 a) Draw a diagram to represent the complex ion formed between a Cr^{3+} ion and three ethanedioate ions $^-O_2C–CO_2^-$.
 b) What is the overall shape of this complex ion?

24 a) Predict an order of stability for the complex ions $[Ni(NH_3)_6]^{2+}$, $[Ni(en)_3]^{2+}$ and $[Ni(edta)]^{2-}$
 b) Explain your prediction.

25 Look at the painting and caption in Figure 11.25. Write the formula of
 a) the hexacyanoferrate(II) ion
 b) iron(III) hexacyanoferrate(II).

Activity

Cis-platin – an important chemotherapy drug

The neutral complex, $PtCl_2(NH_3)_2$, in which Cl^- ions and NH_3 molecules act as ligands, has two isomers. These isomers have different melting temperatures and different chemical properties. One isomer called *cis*-platin is used as a chemotherapy drug in the treatment of certain cancers, whereas the other isomer is ineffective against cancer.

Patients are given an intravenous injection of *cis*-platin which circulates all around the body, including the cancerous area. *Cis*-platin diffuses relatively easily through the tumour cell membrane because it has no overall charge, like the cell membrane.

Once inside the cell, *cis*-platin reacts by exchanging one of its chloride ions for a molecule of water, forming $[Pt(NH_3)_2(Cl)(H_2O)]^+$ which is the 'active principle' (Figure 11.26). This positively charged ion then enters the cell nucleus where it readily bonds with two sites on the DNA. Binding involves co-ordinate bonding from the nitrogen or oxygen atoms in the bases of DNA to the platinum ion.

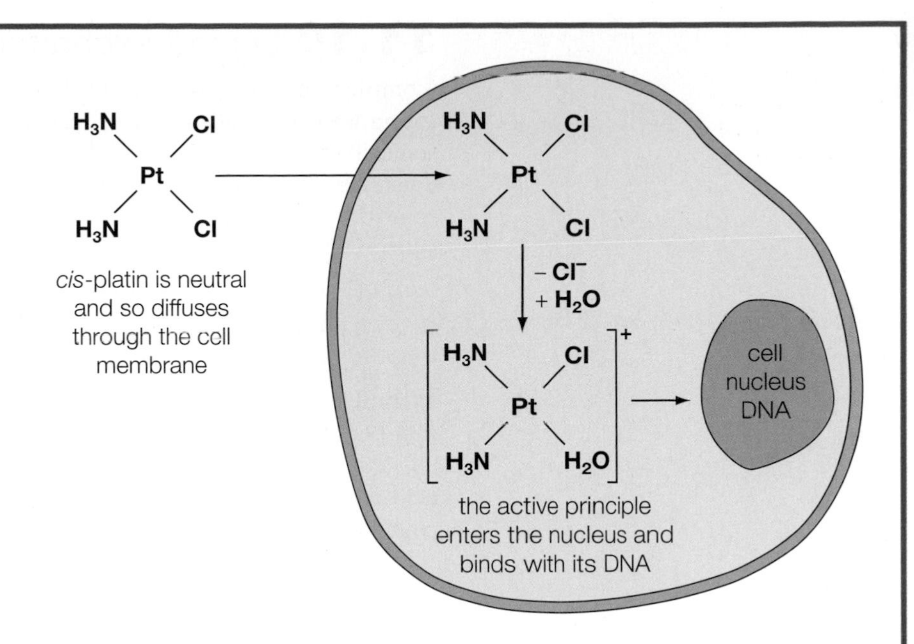

Figure 11.26▲ The action of *cis*-platin.

The *cis*-platin binding changes the overall structure of the DNA helix, pulling it out of shape and shortening the helical turn. The badly shaped DNA can no longer replicate and divide to form new cells, although the affected cells continue to grow. Eventually, the cells die and, if enough of the cancerous cells absorb *cis*-platin, the tumour is destroyed.

Unfortunately, *cis*-platin is not a miracle cure without risks or drawbacks. It is toxic, resulting in unpleasant side-effects, and can cause kidney failure. Clinical trials have, however, led to the discovery of other platinum complexes which cause fewer problems and are already used as anti-cancer drugs.

1 Why is it possible to conclude that *cis*-platin has a square planar rather than a tetrahedral structure?

2 What type of isomerism do *cis*-platin and its isomer show?

3 a) What is the oxidation number of platinum in *cis*-platin?

b) Write the systematic name of *cis*-platin.

c) Draw the structure of *cis*-platin.

4 Why does *cis*-platin diffuse easily through the membrane of cells?

5 What is meant by the term 'active principle' applied to $[Pt(NH_3)_2(Cl)(H_2O)]^+$?

6 When $[Pt(NH_3)_2(Cl)(H_2O)]^+$ has formed inside the cell, it cannot diffuse out through the cell membrane. Why is this?

7 Why is a cell with *cis*-platin binding to DNA unable to replicate?

8 Why is the binding to *cis*-platin from nitrogen and oxygen atoms rather than from carbon and hydrogen atoms in the bases of DNA?

9 Why is *cis*-platin more likely to affect cancerous cells than normal cells?

10 Why is any anti-cancer chemotherapy drug that acts like *cis*-platin likely to have undesirable side-effects?

11.10 Ligand exchange reactions

Complex ions often react by exchanging one ligand for another. These ligand exchange reactions are often reversible and the changes of ligand are sometimes accompanied by colour changes. For example, when excess concentrated ammonia solution is added to pale blue copper(II) sulfate solution, ammonia molecules are exchanged for water molecules around the central Cu^{2+} ion and the colour changes to a deep blue.

$$\underset{\text{pale blue}}{Cu(H_2O_6)^{2+}(aq)} + 4NH_3(aq) \rightleftharpoons \underset{\text{deep blue}}{Cu(NH_3)_4(H_2O)_2{}^{2+}(aq)} + 4H_2O(l)$$

The ligands NH_3 and H_2O are both uncharged and similar in size. This allows exchange reactions between these ligands without a change in co-ordination number of the metal ion.

A similar ligand exchange reaction occurs when concentrated hydrochloric acid is added to copper(II) sulfate solution. This time, the colour changes from pale blue to yellow as Cl^- ions replace water molecules around the Cu^{2+} ion.

$$\underset{\text{pale blue}}{Cu(H_2O)_6{}^{2+}(aq)} + 4Cl^-(aq) \rightleftharpoons \underset{\text{yellow}}{CuCl_4{}^{2-}(aq)} + 6H_2O(l)$$

In this case, however, the ligand exchange involves a change in co-ordination number. Chloride ions are larger than water molecules, so fewer chloride ions can fit round the central Cu^{2+} ion.

Relative stability of complex ions

In aqueous solution, the simple compounds of most transition metals contain complex ions with formulae such as

$$Cu(H_2O)_6{}^{2+},\ Cr(H_2O)_6{}^{3+} \text{ and } Co(H_2O)_6{}^{2+}$$

When solutions containing other ligands, such as Cl^-, are added to aqueous solutions of these hydrated cations, the mixture comes to an equilibrium in which the water molecules of some complexes have been replaced by the added ligands. For example, the equilibrium which results when concentrated sodium chloride solution is added to aqueous copper(II) ions is:

$$Cu(H_2O)_6{}^{2+}(aq) + 4Cl^-(aq) \rightleftharpoons CuCl_4{}^{2-}(aq) + 6H_2O(l)$$

The equilibrium constant, K, for this reaction is:

$$K = \frac{[CuCl_4{}^{2-}(aq)]_{eq}}{[Cu(H_2O)_6{}^{2+}(aq)]_{eq}\,[Cl^-(aq)]^4_{eq}}$$

$[H_2O(l)]$ is constant and therefore it is not included in the equation for K.

Equilibrium constants like this for the formation of complex ions in aqueous solution are called stability constants and the symbol K_{stab} is sometimes used in place of K.

Stability constants enable chemists to compare the stabilities of the complex ions of a cation with different ligands. The larger the stability constant, the more stable is the complex ion compared with that containing water.

Table 11.5 shows the stability constants of three complexes of the copper(II) ion. These show that the relative stabilities of the three copper(II) complexes are:

$$Cu(edta)^{2-} > Cu(NH_3)_4(H_2O)_2{}^{2+} > CuCl_4{}^{2-}$$

Ligand	Complex ion	K
Cl^-	$CuCl_4{}^{2-}$	4.0×10^5
NH_3	$Cu(NH_3)_4(H_2O)_2{}^{2+}$	1.3×10^{13}
edta	$Cu(edta)^{2-}$	6.3×10^{18}

Table 11.5 ◀ The stability constants of three copper(II) complexes.

Complex ions and entropy

When a bidentate ligand, such as 1,2-diaminoethane, replaces a monodentate ligand, such as water, there is an increase in entropy. One molecule of the bidentate ligand replaces two molecules of the monodentate ligand and this results in an increase in the number of product particles. For example, in the reaction

$$Cu(H_2O)_6{}^{2+}(aq) + 3H_2NCH_2CH_2NH_2(aq) \rightarrow Cu(H_2NCH_2CH_2NH_2)_3{}^{2+}(aq) + 6H_2O(l)$$

there are seven product particles but only four reactant particles.

As the entropy of any system depends on the number of particles present, the entropy of the system increases when this reaction occurs. In other words, ΔS_{system} is positive.

When a polydentate ligand, such as edta, replaces a monodentate ligand, an even larger increase occurs in the entropy of the system.

$$Cu(H_2O)_6{}^{2+} + edta^{4-}(aq) \rightarrow Cu(edta)^{2-}(aq) + 6H_2O(l)$$

Because of this increase in the entropy of the system, complexes with polydentate ligands are usually more stable than those with bidentate ligands, which in turn are more stable than complexes with monodentate ligands.

Note

For a spontaneous change to occur, ΔS_{total} must be positive.

$$\Delta S_{total} = \Delta S_{system} + \Delta S_{surroundings}$$

If ΔS_{system} is positive, the products of the system are more thermodynamically stable than the reactants.

Test yourself

26 Write equations for the ligand exchange reactions that occur when:

a) hexaaquacobalt(II) ions react with ammonia molecules to form hexaamminecobalt(II) ions

b) hexaamminecobalt(II) ions react with chloride ions to from tetrachlorocobaltate(II) ions

c) hexaaquairon(II) ions react with cyanide ions to form hexacyanoferrate(II) ions.

27 Explain the following changes with the help of equations.

a) Adding a small amount of ammonia solution to a pale blue solution of hydrated copper(II) ions produces a pale blue precipitate of the hydrated hydroxide.

b) On adding more ammonia solution, the precipitate dissolves to give a deep blue solution.

28 A dilute solution of cobalt(II) chloride is pink because it contains hydrated cobalt(II) ions. The solution turns blue on adding concentrated hydrochloric acid with the formation of tetrachlorocobaltate(II) ions.

a) Write an equation for the reaction that occurs when concentrated HCl is added to dilute cobalt chloride solution and indicate the colour of all species.

b) Explain the chemical basis for the test illustrated in Figure 11.27.

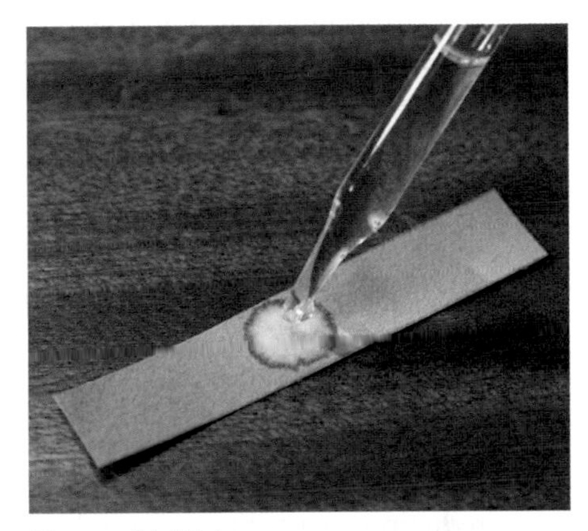

Figure 11.27▲
Filter paper soaked in pink cobalt(II) chloride solution and dried in an oven until it is blue, can be used to test for the presence of water.

29 The stability constants and colour of some cobalt(II) complexes are shown in Table 11.6.

Complex	Stability constant	Colour
$Co(H_2O)_6{}^{2+}$	1.0	Pink
$Co(NH_3)_6{}^{2+}$	3×10^4	Green
$Co(edta)^{2-}$	2×10^{16}	Pink

Table 11.6▲

What would you expect to see when

a) ammonia solution is added to an aqueous solution of cobalt(II) chloride?

b) edta solution is added to a solution of cobalt(II) chloride in aqueous ammonia?

30 Suggest two reasons why the stability constant of $Co(edta)^{2-}$ is so much larger than those of $Co(H_2O)_6{}^{2+}$ and $Co(NH_3)_6{}^{2+}$.

12 Arenes – aromatic hydrocarbons

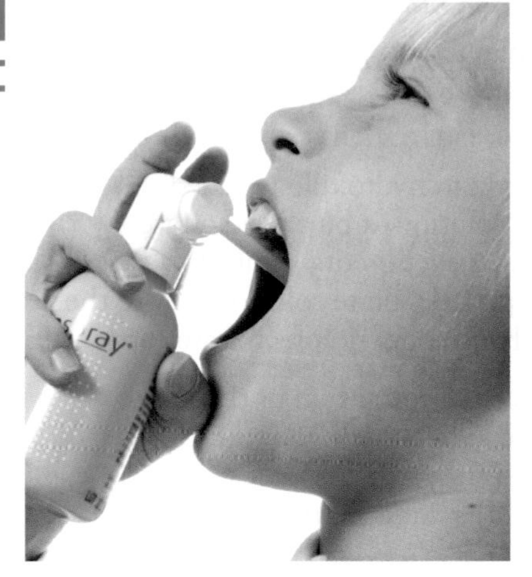

Figure 12.1▲
The antiseptics used in some throat sprays are similar in structure to TCP and Dettol.

Aspirin, paracetamol and ibuprofen are three tried and tested painkillers. Dettol and TCP are two important and effective antiseptics. Terylene and polystyrene are two versatile polymers produced in millions of tonnes every year.

At the heart of all these widely different products is a remarkably stable ring of six carbon atoms – the benzene ring, a constituent of all arenes.

12.1 Arenes

Arenes are hydrocarbons – such as benzene, methylbenzene and naphthalene. They are ring compounds in which there are delocalised electrons. The simplest arene is benzene. Traditionally chemists have called the arenes 'aromatic' ever since the German chemist Friedrich Kekulé was struck by the fragrant smell of oils such as benzene. In their modern name 'arene', the '**ar**-' comes from **ar**omatic and the ending '**-ene**' points to the fact that they are unsaturated hydrocarbons like the alkenes.

Definitions

Arenes are hydrocarbons with a ring or rings of carbon atoms in which there are delocalised electrons.

Delocalised electrons are bonding electrons that are not fixed between two atoms in a bond but shared between three or more atoms. (See also *Edexcel Chemistry for AS*, Sections 4.3 and 7.4).

Figure 12.2►
Benzene is an important and useful chemical. It was first isolated in 1825 by the fractional distillation of whale oil, which was commonly used for lighting homes. Later it was obtained by the fractional distillation of coal tar. Today, it is obtained by the catalytic reforming of fractions from crude oil.

12.2 The structure of benzene

Friedrich Kekulé played a crucial part in our understanding of the structure of benzene as the result of a dream. The dream helped Kekulé to propose a possible structure for benzene which had an empirical formula of CH and a molecular formula of C_6H_6. Kekulé had been working on the problem of the structure of benzene for some time. Then one day in 1865, while dozing in front of the fire, he dreamed of a snake biting its own tail. This inspired him to think of a ring structure for benzene (Figure 12.3).

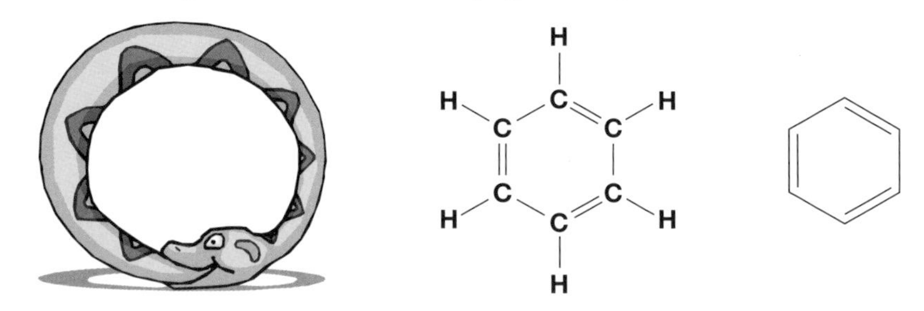

Figure 12.3►
Kekulé's snake and his structural and skeletal formulae for the structure of benzene. Kekulé's formula would have the systematic name cyclohexa-1,3,5-triene.

Kekulé's structure explained many of the properties of benzene. It was acceptable for many years, but still left some problems.

The absence of isomers of 1,2-dichlorobenzene

Kekule's structure suggests that there should be two isomers of 1,2-dichlorobenzene, one in which the chlorine atoms are attached to carbon atoms linked by a single carbon–carbon bond, the other in which the chlorine atoms are attached to carbon atoms linked by a double carbon–carbon bond (Figure 12.4).

Figure 12.4◄
Possible isomers of 1,2-dichlorobenzene.

In practice, it has never been possible to separate two isomers of 1,2-dichlorobenzene or any other 1,2-disubstituted compound of benzene. To get round this problem, Kekulé suggested that benzene molecules might somehow alternate rapidly between the two possible structures, but this failed to satisfy his critics.

The bond lengths in benzene: X-ray diffraction data

The Kekulé structure shows a molecule with alternating single and double bonds. This would imply that three of the bonds are similar in length to the carbon–carbon single bond in alkanes while the other three are similar in length to the carbon–carbon double bond in alkenes. X-ray diffraction studies show that the carbon atoms in a benzene molecule are at the corners of a regular hexagon. All the bonds are the same length, shorter than single bonds but longer than double bonds (Figure 12.5).

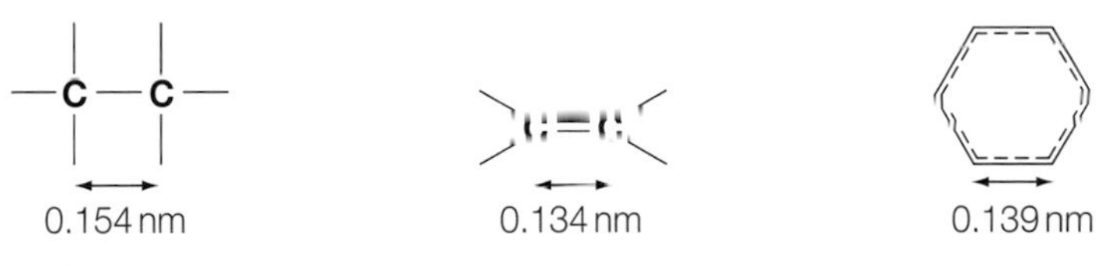

Figure 12.5▲
Carbon–carbon bond lengths in ethane, ethene and benzene.

The resistance to reaction of benzene

An inexperienced chemist looking at the Kekulé structure might expect benzene to behave chemically like a very reactive alkene and to take part in addition reactions with bromine, hydrogen bromide and similar reagents. Benzene does not do this. The compound is much less reactive than alkenes and its characteristic reactions are substitutions not additions.

The stability of benzene: thermochemical data

A study of enthalpy (energy content) changes show that benzene is more stable than expected for a compound with the Kekulé formula. This conclusion is based on a comparison of the enthalpy changes of hydrogenation of benzene and cyclohexene

Cyclohexene is a cyclic hydrocarbon with one double bond. Like other alkenes, it adds hydrogen in the presence of a nickel catalyst at 140 °C to form cyclohexane.

The enthalpy change of the reaction, $\Delta H^{\ominus}$, is $-120\,\text{kJ mol}^{-1}$ (Figure 12.6).

Figure 12.6◄

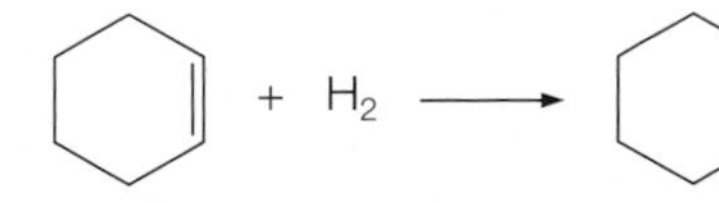

$+ H_2 \longrightarrow$ $\qquad \Delta H^{\ominus} = -120\,\text{kJ mol}^{-1}$

cyclohexene cyclohexane

So, if benzene has three carbon–carbon double bonds as in Kekulé's structure, we might reasonably predict that $\Delta H^{\ominus}$ for the hydrogenation of benzene should be $-360\,\text{kJ mol}^{-1}$. But, when the hydrogenation is carried out, the measured enthalpy change is only $-208\,\text{kJ mol}^{-1}$.

The measured enthalpy change is much less exothermic than the estimated value. This suggests that the addition of hydrogen to benzene does not involve normal double bonds and that benzene is actually much more stable than expected (Figure 12.7).

Figure 12.7▶
Comparing the measured enthalpy change of hydrogenation of benzene with the estimated enthalpy change of hydrogenation for Kekulé's structure.

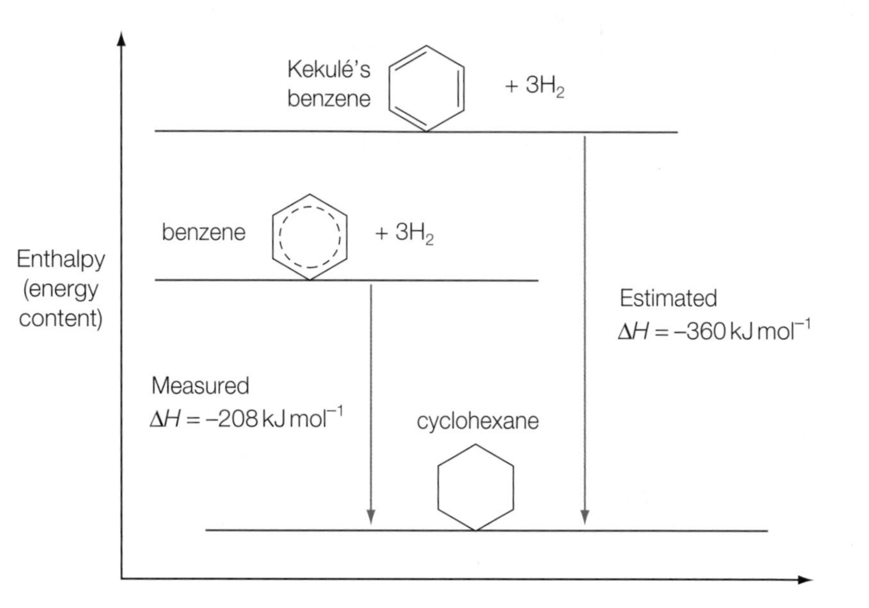

Bonding in benzene: infrared data

The fact that benzene does not have carbon–carbon bonds like ethane or ethene is also reflected in its infrared absorption spectrum. From your studies at AS level, you will know that IR spectra provide unique fingerprints of compounds in which each bond shows characteristic absorptions at specific frequencies in the infrared region of the electromagnetic spectrum. In IR spectra, the frequencies are normally shown as wavenumbers in units of cm^{-1}, i.e. the number of waves in a centimetre rather than the number of waves per second.

Look closely at the infrared spectra of benzene in Figure 12.8 and that of oct-1-ene, $CH_3(CH_2)_5CH{=}CH_2$ in Figure 12.9.

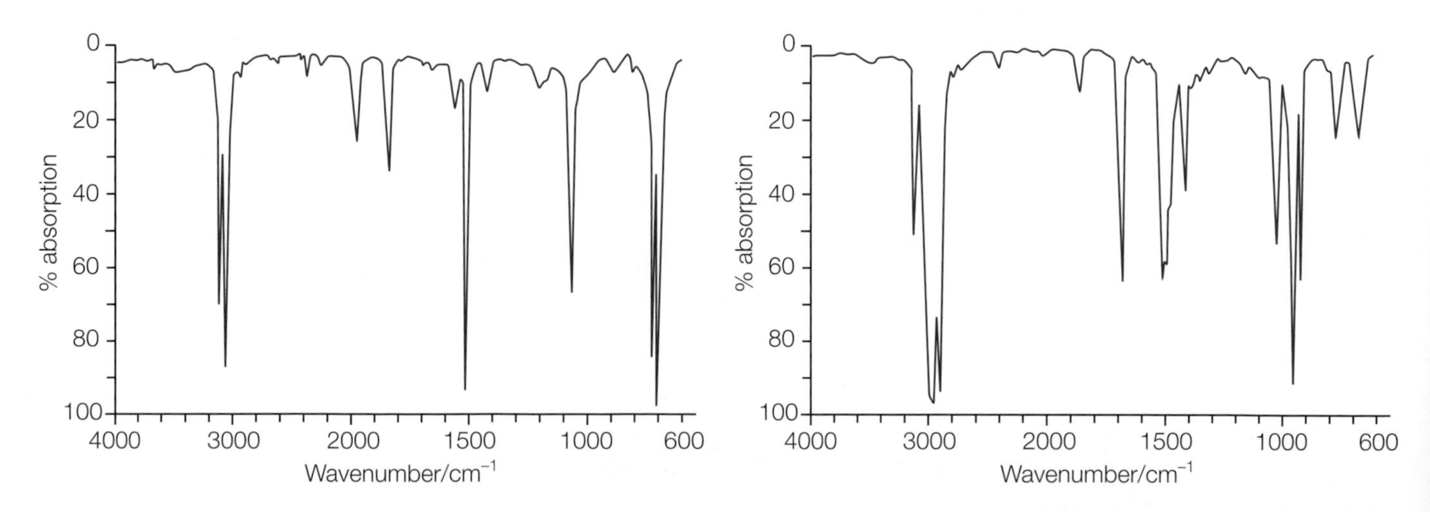

Figure 12.8▲
The infrared absorption spectrum of benzene.

Figure 12.9▲
The infrared absorption spectrum of oct-1-ene.

Notice that benzene does not have the typical strong absorptions of C–H bonds in $-CH_2$ and $-CH_3$ groups in the wavenumber range 2962–2853 cm^{-1}, nor the C=C absorption of an alkene, like oct-1-ene, just below 1700 cm^{-1}.

Instead, and unlike alkanes and alkenes, benzene has strong absorptions at about 3050 cm^{-1} and 750 cm^{-1}. All this provides further evidence that benzene does not have normal C–C or C=C bonds in its structure.

Test yourself

1 Assume that the empirical formula of benzene is CH. What further information is needed to show that its molecular formula is C_6H_6? What methods do chemists use to obtain this information?

2 Draw one possible structure for C_6H_6 that is not a ring. Why does this structure not fit with Kekulé's structure for benzene?

3 An arene consists of 91.3% carbon.
   a) What is the empirical formula of the arene?
   b) What is the molecular formula of the arene if its molar mass is 92 g mol^{-1}?
   c) Draw the structure of the arene.

4 a) Look carefully at Figure 12.7. How much more stable is real benzene than Kekulé's structure for benzene?
   b) Predict the enthalpy change for the complete hydrogenation of cyclohexa-1,3-diene.

12.3 Delocalisation in benzene

The accumulation of the evidence discussed in Section 12.2 led to increased activity in the search for a more accurate model for the structure of benzene. The 'quick fix' was to treat the carbon–carbon bonds in benzene as halfway between single and double bonds and draw them with a full line and a dashed line side-by-side as in Figure 12.5. This model explains the absence of isomers of 1,2-dichlorobenzene, the equal carbon–carbon bond lengths in benzene and also its resistance to reaction. In recent years, the bonding between carbon atoms in benzene has been simplified to a circle inside a hexagon as in Figure 12.10.

Figure 12.10▲
The usual way of representing benzene today.

Although the structure in Figure 12.10 allows an improved understanding of the properties of benzene, a better insight comes from considering its electronic and orbital structure.

Figure 12.11 shows benzene with normal covalent sigma bonds (σ bonds) between its carbon and hydrogen atoms. Each carbon atom uses three of its electrons to form three σ bonds with its three neighbours. This leaves each carbon atom with one electron in an atomic p orbital.

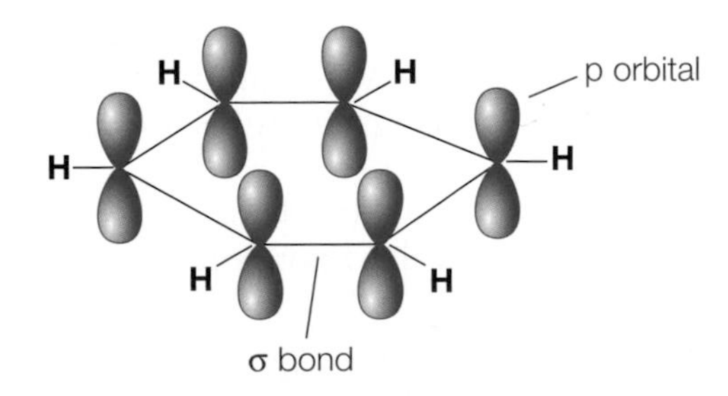

Figure 12.11▲
Sigma bonds in benzene, with one electron per carbon atom remaining in a p orbital.

These six p electrons do not pair up to from three carbon–carbon double bonds (consisting of a σ bond plus a π bond) as in the Kekulé structure. Instead, they are shared evenly between all six carbon atoms, giving rise to circular clouds of negative charge above and below the ring of carbon atoms (Figure 12.12). This is an example of a delocalised π electron system, which occurs in any molecule where the conventional structure shows alternating double and single bonds. Within the π electron system, the electrons are free to move anywhere.

Molecules with delocalised electrons, in which the charge is spread over a larger region than usual, are more stable than might otherwise be expected. In benzene, this accounts for the compound being 152 kJ mol^{-1} more stable than expected for the Kekulé structure.

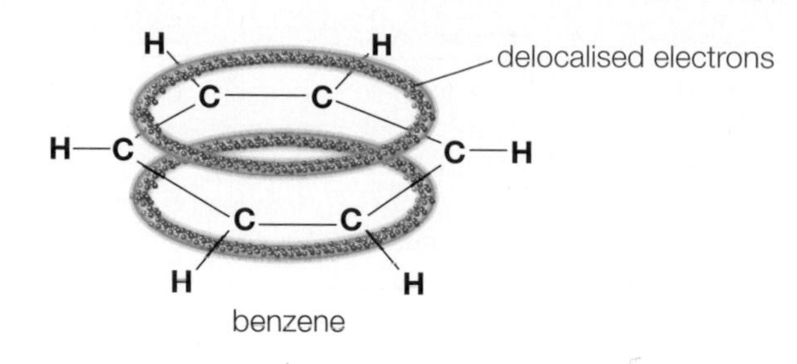

Figure 12.12▶
Representation of the delocalised π bonding in benzene. The circle in a benzene ring represents six delocalised electrons. This way of showing the structure explains the shape and stability of benzene.

The development of ideas concerning the structure of benzene illustrates the way in which theories develop and get modified as new knowledge becomes available.

12.4 Naming arenes

The name 'benzene' comes from *gum benzoin*, a natural product containing benzene derivatives. These derivatives of benzene are named either as substituted products of benzene or as compounds containing the phenyl group, C_6H_5–. The names and structures of some derivatives of benzene are shown in Table 12.1.

Note

The names used for compounds with a benzene ring can be confusing. The phenyl group C_6H_5– is used to name many compounds in which one of the hydrogen atoms in benzene has been replaced by another atom or group. The use of phenyl in this way dates back to the first studies of benzene. At this time 'phene' was suggested as an alternative name for benzene, based on a Greek word for 'giving light'. The name 'phene' was suggested because benzene had been discovered in the tar formed on heating coal to produce gas for lighting.

Systematic name	Substituent group	Structure
Chlorobenzene	Chloro, –Cl	C_6H_5–Cl
Nitrobenzene	Nitro, $-NO_2$	$C_6H_5-NO_2$
Methylbenzene	Methyl, $-CH_3$	$C_6H_5-CH_3$
Phenol	Hydroxy, –OH	C_6H_5–OH
Phenylamine	Amine, $-NH_2$	$C_6H_5-NH_2$

Table 12.1▲
The names and structures of some derivatives of benzene.

When more than one hydrogen atom is substituted, numbers are used to indicate the positions of substituents on the benzene ring (Figure 12.13). The ring is usually numbered clockwise and the numbers used are the lowest ones possible. In some cases, the ring is numbered anticlockwise to get the lowest possible numbers.

In phenyl compounds, such as phenol and phenylamine, the –OH and $-NH_2$ groups are assumed to occupy the 1 position.

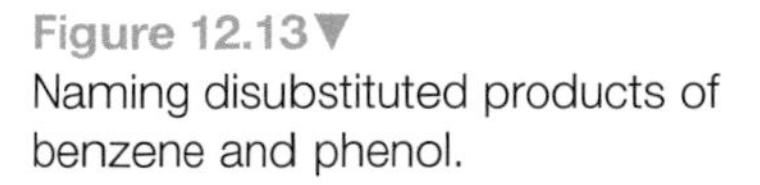
Figure 12.13▼
Naming disubstituted products of benzene and phenol.

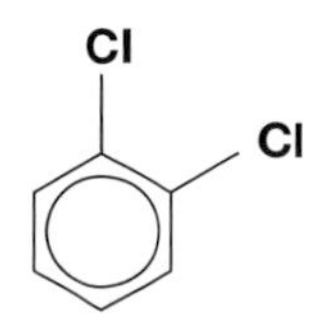

1,2-dichlorobenzene

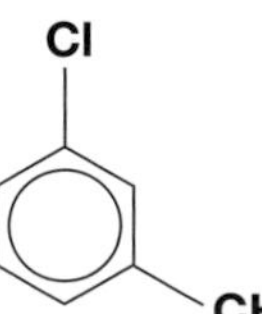
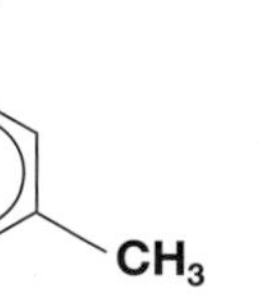

1-chloro-3-methylbenzene

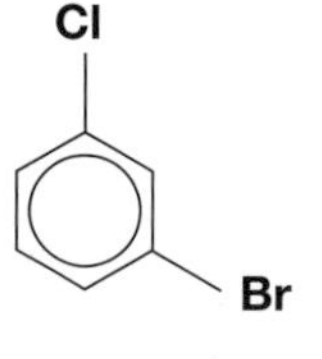

1-bromo-3-chlorobenzene

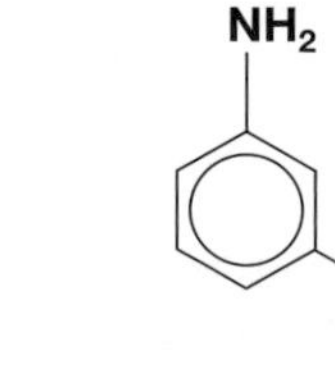

3-nitrophenylamine

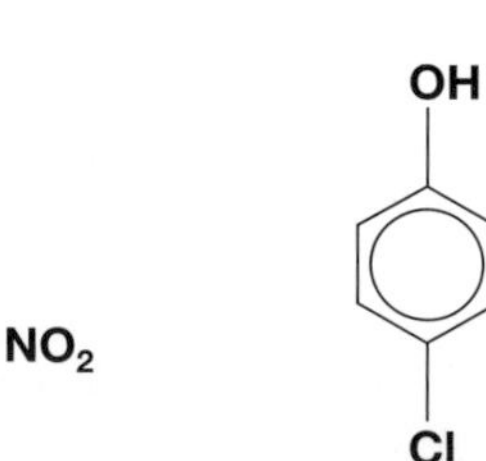

2,4-dichlorophenol

Test yourself

5 How does the model of benzene molecules with delocalised π electrons account for the following?
   a) The benzene ring is a regular hexagon.
   b) There are no isomers of 1,2-dichlorobenzene.
   c) Benzene is less reactive than cycloalkenes.

6 When chemists thought that benzene had alternate single and double bonds, it was sometimes named cyclohexa-1,3,5-triene. Why is this now an unsatisfactory systematic name for benzene?

7 Why is the middle compound in Figure 12.13 named 1-bromo-3-chlorobenzene and not 1-chloro-3-bromobenzene?

12.5 The properties and reactions of benzene and arenes

Arenes are non-polar compounds with weak intermolecular forces between their molecules. The boiling temperatures of arenes depend on the size of the molecules. The bigger the molecules, the higher the boiling temperatures. Benzene and methylbenzene are liquids at room temperature, while naphthalene is a solid (Figure 12.14).

Arenes, like other hydrocarbons, do not mix with water, but they do mix freely with non-polar solvents such as cyclohexane.

Note

Benzene is toxic. It is also a carcinogen. Because of this, benzene is banned from teaching laboratories. Arene reactions may be studied using other compounds such as methylbenzene or methoxybenzene.

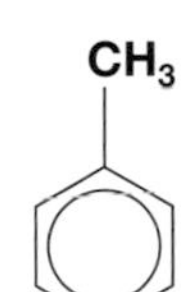

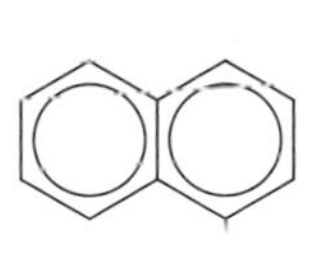

Figure 12.14▲
Three arenes: benzene, methylbenzene and naphthalene.

Test yourself

8 Explain the existence of weak attractive forces between benzene molecules, which are uncharged and non-polar.

9 Explain why benzene does not mix with water.

10 Name a solvent, other than cyclohexane, with which you would expect benzene to mix freely.

The chemical reactions of arenes

Arenes burn in air. Unlike straight-chain alkanes and alkenes of similar molar mass, they burn with a very smoky flame because of the high ratio of carbon to hydrogen in their molecules (Figure 12.15).

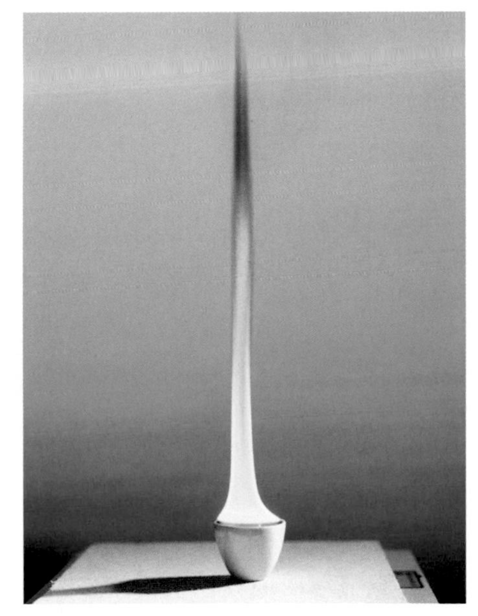

Figure 12.15▲
A sample of methylbenzene burning in air showing the yellow flame and very smoky fumes.

Benzene and other arenes are similar to alkenes in having a prominent and superficial electron-dense region. In arenes this is a delocalised ring of π electrons, while in alkenes it is a single π bond. Because of this similarity, both arenes and alkenes react with electrophiles in many of their reactions. However, the similarity ends there, because the overall reactions of arenes involve substitution, unlike those of alkenes which involve addition.

It is quite easy to see why addition reactions are difficult for arenes. If, for example, benzene reacted with bromine in an addition reaction, the ring of delocalised π electrons would be broken (Figure 12.16). Given the stability associated with the delocalised π system, this would require much more energy than that needed to break the one double bond in ethene.

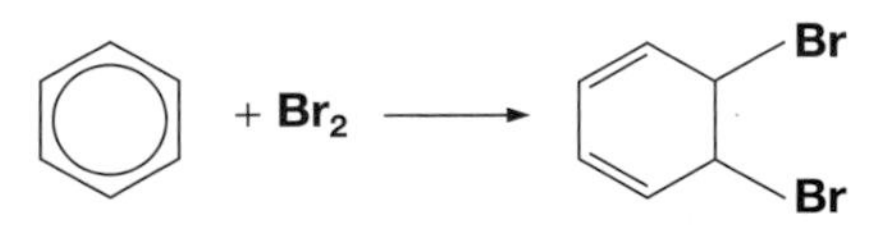

Figure 12.16▲
Benzene does not combine with bromine in an addition reaction because its ring of delocalised π electrons would be broken.

So, arenes tend to retain their delocalised π electrons and undergo substitution rather than addition reactions.

12.6 The substitution reactions of benzene

Halogenation

Benzene will not react with chlorine or bromine in the dark. The halogen molecules are non-polar and have no centre of positive charge to initiate an attack on the delocalised electron system. However, if benzene is warmed with bromine plus a catalyst of iron or iron(III) bromide, a reaction occurs forming bromobenzene (Figure 12.17).

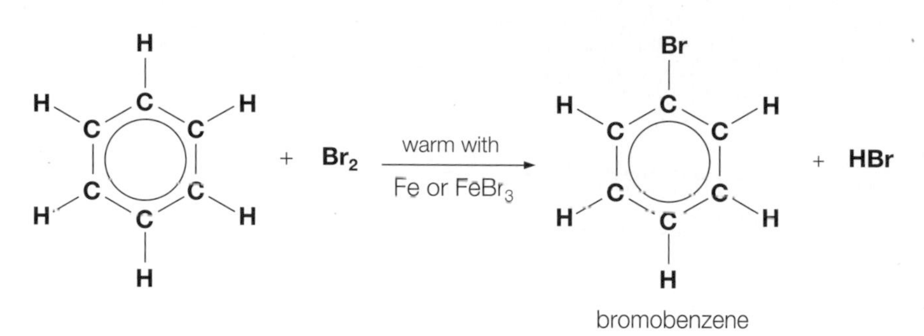

Figure 12.17▶ The reaction of benzene with bromine.

A similar reaction occurs when benzene is warmed with chlorine in the presence of iron, iron(III) chloride or aluminium chloride.

Nitration

When benzene is warmed to about 55 °C with concentrated nitric acid in the presence of concentrated sulfuric acid, the major product is yellow, oily nitrobenzene (Figure 12.18).

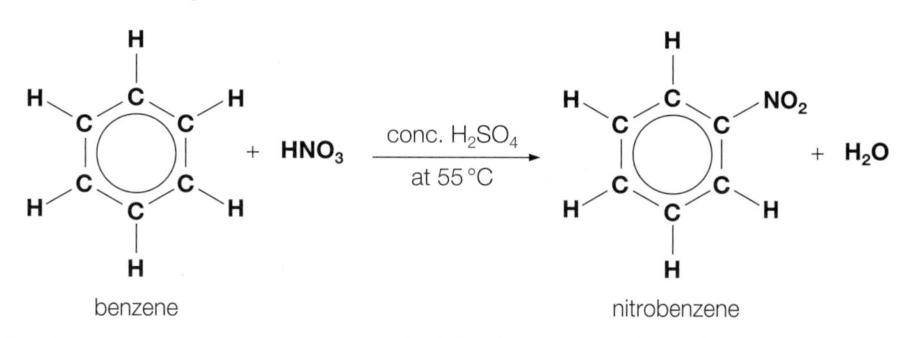

Figure 12.18▶ The reaction of benzene with concentrated nitric acid.

A substitution reaction occurs in which hydrogen is replaced by a nitro group, $-NO_2$. If the reaction mixture is heated above 55 °C, further nitration occurs forming dinitrobenzene.

Figure 12.19▶ Nitrated organic compounds, like TNT (trinitrotoluene) and nitroglycerine, are useful explosives in mining, tunnelling and road building.

Nitration of benzene and other arenes is important because it produces a range of important products including dyes and powerful explosives such as TNT (trinitrotoluene, now called 1-methyl-2,4,6-trinitrobenzene).

Sulfonation

If benzene is treated at room temperature with concentrated sulfuric acid, no reaction occurs. But if 'fuming' sulfuric acid is used, a reaction occurs forming benzenesulfonic acid. The fuming sulfuric acid is a mixture of concentrated sulfuric acid and dissolved sulfur trioxide, SO_3.

The sulfur atoms in SO_3 molecules are strongly polarised δ+. Because of this, they can attack the delocalised electron system in benzene, initially forming benzenesulfonate ions (Figure 12.20) by displacing H^+ ions.

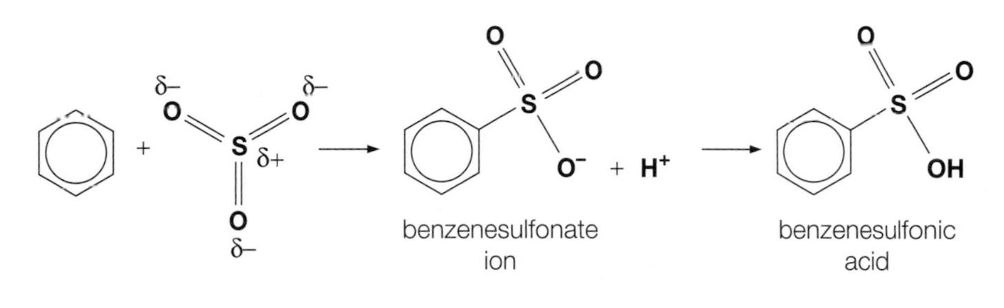

Figure 12.20▲
The reaction of benzene with fuming sulfuric acid.

Alkylation and acylation (Friedel–Crafts reactions)

The Friedel–Crafts reaction is an important method for substituting an alkyl group or an acyl group for a hydrogen atom in an arene. This reaction was discovered and developed jointly by the French organic chemist Charles Friedel (1832–1899) and the American, James Crafts (1839–1917). The reaction is used on both a laboratory and an industrial scale.

In a Friedel–Crafts reaction, a halogenoalkane or an acyl chloride is refluxed with an arene in the presence of aluminium chloride as catalyst. For example, if benzene is refluxed with chloromethane and aluminium chloride, a substitution reaction occurs forming methylbenzene (Figure 12.21).

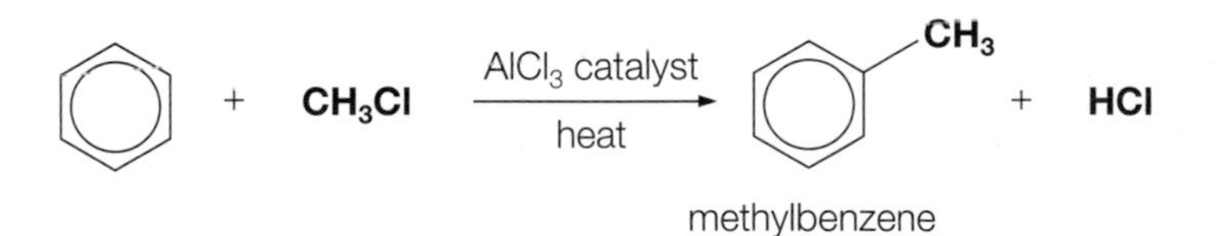

Figure 12.21▲
Friedel–Crafts alkylation of benzene with chloromethane forming methylbenzene.

A similar reaction occurs when benzene is refluxed with the acyl chloride, ethanoyl chloride, plus aluminium chloride as a catalyst. This time, the product is phenylethanone, also known as methylphenylketone (Figure 12.22).

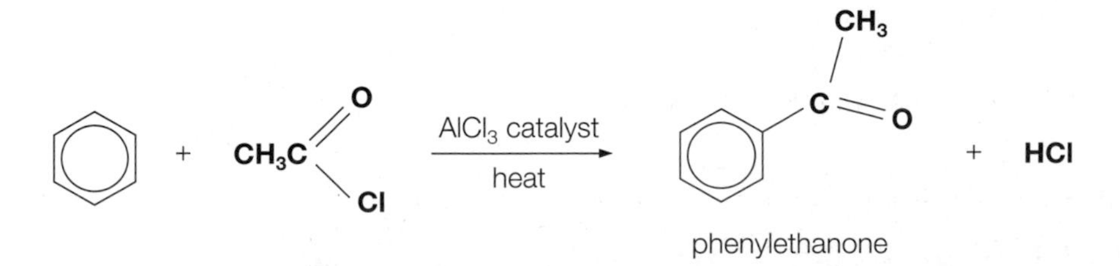

Figure 12.22▲
Friedel–Crafts acylation of benzene with ethanoyl chloride forming phenylethanone.

Test yourself

11 a) Why must the reaction mixture be completely dry during a Friedel–Crafts reaction?

b) Draw the structures of the products of a Friedel–Crafts reaction of benzene with:

i) 2-iodo-2-methylpropane

ii) propanoyl chloride.

c) Suggest a reason for using an iodoalkane instead of a chloroalkane in a Friedel–Crafts reaction.

12 Three possible isomers of dinitrobenzene can be produced. One of these isomers is called 1,2-dinitrobenzene.

a) Draw and name the structures of the other two dinitrobenzenes.

b) Why is there no isomer called 1,6-dinitrobenzene?

13 a) Draw the structure of TNT. (*Hint:* Look closely at its systematic name of 1-methyl-2,4,6-trinitrobenzene.)

b) Why is TNT mixed with a compound containing a high proportion of oxygen, such as potassium nitrate, when it is used as an explosive?

12.7 The addition reactions of benzene with hydrogen

The characteristic reactions of benzene involve substitution because this type of reaction retains the delocalised π electron system with its associated stability. However, addition reactions involving disruption of the π electron system do occur.

Benzene, like alkenes, will undergo addition with hydrogen in the presence of a nickel catalyst, but at considerably higher temperatures (Figure 12.23).

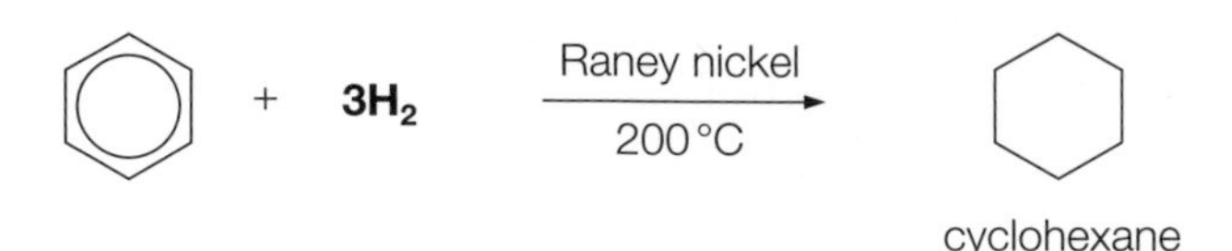

Figure 12.23▲
The addition of hydrogen to benzene forming cyclohexane.

A higher temperature is needed with benzene in order to break up the stable π electron system and allow addition to occur. A special finely divided form of nickel, called Raney nickel, is also used because this has an extremely high surface area.

The addition of hydrogen to benzene is thought to occur in three stages. First, the formation of cyclohexa-1,3-diene, then cyclohexene and finally cyclohexane.

The catalytic hydrogenation of benzene is important industrially in the manufacture of cyclohexane, which is used to make nylon.

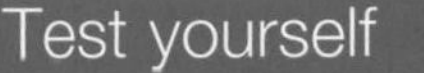

14 a) Why is Raney nickel used in the manufacture of cyclohexane from benzene?

b) Write equations to show the three stages in the hydrogenation of benzene via cyclohexa-1,3-diene and cyclohexene to form cyclohexane.

c) Why do cyclohexa-1,3-diene and cyclohexene react more readily with hydrogen than benzene?

12.8 The mechanism of electrophilic substitution in benzene

Electrophilic substitution is the characteristic reaction of benzene in which the delocalised π electrons are attacked by strong electrophiles.

Nitration

The mixture of concentrated nitric and sulfuric acids that reacts with benzene is called a nitrating mixture. At 55 °C, concentrated nitric acid on its own reacts very slowly with benzene, and concentrated sulfuric acid by itself has practically no effect. However, in a mixture of the two, sulfuric acid reacts with nitric acid to produce nitronium ions, NO_2^+, which are very reactive electrophiles.

$$HNO_3 + H_2SO_4 \rightarrow NO_2^+ + HSO_4^- + H_2O$$

The nitronium ions are formed by removal of OH^- ions from nitric acid by sulfuric acid. In this reaction, HNO_3 is acting as a base and H_2SO_4 is acting as an acid.

The NO_2^+ ion is a reactive electrophile which is strongly attracted to the delocalised electrons in benzene. As it approaches the benzene ring, the NO_2^+ ion forms a covalent bond to one of the carbon atoms using two electrons from the π system (Figure 12.24).

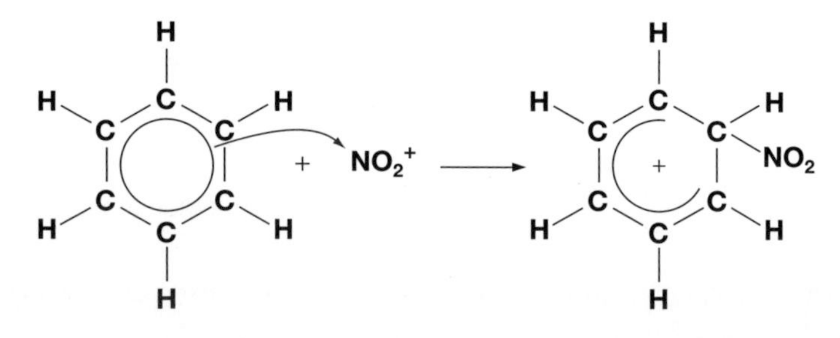

Figure 12.24▶
Electrophilic nitronium ions use two of the delocalised electrons in benzene to form an intermediate cation.

The formation of a covalent bond in an intermediate cation disrupts the delocalised ring. A large input of energy is needed to do this and the reaction has a fairly high activation energy.

The unstable intermediate cation quickly breaks down, producing nitrobenzene. This involves the return of two electrons from a C–H bond to the π electron system. The stability of the delocalised ring is restored and energy is released (Figure 12.25).

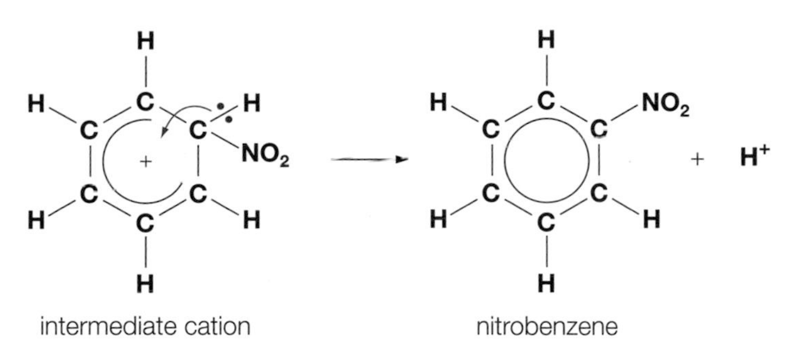

Figure 12.25▲
The intermediate cation breaks down to form nitrobenzene.

Halogenation

When benzene is warmed with bromine in the presence of iron filings, the bromine first reacts with the iron to form iron(III) bromide.

$$2Fe(s) + 3Br_2(l) \rightarrow 2FeBr_3(s)$$

The iron(III) bromide then acts as a catalyst for the reaction of bromine with benzene by polarising further bromine molecules as $Br^{\delta+}$–$Br^{\delta-}$ (Figure 12.26).

$$Br—Br + Fe^{3+}(Br^-)_3 \longrightarrow \overset{\delta+}{Br}—\overset{\delta-}{Br}\cdots\cdots Fe^{3+}(Br^-)_3$$

Figure 12.26▲
Fe^{3+} ions in iron(III) bromide polarise bromine molecules as $Br^{\delta+}$–$Br^{\delta-}$.

The remaining steps in bromination of benzene are similar to nitration. The δ+ bromine atoms in the polarised Br_2 molecules act as electrophiles in a similar way to NO_2^+ ions

An intermediate cation is first produced as the $Br^{\delta+}$ atom forms a covalent bond using two electrons from the delocalised π system leaving a Br^- ion. This intermediate then breaks down to form bromobenzene as two electrons are returned from the C–H bond to the π system and the stable, delocalised ring is restored (Figure 12.27). At the same time, an H^+ ion is released from the intermediate cation. This H^+ ion immediately combines with the Br^- ion released in stage 1 to form hydrogen bromide.

Definition

Chemists sometimes use the term **halogen carrier** to describe substances, such as iron(III) bromide, aluminium chloride and iron, which catalyse the reaction of benzene with chlorine or bromine.

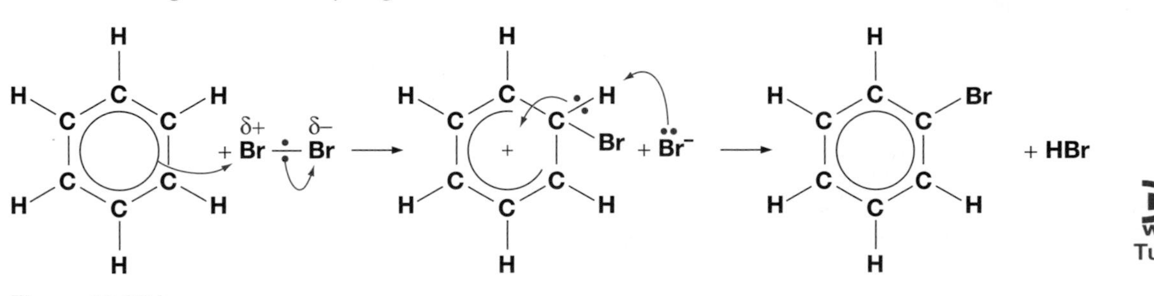

DL www Tutorial

Figure 12.27▲
The mechanism for the bromination of benzene.

Test yourself

15 This question is about the Friedel–Crafts reaction between benzene and ethanoyl chloride, CH_3COCl, in the presence of aluminium chloride, $AlCl_3$, as catalyst.

a) Write an equation to show how $AlCl_3$ molecules react with polarised CH_3COCl molecules to produce reactive ethanoyl electrophiles.

b) Write an equation for the electrophilic substitution of benzene by ethanoyl electrophiles to produce phenylethanone.

c) Write an equation to show how molecules of the aluminium chloride catalyst are regenerated.

Alkylation and acylation (Friedel–Crafts reactions)

In Friedel–Crafts reactions, the aluminium chloride plays an important catalytic role in creating the electrophiles which attack benzene. For example, when chloromethane is mixed with aluminium chloride, $AlCl_3$ molecules remove Cl^- ions from polar $^{\delta+}CH_3$–$^{\delta-}Cl$ molecules allowing reactive CH_3^+ ions to act as electrophiles.

$$^{\delta+}CH_3\text{—}^{\delta-}Cl + AlCl_3 \rightarrow \underset{\text{electrophile}}{^{+}CH_3} + AlCl_4^-$$

These reactive CH_3^+ electrophiles then attack the delocalised π system of benzene molecules to form an intermediate cation, which breaks down producing methylbenzene and H^+ ions (Figure 12.28).

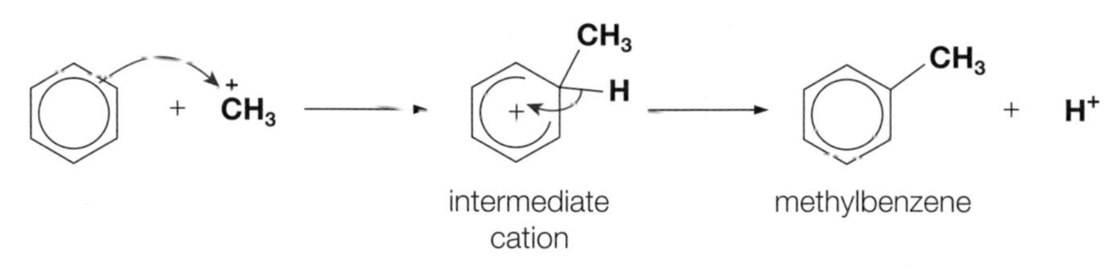

Figure 12.28▲
The reaction of CH_3^+ electrophiles with benzene in the Friedel–Crafts reaction to produce methylbenzene.

Finally, the aluminium chloride catalyst is regenerated as H^+ ions released in the electrophile substitution react with $AlCl_4^-$ ions.

$$H^+ + AlCl_4^- \rightarrow HCl + AlCl_3$$

Activity

Studying the reaction of benzene with chlorine

Figure 12.29 shows the apparatus that might once have been used to prepare chlorobenzene by heating benzene with chlorine gas in the presence of iron filings. This preparation is now banned in teaching laboratories in schools and colleges.

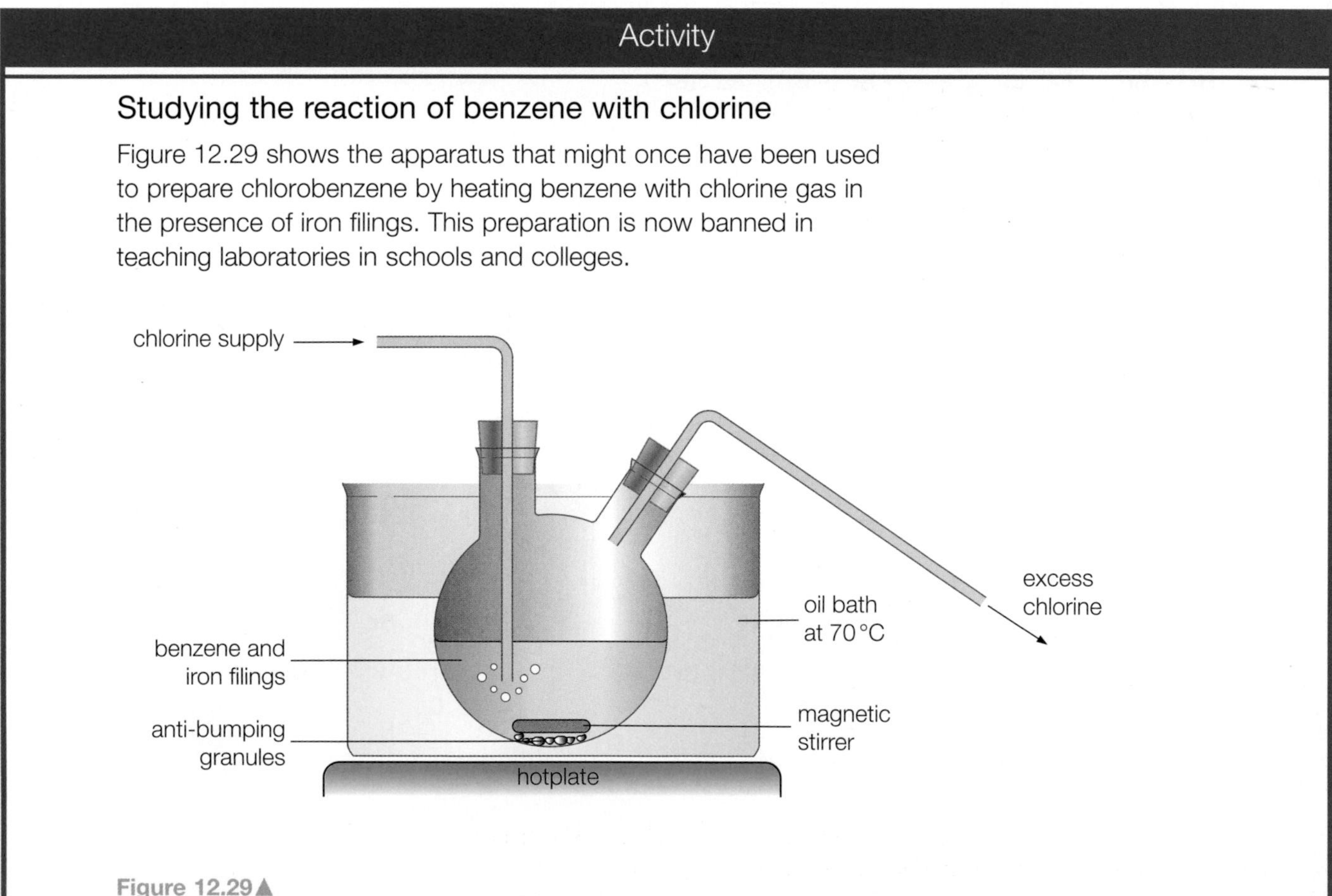

Figure 12.29▲
Preparing chlorobenzene.

1 Why is this reaction banned in the teaching laboratories of schools and universities?

2 a) Why should the reaction be carried out in a fume cupboard?

b) Why is a hotplate used?

c) Why is the oil bath at 70 °C?

3 During the reaction, iron reacts with chlorine to form iron(III) chloride, which then acts as an electron pair acceptor polarising the Cl_2 molecules as $Cl^{\delta+}–Cl^{\delta-}$.

Write equations to show:

a) the formation of iron(III) chloride

b) the polarisation of Cl_2 as $Cl^{\delta+}–Cl^{\delta-}$ by iron(III) chloride.

4 Write a mechanism with two steps for the reaction of polarised $Cl^{\delta+}–Cl^{\delta-}$ molecules with benzene to form an intermediate cation plus Cl^- in the first step, and then chlorobenzene plus HCl in the second step.

5 The reaction shown in Figure 12.29 is just as effective if aluminium chloride or iron(III) chloride are used in place of iron. These three substances (Fe, $FeCl_3$ and $AlCl_3$) are often described as catalysts and halogen carriers.

a) Why are these substances described as halogen carriers for the reaction?

b) Why is it correct to describe aluminium chloride and iron(III) chloride as catalysts?

c) Why is it incorrect to describe iron as a catalyst for the reaction?

6 Aluminium chloride acts as a catalyst for the chlorination of benzene by polarising Cl_2 molecules in the same way as iron(III) chloride.

a) Do you think aluminium chloride will be more effective or less effective than iron(III) chloride?

b) Explain your answer to part **a)**.

12.9 Comparing the reactions of arenes and alkenes

The electrophilic substitution of benzene occurs in three distinct steps whatever the final product.

Step 1 The formation of a reactive electrophile: NO_2^+, $Br^{\delta+}–Br^{\delta-}$, $Cl^{\delta+}–Cl^{\delta-}$, $S^{\delta+}O_3^{\delta-}$, CH_3^+ or CH_3CO^+.

Step 2 The attack of the delocalised π system by the reactive electrophile, forming an unstable intermediate cation.

Step 3 The decomposition of this intermediate cation producing a substituted benzene (nitrobenzene, bromobenzene, methylbenzene, etc.).

Notice that the first two steps of this mechanism (Figure 12.30) are similar to the first two steps in the electrophilic addition of electrophiles, such as chlorine, bromine and hydrogen bromide, to alkenes.

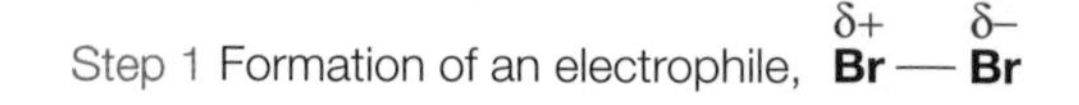

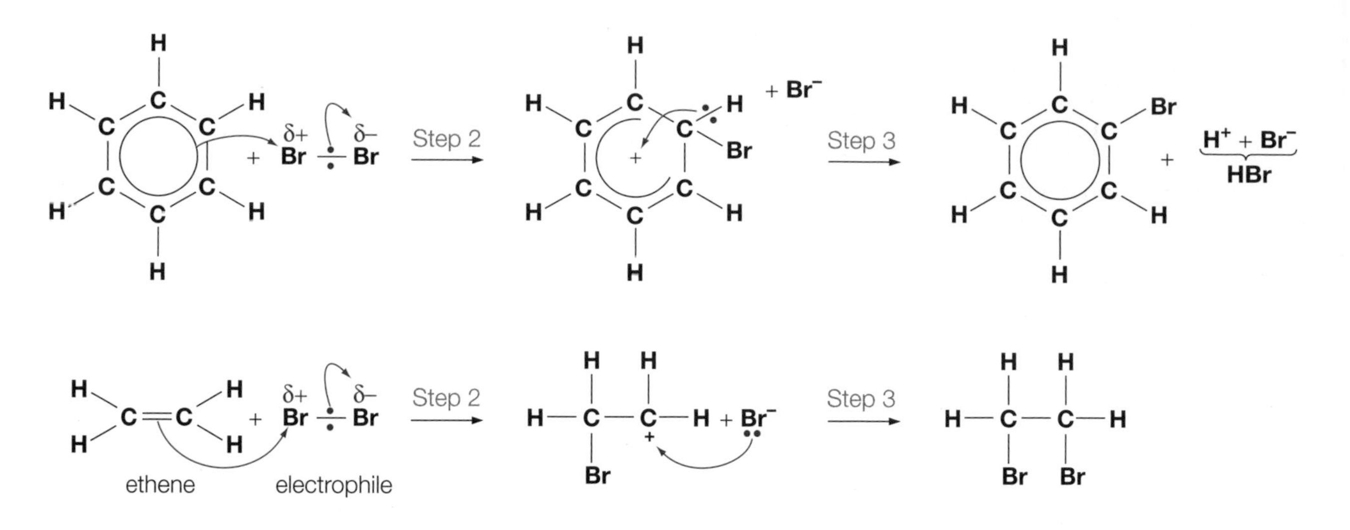

Figure 12.30▲
Comparing the mechanisms of the reactions of bromine with benzene and ethene.

The third step of the mechanism is, however, very different. With benzene, the intermediate cation loses its charge by capturing two electrons from a C–H bond and releasing an H^+ ion. The delocalised, stable aromatic system is restored and a *substitution* reaction occurs.

With alkenes, there is no delocalised system or its associated stability. So, in this case, the intermediate cation combines with an anion such as Br^- or Cl^- and an *addition* reaction occurs.

These reactions of arenes and alkenes with electrophiles also differ in another important way – their relative reactivities.

Benzene and arenes are resistant to halogenation, nitration, alkylation and acylation because of the stability of their delocalised π electrons. They must be heated and, in the cases of halogenation, alkylation and acylation, catalysed as well before a reaction occurs.

Alkenes, however, will react with halogens and hydrogen halides rapidly at room temperature because of the localised electron density of the C=C bond. Electrons in the π bond of ethene and other alkenes are readily available to any attacking electrophile.

Test yourself

16 a) Describe briefly how you would prepare bromobenzene from benzene.

b) Why does benzene not react with bromine unless a halogen carrier is present?

17 Iodine will not react with hot benzene even in the presence of iron, but a good yield of iodobenzene can be produced by reacting benzene with iodine(I) chloride.

a) Why do you think iodine will not react with hot benzene even in the presence of iron?

b) How is iodine(I) chloride polarised?

c) Using your answer to part **b)**, explain why iodine(I) chloride reacts with benzene to give good yields of iodobenzene.

12.10 Phenol

Phenol is an example of a compound with a functional group directly attached to a benzene ring. In phenol, the functional group is –OH. Experiments show that the –OH group affects the behaviour of the benzene ring while the benzene ring modifies the properties of the –OH group. As a result of this, phenol has some distinctive and useful properties.

As expected, the –OH group gives rise to hydrogen bonding in phenol and therefore much stronger intermolecular forces than in benzene. This results in phenol being a solid at room temperature (Figure 12.31).

Figure 12.31▲
Crystals of phenol.

The –OH group in phenol also allows it to hydrogen bond with water. As a result of this, phenol dissolves slightly in water and much more readily in alkalis, such as sodium hydroxide solution (Figure 12.32) with which it forms soluble ionic compounds. In this reaction, phenol is behaving as an acid. However, phenol does not ionise significantly in water and it does not react with carbonates to produce carbon dioxide.

Figure 12.32◀
Phenol reacts with aqueous sodium hydroxide to form a colourless solution of sodium phenoxide.

The derivatives of phenol are named in a similar fashion to those of benzene, by numbering the carbon atoms in the benzene ring starting from the –OH group. The numbering runs clockwise or anticlockwise to give the lowest possible numbers for the substituted groups (Figure 12.33).

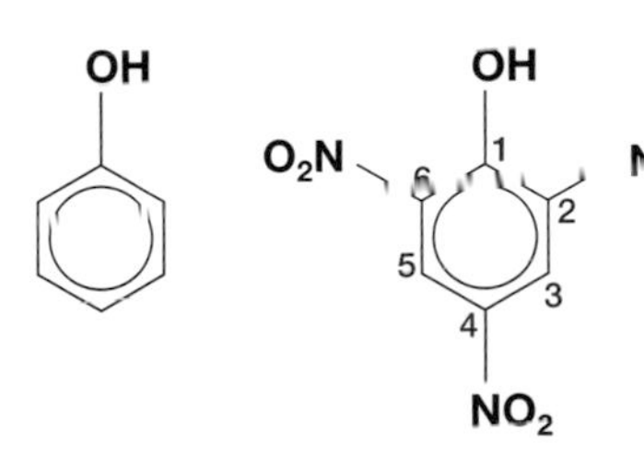

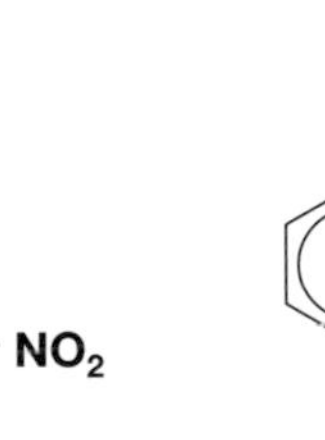

OH

Br

phenol | 2,4,6-trinitrophenol | 2-methyl-5-nitrophenol (not 6-methyl-3-nitrophenol) | 4-bromophenol

Figure 12.33▲
The structure of phenol and other phenols.

Test yourself

18 Explain in terms of intermolecular forces why:
  a) phenol is a solid while benzene is a liquid at room temperature
  b) phenol, unlike benzene, is slightly soluble in water
  c) phenol does not mix with water as freely as ethanol.

19 What would you expect to observe on heating phenol until it burns?

20 Identify one way in which the –OH group behaves similarly in phenol and ethanol, and one way in which it behaves differently.

21 a) What would you expect to observe if you added enough dilute hydrochloric acid to a solution of phenol in sodium hydroxide to make the mixture acidic?
  b) Explain the reaction that occurs.

12.11 Reactions of the benzene ring in phenol

Comparing the reactivity of benzene and phenol

The –OH group in phenol activates its benzene ring and makes it more reactive than benzene itself. A lone pair of electrons on the –OH group interacts with the delocalised electrons in the benzene ring, releasing electrons into the ring and making electrophilic attack easier. As a result, electrophilic substitution takes place under much milder conditions with phenol than with benzene.

Reaction with bromine

An aqueous solution of phenol reacts readily with bromine water to produce an immediate white precipitate of 2,4,6-tribromophenol as the orange/yellow bromine colour fades. The reaction is rapid at room temperature with bromine water. There is no need to heat the mixture or use a catalyst (Figure 12.34).

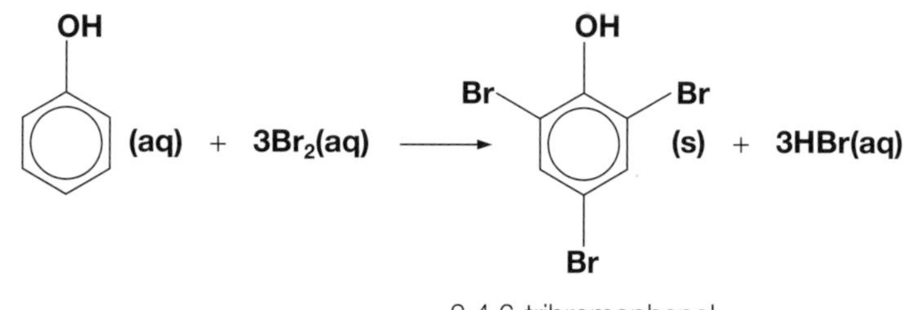

Figure 12.34▶
The reaction of phenol with bromine.

Reaction with nitric acid

Dilute nitric acid reacts rapidly with phenol at room temperature to form a brown mixture. The main products of the reaction are 2-nitrophenol and 4-nitrophenol (Figure 12.35). Notice how the conditions for nitrating phenol are so much milder than those needed to nitrate benzene.

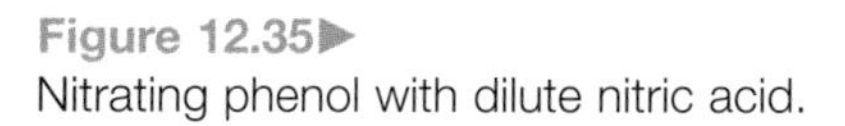

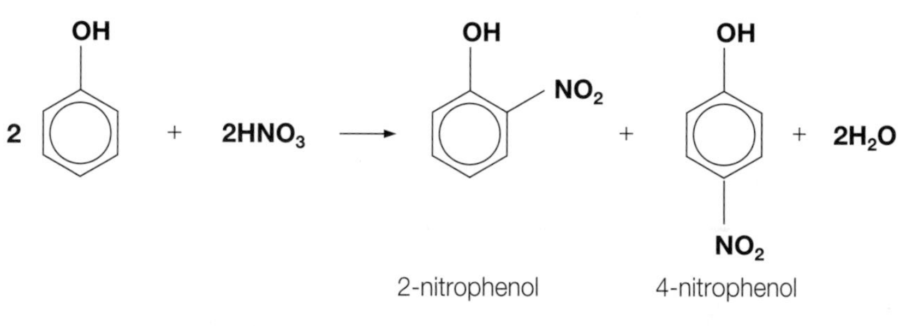

Figure 12.35▶
Nitrating phenol with dilute nitric acid.

Activity

Manufacturing phenol

Phenol is manufactured from benzene, propene and oxygen in two stages. The process is known as the cumene process (Figure 12.36).

The first stage of the process involves the acid-catalysed electrophilic substitution of benzene with propene to form cumene.

The second stage involves the air oxidation of cumene. This produces equimolar amounts of phenol and propanone, a valuable co-product. About 100 000 tonnes of phenol are manufactured each year in the UK using the cumene process.

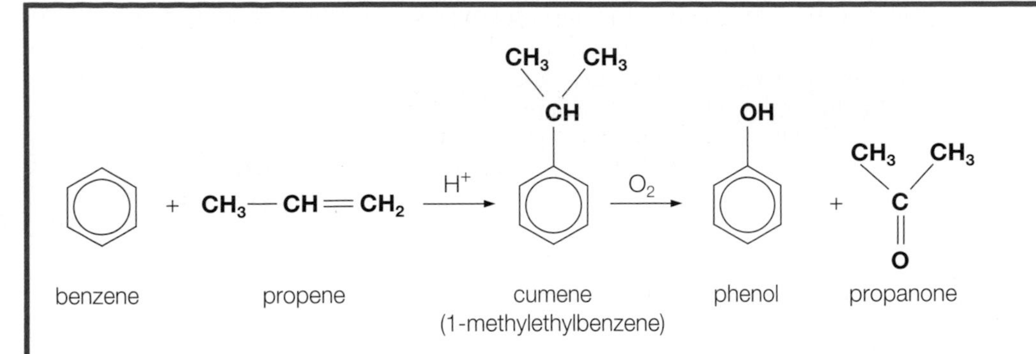

Figure 12.36 The manufacture of phenol by the cumene process.

1 Benzene and propene are obtained for the cumene process from crude oil. What processes, starting with crude oil, are used to produce:

a) benzene **b)** propene?

2 In the first stage of the cumene process, H^+ ions react with propene to produce electrophiles.

a) Write the formulae of two possible electrophiles produced when H^+ ions react with propene.

b) Explain why one of these electrophiles is more stable than the other.

c) Name and draw the structure of a second possible product of this first stage besides cumene.

3 a) Write an equation for the reaction of the stable electrophile, identified in question **2b)**, with benzene to produce cumene.

b) Why is this reaction described as acid-catalysed?

4 Write an equation for the second stage of the process in which cumene is oxidised to phenol and propanone.

5 The actual yield in the cumene process is 85%. Calculate the mass of benzene required to manufacture 1 tonne of phenol and the mass of propanone formed at the same time.

12.12 The uses and importance of phenol

The main uses of phenol are in the production of:

- antiseptics and disinfectants
- thermosetting plastics.

Antiseptics and disinfectants

Phenol is a powerful disinfectant that has been used to kill germs ever since it was isolated from coal tar in the nineteenth century. For a time, phenol was famous as a chemical that made surgery safe. In 1857, a young surgeon called Joseph Lister working in Glasgow read Louis Pasteur's papers about the germ theory of disease. Lister realised that he could use a disinfectant to prevent infections after serious operations. He developed a technique of spraying a solution of phenol over the open wounds during operations. As a result of this, his patients had a much better chance of their wounds healing without becoming infected.

Unfortunately, there are also risks in the use of phenol. It is an unpleasant chemical that burns the skin and is therefore unsuitable as an antiseptic.

Definitions

Disinfectants are chemicals that destroy microorganisms. Phenol is a disinfectant and so is chlorine. Unlike antiseptics, disinfectants are such powerful reagents that they are no longer used on skin and other living tissues.

Antiseptics are chemicals that kill microorganisms but, unlike disinfectants, they can be used safely on the skin.

Since Lister's time, chemists have discovered substituted phenols that are suitable as antiseptics. These include Dettol and TCP (Figure 12.37) in which chlorination increases the antibacterial properties of phenol and also improves its healing properties as an antiseptic.

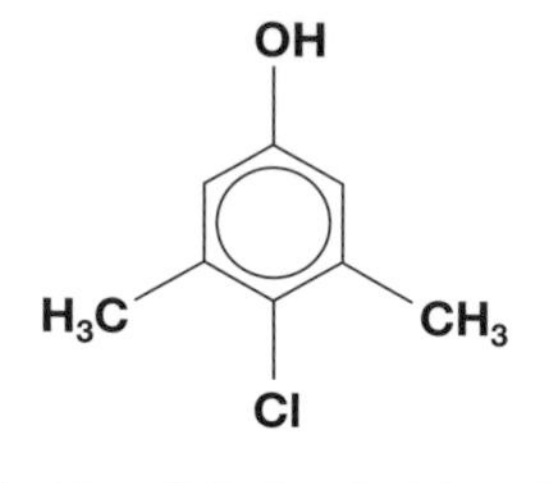

4-chloro-3,5-dimethylphenol
(Dettol)

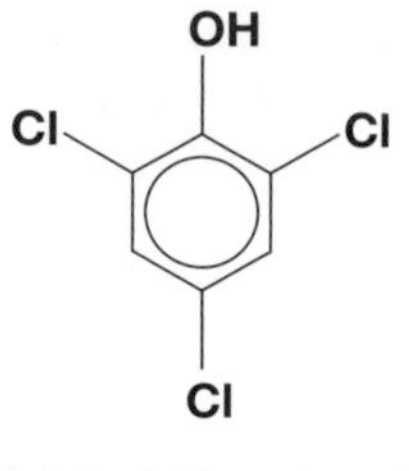

2,4,6-trichlorophenol
(TCP)

Figure 12.37▶ The structure of Dettol and TCP.

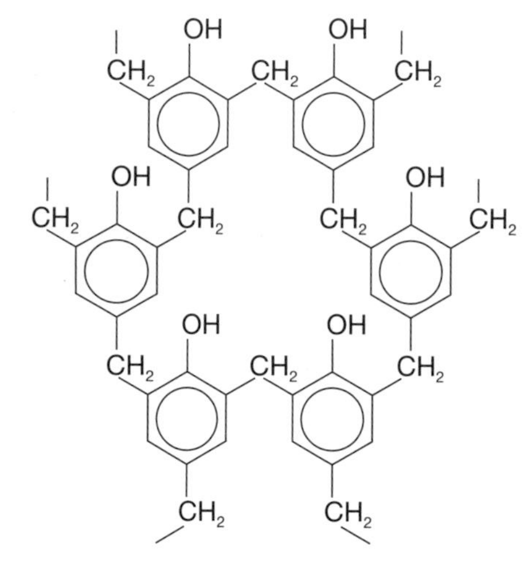

Figure 12.38▲ The thermosetting plastic produced from phenol and methanal.

Thermosetting plastics (resins)

When phenol is heated with methanal in the presence of an acid or an alkali as catalyst, a hard, brittle plastic is formed. Initially, the reaction links phenol molecules through their 2 and 6 positions to form a linear polymer. These linear polymers then react further with methanal molecules to form links through their 4-positions (Figure 12.38).

The final product is a molecular network in which all the benzene rings are substituted in the 2, 4 and 6 positions. This polymer is very hard because of its extensively cross-linked network. The hard, dark brown material was discovered in 1910 in the USA by Leo Baekeland and named 'Bakelite'. Bakelite is an excellent thermal and electrical insulator that is still widely used for heat-resistant handles and knobs as well as electrical plugs and fittings.

Bakelite has the disadvantage that it sets hard as it forms and cannot be remelted or softened. This makes it difficult to mould. Polymers of this kind are described as thermosetting and are sometimes called resins. In contrast, polymers such as polythene and pvc soften on heating and can be moulded into different shapes which they retain on cooling. These polymers are described as thermoplastic.

> **Definitions**
>
> **Thermoplastic polymers** soften on heating and can be moulded into different shapes which are retained on cooling.
>
> **Thermosetting polymers** (or resins) set hard on formation and cannot be softened or moulded into different shapes.

Figure 12.39▶ An advert for Bakelite in the 1950s extolling its properties. Bakelite was the first commercially successful plastic.

REVIEW QUESTIONS

Extension questions

1 a) Describe the structure and bonding in benzene and explain why benzene is less reactive with electrophiles than alkenes. (8)

b) Describe the conditions needed for the nitration of benzene to form nitrobenzene, and outline the mechanism of the reaction using curly arrows where appropriate. (5)

2 a) Chlorobenzene can be produced from benzene and chlorine with a suitable catalyst.

i) Name the catalyst. (1)

ii) Describe briefly how chlorobenzene could be prepared. (3)

b) Under suitable conditions benzene can be used to make the halogenoalkane shown below.

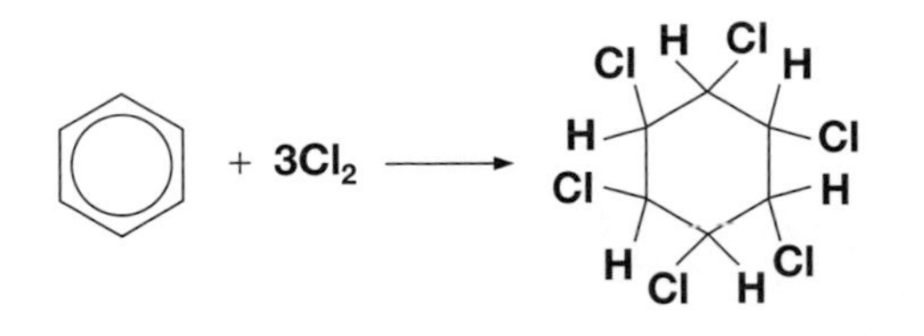

i) What type of reaction has occurred? (1)

ii) What is the name of the halogenoalkane produced? (2)

c) The halogenoalkane from part b) reacts on warming with excess aqueous sodium hydroxide.

i) Draw the structure of the final product with excess sodium hydroxide solution. (1)

ii) What type of reaction has occurred? (2)

3 Naphthalene is an arene containing two fused benzene rings. Its skeletal formula is

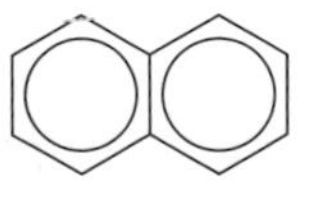

a) What is i) the molecular formula ii) the empirical formula of naphthalene? (2)

b) Describe the delocalisation of electrons in naphthalene, including the shape of the delocalised system and the number of delocalised electrons. (5)

c) Naphthalene reacts with nitric acid in the presence of sulfuric acid to form

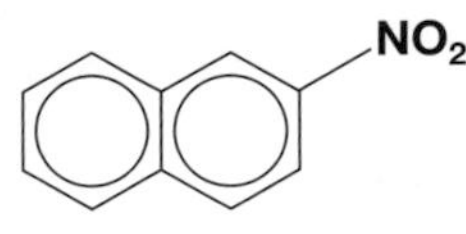

i) Describe the conditions required for the reaction. (2)

ii) What type of reaction occurs and what is the reactive species? (2)

d) Copy, complete and balance the following equation for the reaction of naphthalene with iodine(I) chloride.

+ ICl ⟶ (2)

4 Three reactions of phenol are summarised in the flow diagram below.

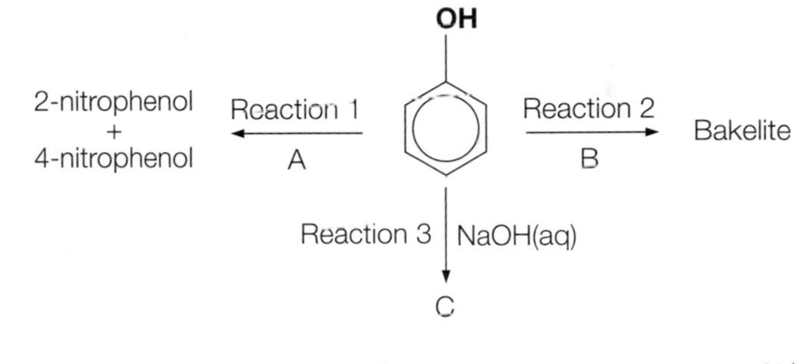

a) Name reagents A and B. (2)

b) Draw the structural formula of C. (2)

c) Under suitable conditions, nitrophenols can be converted to dinitrophenols.

i) Suggest the reaction conditions for converting nitrophenols to dinitrophenols. (2)

ii) One possible dinitrophenol is 2,3-dinitrophenol. Write the names of all the other possible dinitrophenols. (3)

d) Draw the structures of the molecules produced when two molecules of phenol react with one molecule of reagent B. (2)

13 Amines and amides – organic nitrogen compounds

Figure 13.1▲
The active constituent of asthma inhalers is salbutamol which contains the amine functional group.

Amines are nitrogen compounds in which the hydrogen atoms in ammonia, NH_3, are replaced by alkyl or aryl groups. Amines smell like ammonia, but with a distinctly fishy character. The importance of amines, however, is not their smell, but their role in biochemistry and medicine. The amine group is present in amino acids, the monomers for proteins (Topic 14). As a result of this, the amine group plays an important part in metabolism and it is part of the structure of many medicinal drugs.

Amides are nitrogen compounds derived from carboxylic acids in which an $-NH_2$ group replaces the –OH group. The chemistry of amides is important because amide groups link up the monomers in both proteins and in synthetic polymers such as nylon and Kevlar.

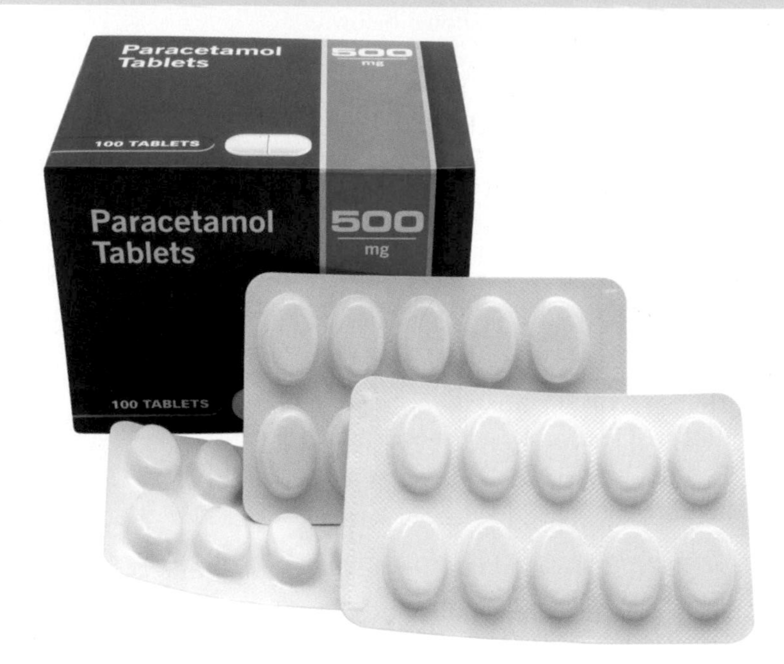

Figure 13.2▲

Paracetamol in Panadol tablets contains the amide group, $-\underset{\underset{O}{\|}}{C}-\underset{\underset{H}{|}}{N}-$

13.1 The structures and names of amines

Note

Notice that the terms primary, secondary and tertiary do not have the same meaning with amines as they do with alcohols.

Amines are nitrogen compounds in which one or more of the hydrogen atoms in ammonia, NH_3, has been replaced by an alkyl or an aryl group. The number of these groups determines whether the compound is a primary amine, a secondary amine or a tertiary amine. If one H atom in ammonia is replaced by an alkyl or aryl group, the compound is a primary amine. If two H atoms in ammonia are replaced, the compound is a secondary amine and if all three H atoms in ammonia are replaced, the compound is a tertiary amine.

Chemists have two systems for naming amines.

Simple amines

Simple amines are treated as a combination of the alkyl or aryl group followed by the ending **-amine**. So, $CH_3CH_2NH_2$ is ethylamine, $C_6H_5NH_2$ is phenylamine and $CH_3CH_2NHCH_3$ is ethylmethylamine. The prefixes di- and tri- are used when there are two or three of the same alkyl or aryl group (Figure 13.3).

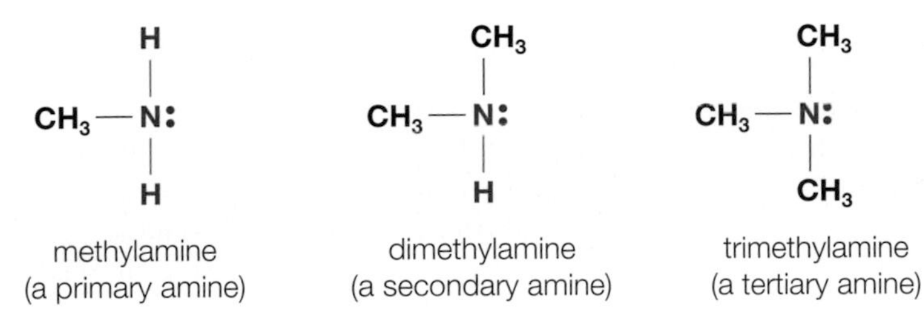

Figure 13.3◀
The structures and names of primary, secondary and tertiary amines containing the methyl group.

More complex amines

The prefix 'amino' is used in compounds that have a second functional group, such as amino acids. So the systematic name for NH_2CH_2COOH is aminoethanoic acid. The prefix 'diamino' is used for compounds containing two amino groups such as 1,2-diaminoethane, $H_2NCH_2CH_2NH_2$.

Test yourself

1 Draw the structures of:
 a) diethylamine b) ethylmethylpropylamine c) 1,6-diaminohexane
 d) 1,2-diaminopentane which contributes to the smell of rotting flesh and has the common name cadaverine
 e) 1-phenyl-2-aminopropane, an amphetamine which is an addictive stimulant.

2 Salbutamol is the active ingredient in asthma inhalers.

Figure 13.4▶
The structure of salbutamol.

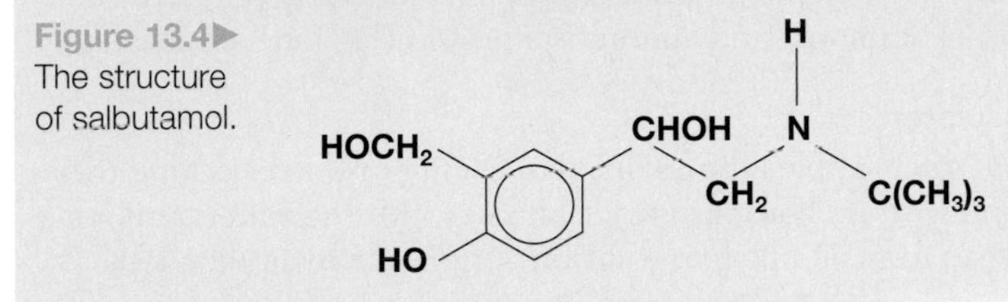

 a) Is its amine group primary, secondary or tertiary?
 b) What other functional groups does salbutamol contain?

13.2 The structures and names of amides

Amides have the general structure $R{-}\underset{\underset{O}{\|}}{C}{-}NH_2$

and the amide functional group is $-\underset{\underset{O}{\|}}{C}-\underset{\underset{H}{|}}{N}-$

Amides are named using the suffix **-amide** after a stem that indicates the number of carbon atoms in the molecule including that in the C=O group (Figure 13.5).

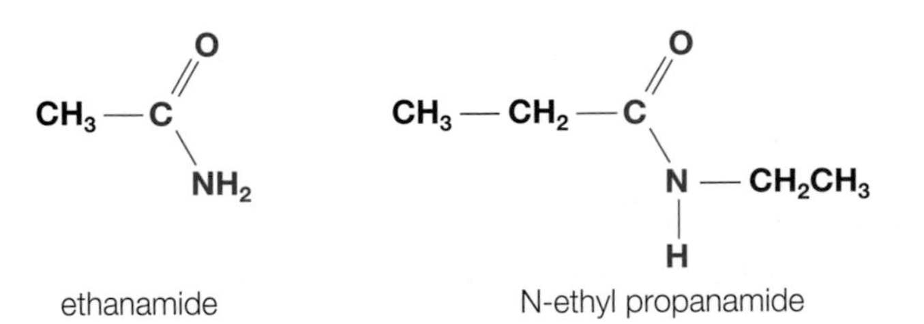

Figure 13.5▲
The structures and names of amides. Notice that N-ethyl propanamide has an ethyl group substituted for one of the hydrogen atoms of the $-NH_2$ group.

Figure 13.6▲
The smell of fish is partly due to ethylamine.

13.3 The properties and reactions of amines

The physical and chemical properties of the simplest amines are similar to those of ammonia. So, methylamine and ethylamine are gases at room temperature and they smell like ammonia, though with a fishy character.

Alkyl amines with short hydrocarbon chains are freely soluble in water, like ammonia. Phenylamine, with its large non-polar benzene ring, is only slightly soluble in water.

Test yourself

3 Methylamine, like ammonia, will mix with and dissolve in water whatever proportions of the two are mixed together. Why is this?

4 Ethane (boiling temperature −89 °C) and methylamine (boiling temperature −6 °C) have very similar molar masses, but very different boiling temperatures. Why is this?

5 Look at the boiling temperatures of methylamine, dimethylamine and trimethylamine on the data sheets on the Dynamic Learning Student website. Why do you think the boiling temperature of trimethylamine, $(CH_3)_3N$, is lower than that of dimethylamine?

Figure 13.7▲
Comparing the basic character of methylamine and ammonia.

Amines as bases

Primary amines, like ammonia, can act as bases. The lone pair of electrons on the nitrogen atom of ammonia and amines is a proton (H^+ ion) acceptor.

Reaction with water

Methylamine and other simple amines dissolve readily in water because they can hydrogen bond with it. They also react as bases with the water, removing H^+ ions (protons) to form an alkaline solution containing hydroxide ions.

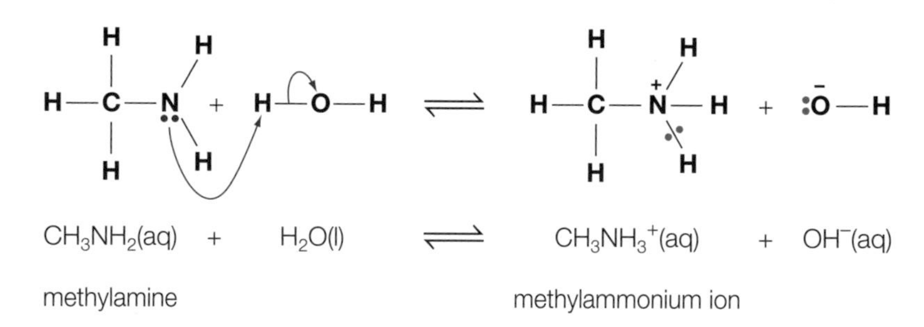

Figure 13.8▲
The reaction of methylamine with water.

The reactions of simple amines, like methylamine, with water are very similar to that of ammonia with water.

$$NH_3(aq) + H_2O(l) \rightleftharpoons NH_4^+(aq) + OH^-(aq)$$

Reaction with acids – formation of salts

Amines react even more readily with acids than they do with water. The lone pair on the nitrogen atom rapidly accepts an H^+ ion (proton) from the acid to form a substituted ammonium salt.

When ethylamine vapour reacts with hydrogen chloride gas, the product is ethylammonium chloride. This forms as a white smoke, which settles as a white solid (Figure 13.9).

$$\underset{\text{ethylamine}}{CH_3CH_2NH_2(g)} + HCl(g) \rightarrow \underset{\text{ethylammonium chloride}}{CH_3CH_2NH_3^+Cl^-(s)}$$

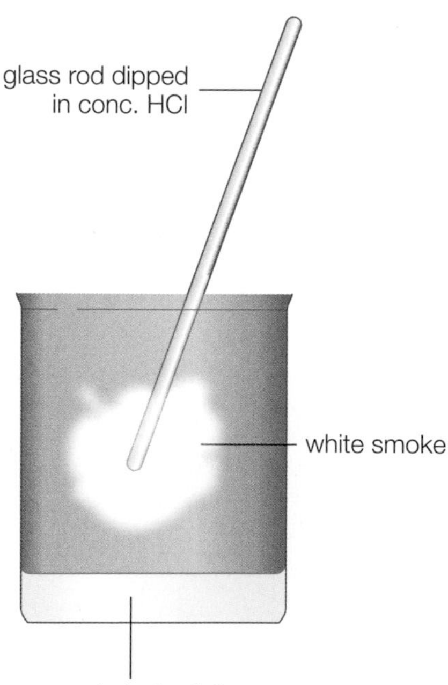

Figure 13.9▲
The vapours from ethylamine solution and concentrated hydrochloric acid react to form a white smoke.

This reaction is very similar to that of ammonia with hydrogen chloride to form ammonium chloride.

$$NH_3(g) + HCl(g) \rightarrow NH_4^+Cl^-(s)$$

Phenylamine, $C_6H_5NH_2$, is only slightly soluble in water, but it dissolves in concentrated hydrochloric acid very easily. This is because it reacts with H^+ ions in the acid to form phenylammonium ions which are soluble in the aqueous mixture.

$$C_6H_5NH_2(l) + H^+(aq) \rightarrow C_6H_5NH_3^+(aq)$$

If a strong base, such as sodium hydroxide, is added to the aqueous phenylammonium ions, H^+ ions are removed from the phenylammonium ions and yellow, oily phenylamine reforms.

$$C_6H_5NH_3^+(aq) + OH^-(aq) \rightarrow \underset{\text{phenylamine}}{C_6H_5NH_2(l)} + H_2O(l)$$

Amines as nucleophiles

Reaction with halogenoalkanes

Amines are strong nucleophiles as well as strong bases, just like ammonia. As nucleophiles, their lone pair of electrons is attracted to any positive ion or positive centre in a molecule.

So, amines will react with the δ+ carbon atoms in the C–Hal bond of halogenoalkanes (Figure 13.10).

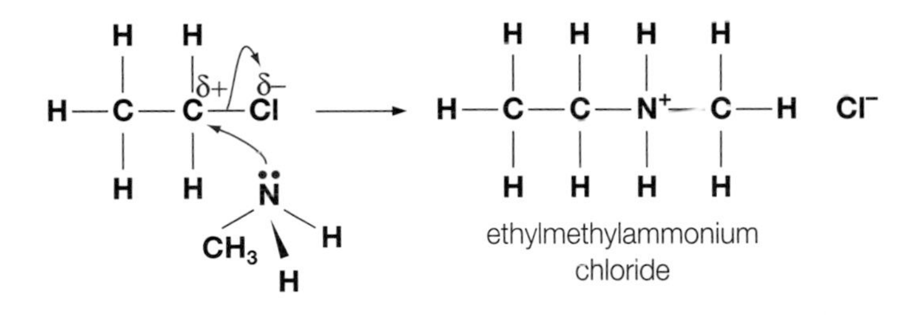

Figure 13.10▲
The reaction of methylamine with chloroethane.

Reaction with acyl chlorides

Amines will also react in a similar fashion with the δ+ carbon atoms in the $-\underset{\underset{O}{\|}}{C}-Cl$ group of acyl chlorides such as ethanoyl chloride (Figure 13.11).

A reaction of this kind involving the ethanoyl group is involved in the manufacture of paracetamol and this is discussed in the following activity.

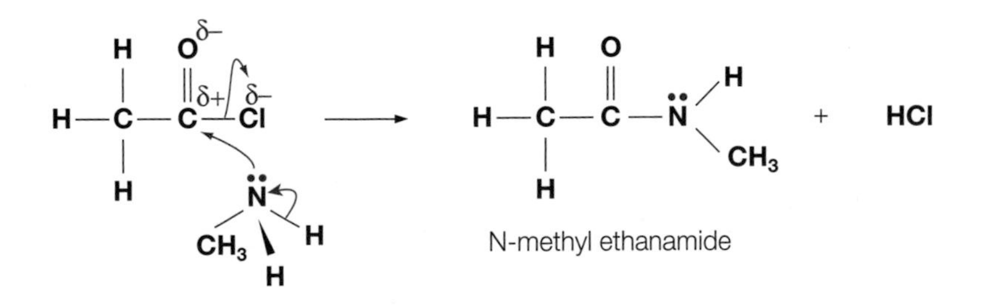

Figure 13.11▲
The reaction of methylamine with ethanoyl chloride.

> **Test yourself**
>
> 6 a) Write an equation to show the formation of a salt when propylamine vapour reacts with hydrogen bromide gas.
>   b) Explain why the reactants are both gases, but the product is a solid.
>
> 7 Write equations for the following reactions and name the products:
>   a) methylamine with concentrated sulfuric acid
>   b) dimethylamine with concentrated sulfuric acid
>   c) chloromethane with ethylamine
>   d) propanoyl chloride with ethylamine.

Activity

Paracetamol – an alternative to aspirin

Aspirin and paracetamol (Figure 13.12) are by far the largest selling pain relievers available without a doctor's prescription. However, aspirin is not without hazards. Every year, about 200 people die of aspirin poisoning due to deliberate or accidental overdose. A large proportion of these deaths are of young children who die accidentally after eating the tablets. Aspirin is also recognised as a cause of internal bleeding and gastric ulcers.

Paracetamol has been produced and marketed as a safer alternative to aspirin. It is a good analgesic (pain reliever) without the harmful side-effects of aspirin.

The compound from which paracetamol is produced and the physiologically active compound that paracetamol produces in the body is 4-aminophenol. Unfortunately, this is toxic, so its harmful effect is reduced by conversion to its ethanoyl derivative.

Paracetamol can be produced by reacting 4-aminophenol with ethanoyl chloride. In industry, however, ethanoic anhydride, $(CH_3CO)_2O$, is used in preference to ethanoyl chloride because it is cheaper and less vigorous in its reactions.

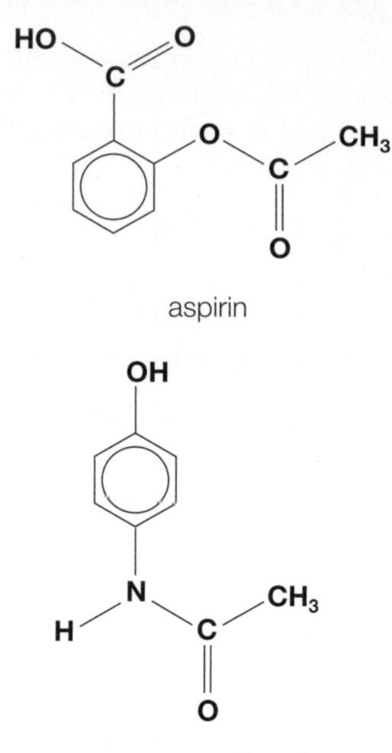

Figure 13.12▲
The structural formulae of aspirin and paracetamol.

1 Draw the structure of 4-aminophenol.

2 a) Write an equation for the reaction of 4-aminophenol with ethanoyl chloride to produce paracetamol.

b) Why does ethanoyl chloride react in this way with 4-aminophenol?

3 When ethanoyl chloride reacts with 4-aminophenol, the –OH group in 4-aminophenol is susceptible to attack as well as the $-NH_2$ group.

a) Why is the –OH group in 4-aminophenol also susceptible to reaction with ethanoyl chloride?

b) Write an equation for the reaction of ethanoyl chloride with the –OH group in 4-aminophenol.

4 Fortunately, the $-NH_2$ group in 4-aminophenol is more reactive than the –OH group. So, in industry the reaction conditions can be carefully chosen so that only the $-NH_2$ group is ethanoylated using ethanoic anhydride. Write an equation for the reaction of 4-aminophenol with ethanoic anhydride to produce paracetamol.

5 Aspirin, like paracetamol, is manufactured by ethanoylation using ethanoic anhydride.

a) What do you understand by the term 'ethanoylation'?

b) Draw the structure of the compound that is ethanoylated to produce aspirin.

6 a) What is the main benefit of aspirin tablets?

b) Summarise the risks posed by aspirin tablets.

c) What are the advantages of paracetamol over aspirin?

13.4 The preparation of amines

Two key reactions are used in the preparation of amines: one for aliphatic amines, the other for aromatic (aryl) amines.

Preparing aliphatic amines

Aliphatic amines can be prepared by heating the corresponding halogenoalkanes in a sealed flask with excess ammonia in ethanol.

Ammonia acts as a nucleophile during the reaction. The initial product is a salt of the amine. The free amine can be liberated from its salt by adding dilute sodium hydroxide solution (Figure13.13).

Figure 13.13▼
The preparation of butylamine from 1-bromobutane and excess concentrated ammonia in ethanol.

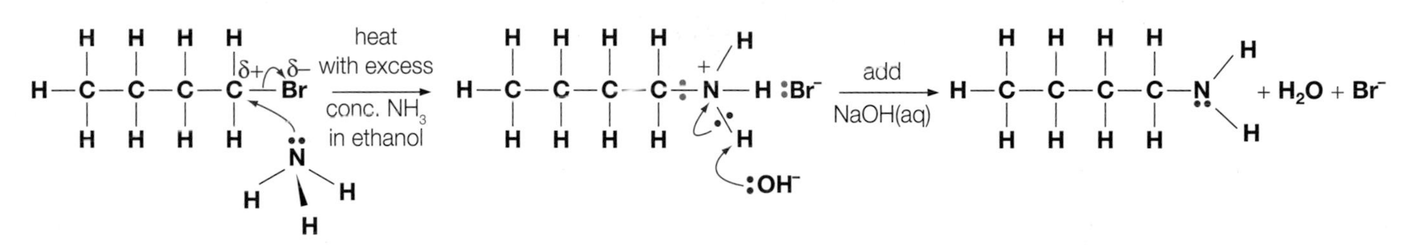

Preparing aromatic amines

The usual laboratory method for introducing an amine group into an aromatic compound is a two-step process – first nitration to make a nitro compound and then reduction (Figure 13.14). The reduction of the aromatic nitro-compound is achieved by boiling under reflux with tin and concentrated hydrochloric acid (Figure 13.15). The aromatic amine dissolves in excess concentrated hydrochloric acid, forming a salt. The free amine can be liberated from the solution by adding sodium hydroxide solution. It is then separated from the mixture by steam distillation.

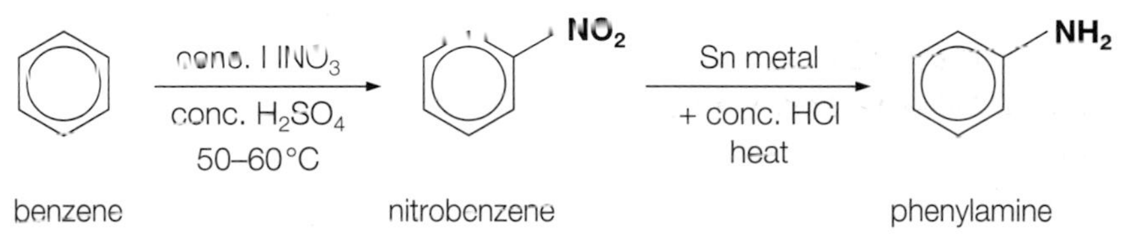

Figure 13.14▲
The two-step preparation of phenylamine from benzene.

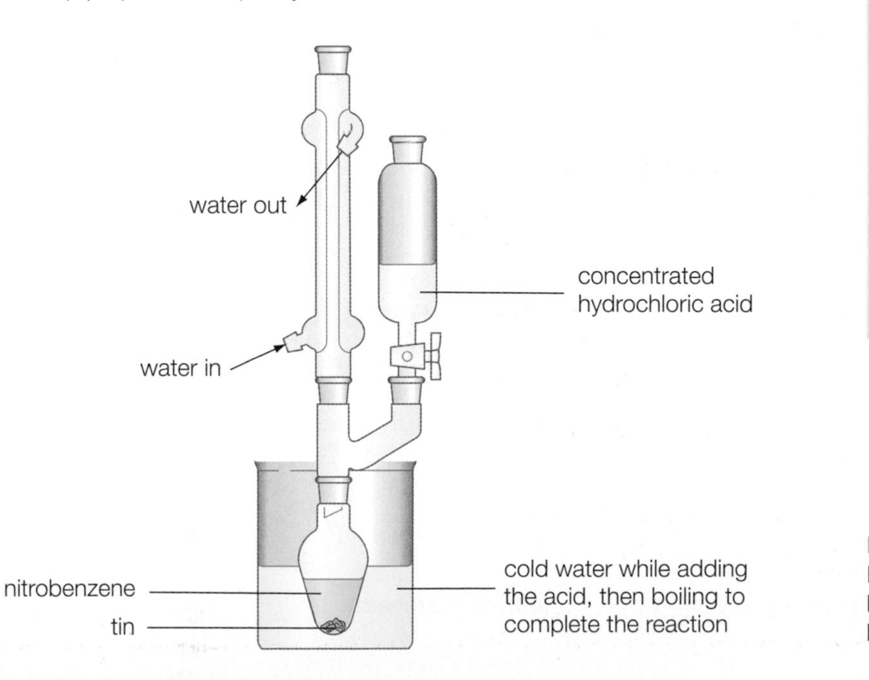

Figure 13.15◀
Reducing nitrobenzene to phenylamine by refluxing with tin and hot concentrated hydrochloric acid.

Test yourself

8 Describe the mechanism, using appropriate curly arrows, for the nucleophilic substitution of 1-chloropropane with ammonia.

9 When preparing primary alkylamines from halogenoalkanes:

a) excess ammonia is used. Why is this? (*Hint:* How would the alkylamine which forms react with any excess halogenoalkane?)

b) a solution of ammonia in ethanol is used rather than ammonia in water. Why is this? (*Hint:* ammonia reacts with water to produce an alkaline solution containing OH^- ions.)

13.5 Diazonium salts

Alkyl and aromatic amines react with nitrous acid (nitric(III) acid), HNO_2, below 10 °C to produce diazonium salts containing the diazo group, $-N^+\equiv N$. The diazonium salts of aromatic amines are important because they are intermediates in the manufacture of azo dyes.

Diazonium salts are unstable. However, the diazonium salts of aromatic amines are stabilised by delocalisation and they are useful reagents if kept cool. Benzenediazonium chloride is prepared by adding a cold solution of sodium nitrite (sodium nitrate(III)), $NaNO_2$, to a solution of phenylamine in concentrated hydrochloric acid at 5 °C.

Initially, sodium nitrite reacts with the hydrochloric acid to form nitrous acid. This then reacts with phenylamine and more hydrochloric acid to produce benzenediazonium chloride.

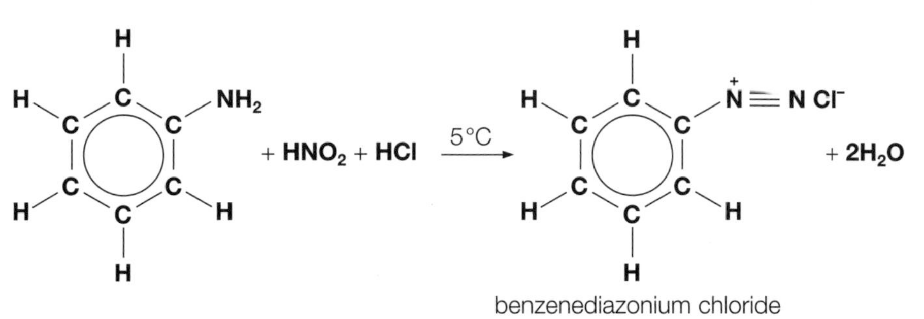

Figure 13.16▲
Preparing benzenediazonium chloride from phenylamine.

The diazonium salts of alkyl amines are much more unstable and decompose immediately to form the corresponding alcohol and nitrogen gas, even at 5 °C.

$$CH_3CH_2{-}NH_2 + HNO_2 + HCl \xrightarrow{5\,°C} CH_3CH_2{-}N^+\equiv N\ Cl^- + 2H_2O$$

then $CH_3CH_2{-}N^+\equiv N\ Cl^- + H_2O \rightarrow CH_3CH_2OH + N_2 + HCl$

Nitrous acid itself is also unstable. It is a weak acid which looks very pale blue in aqueous solution. At room temperature, nitrous acid starts to decompose forming water, nitrogen monoxide and nitrogen dioxide.

$$2HNO_2(aq) \rightarrow H_2O(l) + NO(g) + NO_2(g)$$

The nitrogen dioxide is soluble in water, but the nitrogen monoxide escapes into the air and turns brown as it reacts with oxygen to form nitrogen dioxide.

$$2NO(g) + O_2(g) \rightarrow 2NO_2(g)$$

As the nitrous acid is unstable, it is usually prepared as and when needed by adding hydrochloric acid to sodium nitrite. The nitrous acid is said to be generated *in situ*.

Test yourself

10 a) Write an equation for the formation of nitrous acid, HNO_2, 'in situ' from sodium nitrite, $NaNO_2$, solution and hydrochloric acid.
b) Write an ionic equation for the formation of nitrous acid, excluding spectator ions and showing only the ions involved.
c) The structural formula of nitrous acid is H–O–N=O. Draw a dot-and-cross diagram for nitrous acid showing outer shell electrons only.

11 When nitrous acid decomposes, it disproportionates (gets oxidised and reduced at the same time).
a) Write an equation for the decomposition of nitrous acid.
b) Calculate the oxidation numbers of the different atoms before and after decomposition, and explain why disproportionation has occurred.

12 A solution of benzenediazonium chloride decomposes forming phenol above 10 °C.
a) What are the other products of the decomposition?
b) Write a balanced equation with state symbols for the decomposition.

13 What kind of reactants would you expect diazonium ions with the diazo group, $-N^+\equiv N$, to be?

13.6 Coupling reactions to form azo dyes

The positive charge on the $-N^+\equiv N$ group means that diazonium ions are strong electrophiles. So, we would expect them to attack aromatic compounds with delocalised π electrons, particularly those that have an electron-donating group, such as –OH in phenols (Figure 13.17) and $-NH_2$ in aromatic amines.

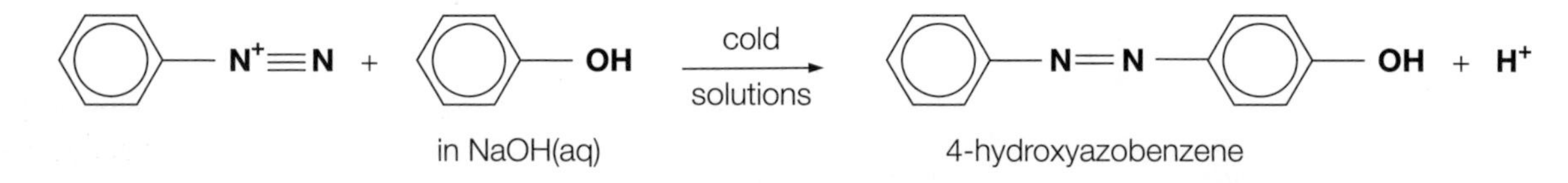

Figure 13.17▼
The reaction between benzenediazonium ions and phenol.

Reactions like this between diazonium ions and phenols or aromatic amines are called coupling reactions. If a cold solution of benzenediazonium chloride is added to a cold solution of phenol in sodium hydroxide, an orange precipitate forms. The precipitate is 4-hydroxyazobenzene which is an azo dye.

The commercial importance of diazonium salts is based on their coupling reactions with phenols and aromatic amines to form azo dyes. Most of these azo dyes are red, orange or yellow. As we have already seen, benzenediazonium chloride reacts with phenol to give an orange dye. With phenylamine it produces a yellow dye called 4-aminoazobenzene (Figure 13.18).

Unlike diazonium compounds, azo compounds are very stable and unreactive.

Figure 13.18▼
The reaction of benzenediazonium chloride with phenylamine to form the yellow dye, 4-aminoazobenzene.

The bright colours of azo compounds result from the extended delocalised electron systems that spread across the whole molecule through the azo group, –N=N–. These delocalised azo systems absorb light in the blue region of the spectrum, which results in the yellow, orange and red dyes. When azo dyes were first discovered in the late nineteenth century, they heralded a new regime for the dyeing of different fabrics. Vegetable dyes, which had been used in the past, faded easily. Azo dyes fade much more slowly as a result of atmospheric oxidation and are not removed by water, soap and other cleaning agents, because they attach themselves more firmly to fabrics.

Figure 13.19▲
This shirt is dyed with azo dyes.

Test yourself

14 a) Write an equation for the coupling reaction between benzenediazonium chloride and naphthalen-2-ol (Figure13.20) to form an azo dye.

b) Why is the reaction usually carried out with the naphthalen-2-ol dissolved in sodium hydroxide solution?

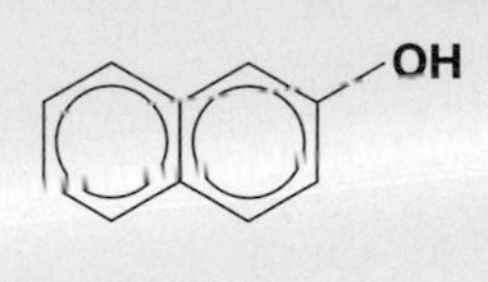

Figure 13.20▲

Azo dyes are also used in some foods (Figure 13.21) in spite of their toxicity. Fortunately, azo dyes are so strongly coloured that the quantities used amount to only milligrams per kilogram of food. Even so, some azo dyes have been banned from use in food. The toxicity arises not from the azo dyes themselves, but when they are metabolised and broken down in the body. Their breakdown produces aromatic amines, some of which are carcinogenic.

Figure 13.21▲
Azo dyes are used to colour foods such as sweets and fruit drinks.

13.7 The preparation of amides

Amides form rapidly at room temperature when acyl chlorides, such as ethanoyl chloride, react with ammonia or with amines. For example, when ethanoyl chloride is carefully added to a concentrated aqueous solution of ammonia, a vigorous reaction takes place producing fumes of hydrogen chloride and ammonium chloride plus a residue of ethanamide.

$$CH_3COCl(l) + NH_3(aq) \rightarrow \underset{\text{ethanamide}}{CH_3CONH_2(s)} + HCl(g)$$

$$HCl(g) + NH_3(g) \rightarrow \underset{\text{ammonium chloride}}{NH_4Cl(s)}$$

The preparation of paracetamol discussed in the Activity in Section 13.3 involves the synthesis of a substituted amide.

REVIEW QUESTIONS

Extension questions

1 Diazonium compounds are important intermediates in the manufacture of synthetic azo dyes.

a) What reaction conditions and reagents are used to make an aqueous solution of the diazonium ion, $C_6H_5–N^+≡N\ Cl^-$, from phenylamine? (3)

b) Explain the following, including appropriate equations in your answer.

i) Solid benzenediazonium chloride, $C_6H_5–N^+≡NCl^-$, is not usually isolated because it is explosive. (3)

ii) Stable solutions of diazonium ions cannot be obtained from aliphatic primary amines like propylamine. (3)

c) The benzenediazonium ion reacts with phenol in alkaline solution to form the azo dye, 4-hydroxyazobenzene (Figure 13.22).

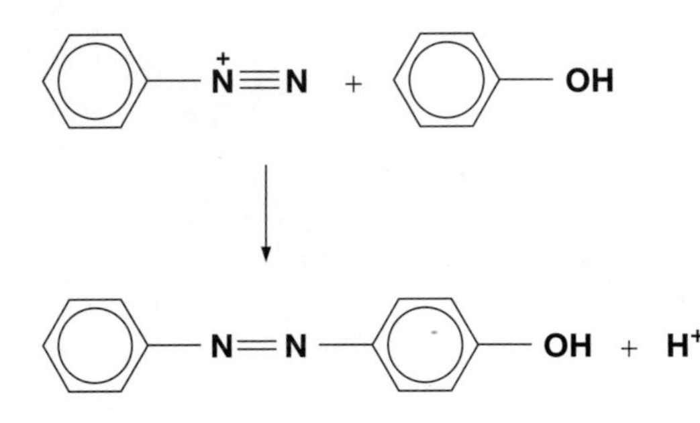

Figure 13.22▲

i) What type of reagent is the diazonium ion in this reaction? (1)

ii) Suggest two reasons why the reaction is carried out in alkaline solution. (2)

2 Describe the main reactions of amines, pointing out the similarities and differences between aliphatic amines, such as butylamine, and aromatic amines, such as phenylamine.

Write balanced equations where appropriate. (14)

3 Figure 13.23 shows a series of reactions beginning with the amine cadaverine. Cadaverine is formed when proteins decompose.

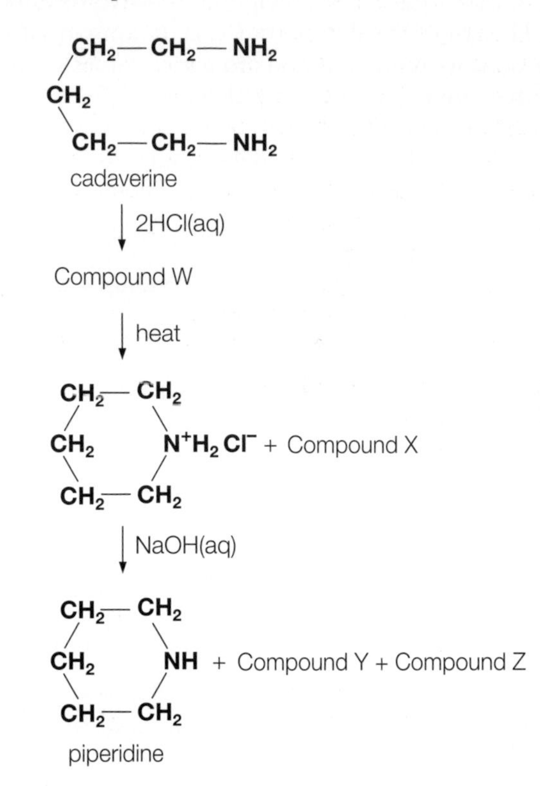

Figure 13.23▲

a) i) What characteristic physical property of cadaverine would you expect to notice if you were provided with a sample of it? (1)

ii) What is the systematic name of cadaverine? (1)

iii) Draw the structural formula of compound W. (1)

iv) Write the name and formula of compound X. (2)

v) Write the formulae of compounds Y and Z. (2)

b) Amines are classed as primary, secondary and tertiary.

i) Explain the difference in structure between the three types of amine. (3)

ii) Which type(s) do cadaverine and piperidine belong to? (2)

c) How will the infrared spectrum of cadaverine compare with that of piperidine? Explain your answer. (2)

14 Amino acids, proteins and polymers

Amino acids are the monomers that make up proteins – naturally occurring polymers which comprise 15% of the human body. There are many different protein molecules in our bodies, each able to do its own special job (Figure 14.1). Muscles, hair, enzymes and hormones all consist of proteins. Other natural polymers include rubber and carbohydrates such as starch and cellulose.

In the last 60 years, an increasing number of synthetic polymers have been produced and manufactured in ever-increasing amounts. These synthetic polymers include polythene and pvc produced by the addition polymerisation of compounds with carbon–carbon double bonds, and others such as polyesters and polyamides formed by condensation polymerisation.

In many respects, polymers have changed the way we live, but the big disadvantage of most of them is that they are not biodegradable.

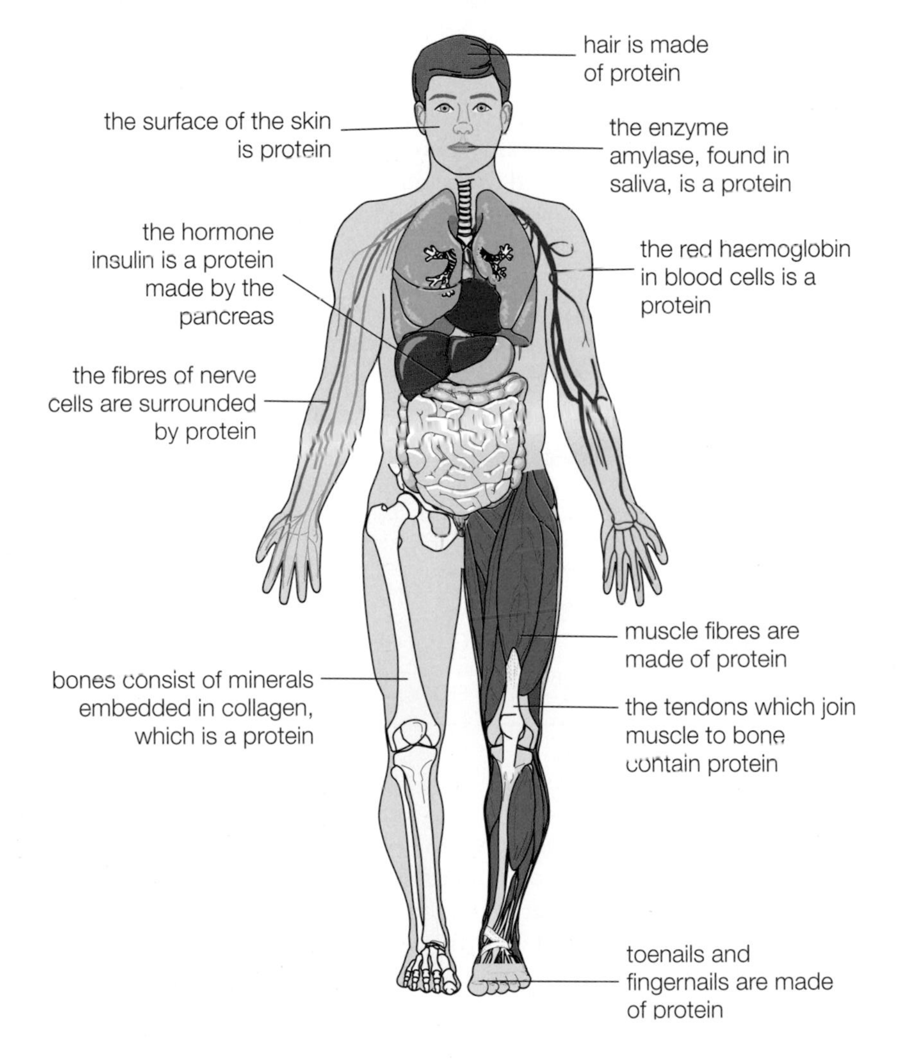

Figure 14.1 ◀ Proteins in the human body.

14.1 Amino acids

Amino acids are the compounds that join together in long chains to make proteins. They are compounds with two functional groups – the amino group, $-NH_2$, and the carboxylic acid group, $-COOH$. About 20 different amino acids are found widely in naturally occurring proteins. Some proteins contain thousands of amino acid units.

Names and formulae

The names and formulae of six naturally occurring amino acids are shown in Figure 14.2. The simplest amino acid is glycine, $H_2N–CH_2–COOH$.

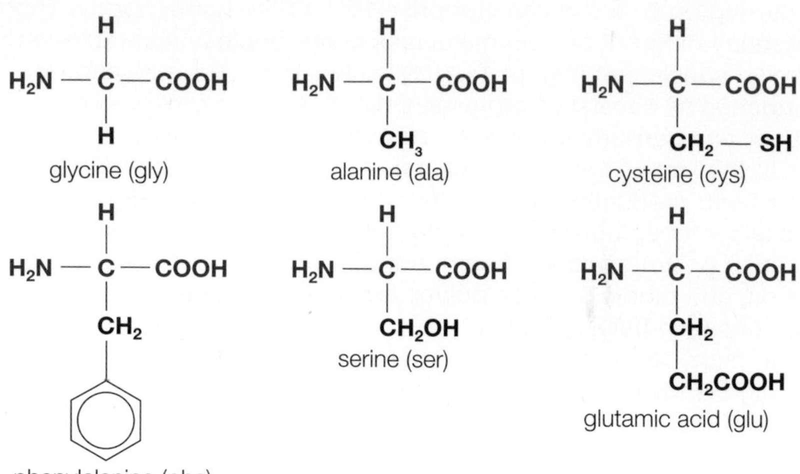

Figure 14.2▶ Six of the amino acids that occur in proteins.

Notice in Figure 14.2 that all six formulae have the amino group attached to the carbon atom next to the carboxylic acid group. This is the case with all the amino acids that occur naturally. The carbon atom next to the carboxylic acid group is sometimes described as the alpha (α) carbon atom or the 2-carbon atom in systematic names. So, all the amino acids in proteins are α-amino acids (2-amino acids) and their general formula can be written as $RCH(NH_2)COOH$.

R stands for the sidegroups in different amino acids (Table 14.1). The common names and R sidegroups of several other amino acids are shown on a data sheet on the Dynamic Learning Student website.

DL www Data

Common name	Abbreviated name	R sidegroup
Glycine	gly	H–
Alanine	ala	CH_3–
Cysteine	cys	HS–CH_2–
Phenylalanine	phe	C_6H_5–CH_2–
Aspartic acid	asp	HOOC–CH_2–

Table 14.1▲ The R sidegroups in some amino acids.

Many of the natural amino acids have complex structures, so it is simpler and more convenient to use their common names rather than systematic names. These common names are sometimes abbreviated to a 'three-letter code' which is usually the first three letters in the name. So, H_2NCH_2COOH is normally called 'glycine' rather than 2-aminoethanoic acid and its abbreviated name is 'gly'.

Amino acid structures

All the amino acids that occur in proteins, except glycine, have a central carbon atom attached to four different groups. This is shown clearly in Figure 14.2. So, except for glycine, all these amino acids have chiral molecules which can exist as mirror images (Section 6.2). The mirror image forms of the amino acid alanine are shown in Figure 14.3.

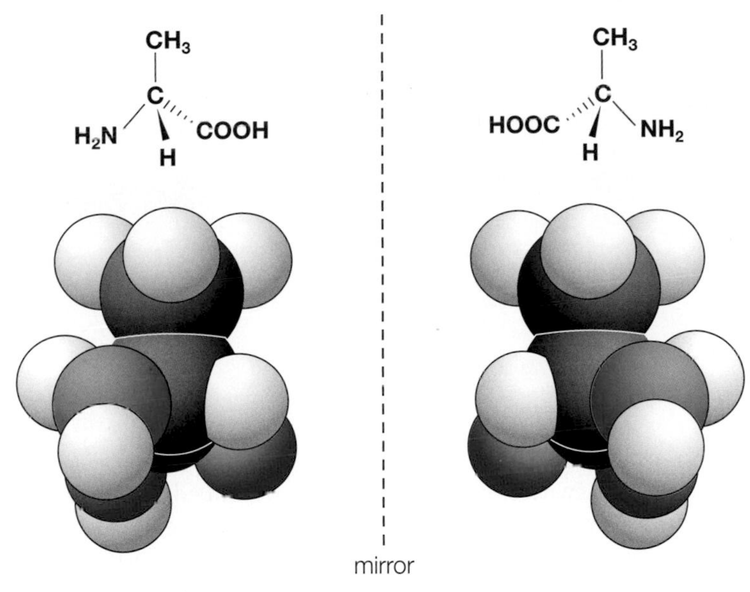

Figure 14.3▲
The mirror image forms of the amino acid alanine. The mirror images are chiral and cannot be superimposed.

As a result of their chirality, the separate (+) and (–) isomers of all naturally occurring amino acids, except glycine, can rotate the plane of plane-polarised light.

Test yourself

1. What is the systematic name for:
   a) alanine
   b) phenylalanine
   c) serine?
2. A dipeptide contains two amino acids linked together. How many different dipeptides can be formed from the naturally occurring amino acids? (Assume there are 20 different naturally occurring amino acids.)
3. Explain why the amino acid glycine is not chiral.
4. How could you distinguish between samples of the two mirror image forms of an amino acid by experiment?

14.2 The acid–base properties of amino acids

As amino acids carry an amino group, $-NH_2$, and a carboxylic acid group, –COOH, they show both the basic properties of primary amines and the acidic properties of carboxylic acids.

The formation of zwitterions

In aqueous solution, carboxylic acid groups ionise producing hydrogen ions, $H^+(aq)$, whereas amino groups are basic and attract hydrogen ions. As a result of this, amino acids form ions in aqueous solution because the amino groups accept hydrogen ions (protons) and the acid groups give them away (Figure 14.4). The ions formed are, however, unusual in that they have both positive and negative charges. Chemists call them zwitterions, from a German word 'zwei' meaning two.

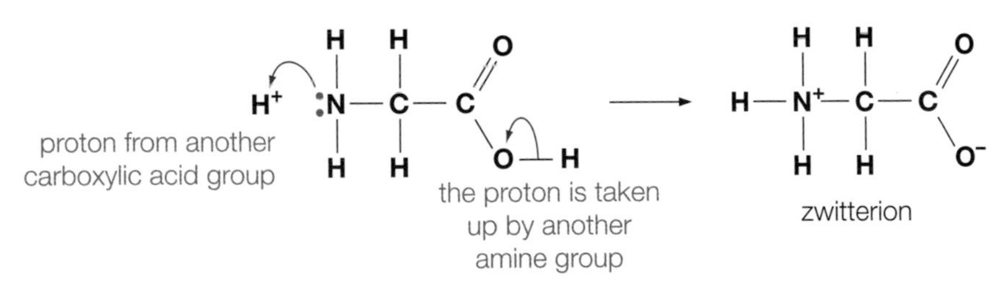

Figure 14.4◀
Glycine forming a zwitterion.

An amino acid will only form zwitterions at a particular pH. If the pH is too high, the solution is too alkaline and in these conditions OH^- ions remove H^+ ions from the zwitterions, forming negative ions (Figure 14.5). On the other hand, if the pH is too low, the solution is too acidic. In this case, H^+ ions react with the zwitterions, producing positive ions (Figure 14.5). Amino acids can therefore exist in three forms depending on the pH – a cation form, a zwitterion and an anion form. However, at one particular pH, molecules of the amino acid will be in the zwitterion form and this pH value is called the isoelectric point.

Definitions

A **zwitterion** is an ion with both a positive and a negative charge.

The **isoelectric point** of an amino acid is the pH value at which it exists as a zwitterion.

Notice from Figure 14.5 that the net charge on an amino acid molecule will vary with the pH. The net charge will be positive in acid solutions and negative in alkaline solutions. At the isoelectric point, the positive and negative charges balance and the net charge on the zwitterion is zero.

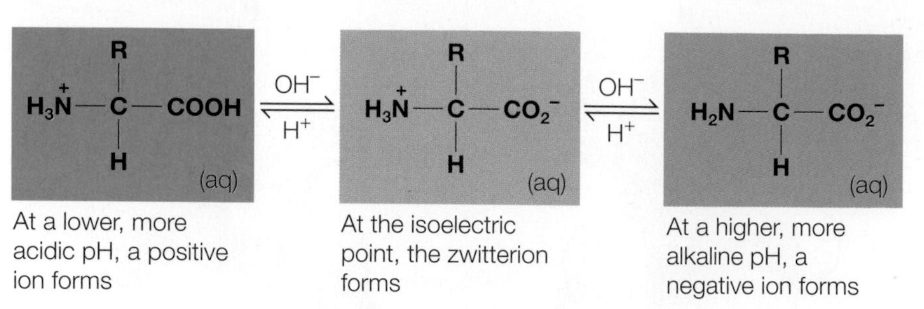

Figure 14.5▶ The ions formed by an amino acid at different pH values.

All amino acids form zwitterions along the lines described above, but their isoelectric points may differ because of the different character of their R groups. In fact, some amino acids, like glutamic acid and aspartic acid, have two –COOH groups and others have two $–NH_2$ groups which influences their isoelectric point significantly.

The movement of H^+ ions from the –COOH group of an amino acid to its $–NH_2$ group will occur in solution before the solid amino acid crystallises out. This means that amino acids also exist as zwitterions in the solid state. This ionic character of amino acids accounts for their high solubility in water and their high melting temperatures.

Test yourself

5 The relative molecular masses of butylamine, $CH_3(CH_2)_3NH_2$, propanoic acid, CH_3CH_2COOH, and glycine, H_2NCH_2COOH are very similar. But glycine (melting temperature 262 °C) is a solid at room temperature, whereas butylamine (melting temperature –49 °C) and propanoic acid (melting temperature –21 °C) are liquids. Why is this?

6 a) Write equations to show the reactions of alanine with:
   i) dilute hydrochloric acid
   ii) aqueous sodium hydroxide.
  b) How do the products from alanine of these two reactions differ from the zwitterions of alanine at its isoelectric point?

7 Why do zwitterions of amino acids exist just as readily in the solid state as they do in aqueous solution?

14.3 From amino acids to peptides and proteins

Peptides are compounds made by linking amino acids together in chains. The simplest example is a dipeptide with just two amino acids linked together by a peptide bond. Figure 14.6 shows the formation of a peptide bond between alanine and glycine to form the dipeptide, 'ala–gly'.

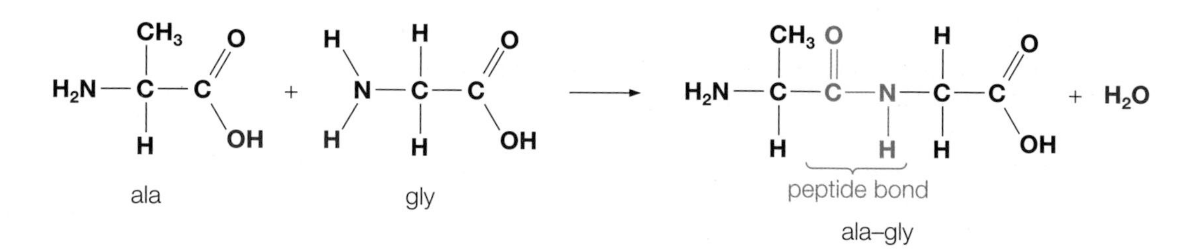

Figure 14.6▲ The formation of a peptide bond between two amino acids.

For chemists, the peptide bond, $-\underset{\underset{O}{\|}}{C}-\underset{\underset{H}{|}}{N}-$, is simply an example of the amide bond (Section 13.2). However, the tradition in biochemistry is to call it a 'peptide bond'.

Notice in Figure 14.6 that when a peptide bond forms between two amino acid molecules, a molecule of water is eliminated at the same time. This is an example of a condensation reaction.

Further condensation reactions can occur between the dipeptide and other amino acid molecules to produce polypeptides and eventually proteins. This is what happens when proteins are synthesised from amino acids in our bodies. The overall process is an example of condensation polymerisation (Section 14.7).

Polypeptides are long-chain peptides. There is no clear dividing line between peptides and polypeptides or between polypeptides and proteins. Some chemists do, however, make a distinction between polypeptides and the longer amino acid chains in proteins. They restrict the definition of polypeptides to chains with 10 to 50 or so amino acids.

Definitions

A **condensation reaction** is one in which molecules join together by splitting off a small molecule, such as water or hydrogen chloride.

Condensation polymerisation involves a series of condensation reactions between the functional groups of monomers to produce a polymer.

The hydrolysis of peptides and proteins

Digestive enzymes in the stomach and small intestine catalyse the hydrolysis of peptide bonds, splitting proteins into polypeptides and then polypeptides into amino acids. Chemists can achieve the same result and hydrolyse the peptide bond by treating proteins and peptides with suitable enzymes, or by heating in acidic or alkaline solution. Heating alone will start to hydrolyse some of the peptide links.

When proteins and peptides are hydrolysed by refluxing with concentrated hydrochloric acid, the product contains the cation forms of the α-amino acids. These are converted to the α-amino acids on dilution with water (Figure14.7).

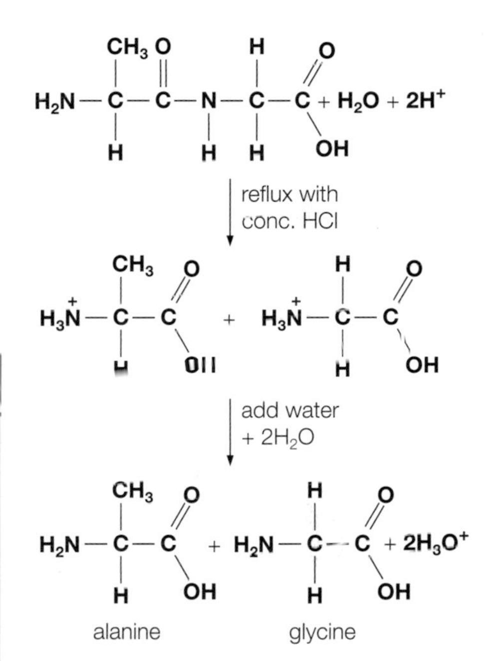

Figure 14.7▲ Hydrolysing a peptide with acid to produce α-amino acids.

Test yourself

8 Draw the structures of the two dipeptides that can be produced from serine and phenylalanine

9 Show that splitting a dipeptide into two amino acids is an example of hydrolysis.

10 a) Identify the functional groups in the sweetener, aspartame (Figure14.8).

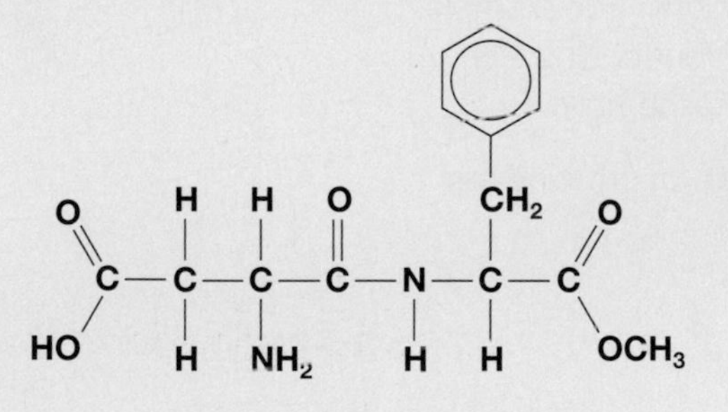

Figure 14.8▲ The structure of aspartame (Nutrasweet)

b) How does aspartame differ from a dipeptide?

c) Suggest a reason why aspartame cannot be used to sweeten food that will be cooked.

d) Why do you think that soft drinks sweetened with aspartame carry a warning for people with the genetic disorder that means that they must not eat phenylalanine?

Data

Tutorial

Activity

The structure of proteins

A protein molecule consists of one or more polypeptide chains. Chemists describe the structure of proteins at four different levels.

- The **primary structure** of a protein is the sequence of amino acids in the polypeptide chain or chains.
- The **secondary structure** describes the repeating patterns in the structure of sections of the polypeptide chains. X-ray diffraction methods have shown that helices and pleated sheets are common repeating structures in proteins. Fibrous proteins such as α-keratin in hair and wool have helical chains of amino acids held together by hydrogen bonds (Figure 14.9).

Stringy α-keratin in silk fibres forms pleated sheets of parallel polypeptide chains held together side-by-side with hydrogen bonds.

- The **tertiary structure** describes the overall three-dimensional folding and shape of a protein. This is held together by hydrogen bonds and other weak interactions between the R groups. Proteins tend to fall into two groups in terms of their tertiary structure:
  - fibrous proteins; long molecules forming fibres of structural material such as α-keratin in hair and collagen in muscle fibres
  - globular proteins; compact, well folded molecules such as enzymes and protein hormones.
- The **quaternary structure** describes the linking between chains in proteins with two or more polypeptide chains. For example, haemoglobin molecules in blood consist of four chains fitting together tightly to form a compact globular assembly.

These four levels of description for the structure of proteins are illustrated in Figure 14.10.

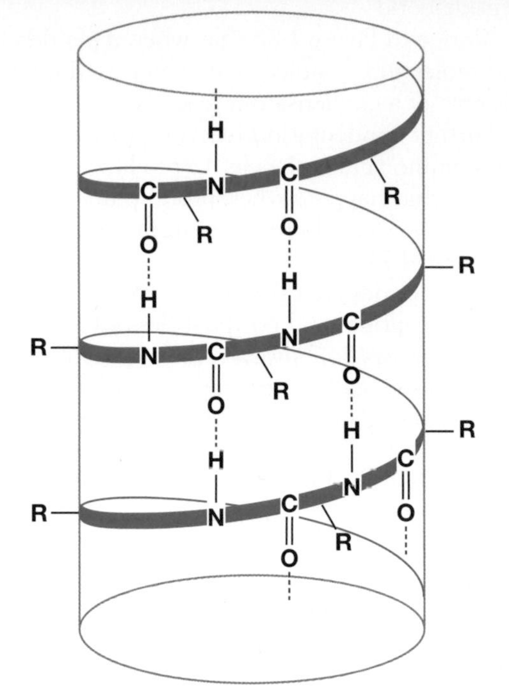

Figure 14.9▲
The alpha helix is common in the secondary structure of many proteins.

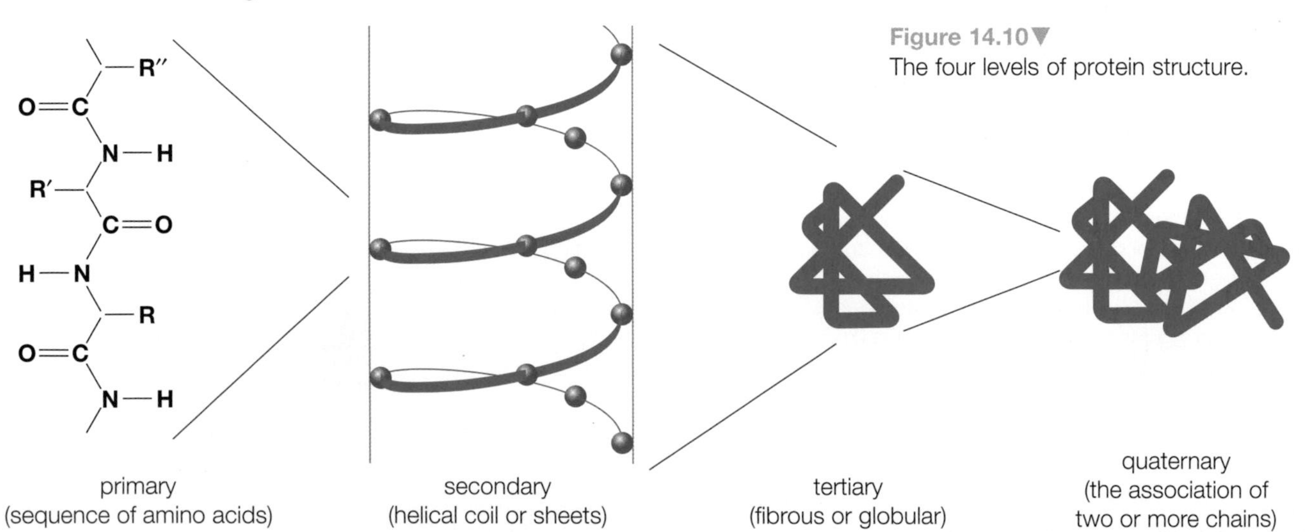

Figure 14.10▼
The four levels of protein structure.

1 The structure in Figure 14.11 shows a short section of a protein molecule.

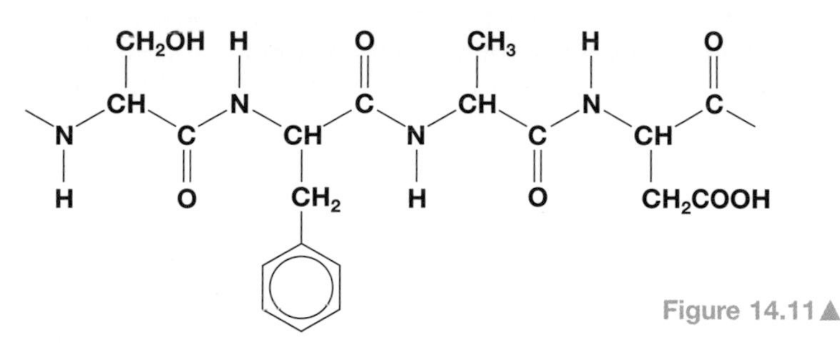

Figure 14.11▲

a) Identify the amino acids that are joined to make this section of a protein.

b) Which of these amino acids have side chains that:

i) are non-polar **ii)** are polar **iii)** can ionise?

www Data

2 Describe the way in which hydrogen bonds hold the helices together in fibrous proteins such as α-keratin.

3 Haemoglobin, the oxygen carrier in blood, is a protein with a relative molecular mass of 66 000.

a) Write one word to describe the tertiary structure of haemoglobin.

b) Assuming that the average mass of an amino acid unit in proteins is equal to that of an aspartic acid unit, calculate the approximate number of amino acid molecules that are needed to produce one molecule of haemoglobin.

4 How does hydrogen bonding explain:

a) the solubility of many proteins in water

b) the precise three-dimensional structure of those enzymes which are proteins

c) the elasticity of natural protein fibres such as wool and silk?

5 Biochemists talk about enzymes being 'denatured' by strong acids, strong bases or by a rise in temperature.

a) What do you think the term 'denatured' means?

b) Suggest a reason for the loss of catalytic activity when an enzyme is denatured by acids, bases or a rise in temperature.

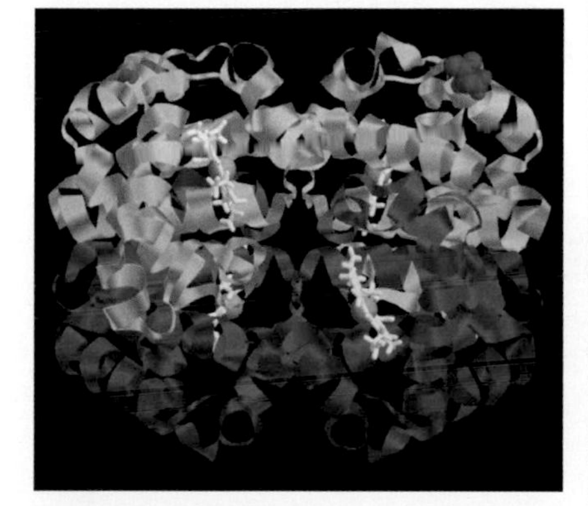

Figure 14.12▲
A computer graphic of a haemoglobin molecule showing the quaternary structure. Each one of the four protein chains (blue and yellow) also carries a haem group (white).

14.4 Investigating the amino acids in peptides and proteins

The amino acids in peptides and proteins can be investigated by hydrolysing the peptides and proteins with concentrated hydrochloric acid, then separating and identifying the amino acids produced using either thin-layer or paper chromatography.

Separating amino acids by chromatography

Thin-layer chromatography (TLC) is quick, cheap and requires only a very small sample for analysis. The technique is widely used both in research laboratories and in industry. In this type of chromatography, the stationary phase is a thin layer of solid supported on a glass or plastic plate.

Definition

Chromatography involves the separation and identification of compounds using a stationary phase (solid) and a mobile phase (liquid or gas). The components of a mixture separate as the mobile phase moves through the stationary phase.

Components that tend to mix with or dissolve in the mobile phase move faster. Components that tend to adsorb on the stationary phase move slower.

In paper chromatography, the stationary phase is absorbent paper like high-quality filter paper. In each of these forms of chromatography, the mobile phase is a liquid. The rate at which a compound moves up the TLC plate or the paper depends on the equilibrium between its adsorption on the solid and its solubility in the liquid solvent. The position of this equilibrium varies from one compound to another, so the components of the mixture separate (Figure 14.13).

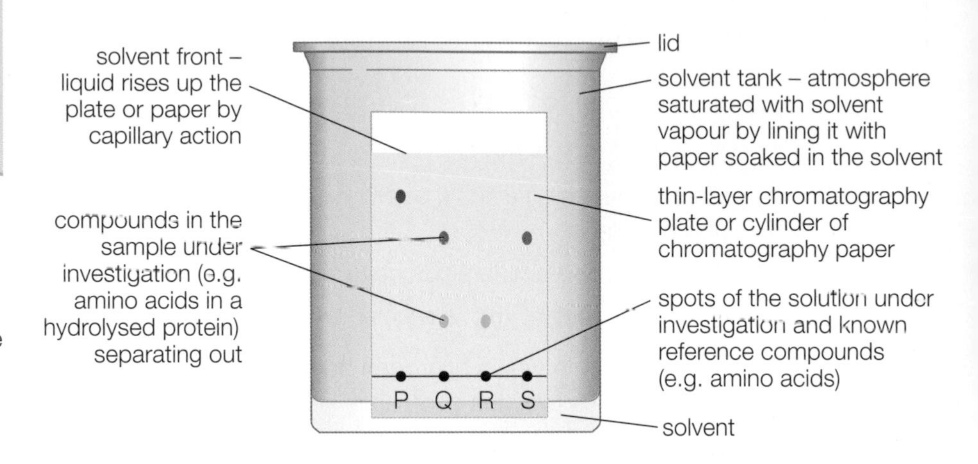

Figure 14.13▶
Thin-layer chromatography or paper chromatography can be used to separate and identify amino acids after a protein has been hydrolysed. The sample under investigation is spotted at Q and known amino acids are spotted at P, R and S.

Identifying amino acids – ninhydrin and R_f values

After chromatography, it is easy to see coloured compounds on the TLC plate or chromatography paper. With colourless compounds, such as amino acids, it is necessary to use a locating agent. Locating agents are sprayed on the plate or paper, where they react with the separated colourless compounds to form coloured products. The locating agent for amino acids is ninhydrin.

When ninhydrin is sprayed on the chromatography plate or paper and then heated in an oven at about 100 °C, it reacts with any amino acids to form purple spots which fade and turn brown with time.

R_f values are used to identify the different separated spots from the sample under investigation with those from the known reference compounds. R_f stands for 'relative to the solvent front'.

$$R_f = \frac{\text{distance moved by a reference compound}}{\text{distance moved by solvent front}}$$

Test yourself

11 What can you conclude about the composition of Q in Figure 14.13?
12 Suggest two factors that will change the R_f value of a particular amino acid.

14.5 Polymer chemistry

Polymer chemistry is the study of the synthesis, structure and properties of polymers. Polymers are long-chain molecules. Natural polymers include proteins, rubber and carbohydrates such as starch and cellulose. Synthetic polymers include polythene and pvc produced by the addition polymerisation of compounds with carbon–carbon double bonds, and others such as polyesters and polyamides formed by condensation polymerisation.

The first synthetic polymers were produced more by good luck than good management. Nowadays, chemists are capable of developing new polymers with specific properties by applying the theories of bonding and structure.

In 1922, the German chemist Hermann Staudinger published his theory about large molecules in which he suggested that substances like rubber and cellulose consisted of long-chain molecules. Staudinger had to fight hard to persuade other chemists to accept his ideas, which are now taken for granted.

One person who was convinced by Staudinger's theory was the American industrial chemist Wallace Carothers. In 1931, Carothers wrote an article introducing the terms addition polymerisation and condensation polymerisation.

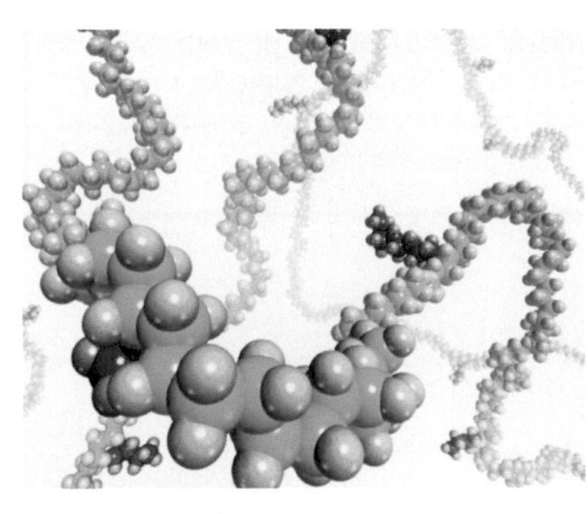

Figure 14.14▲
A false colour computer graphic of low density polythene showing the branches which prevent the polymer chains from packing close together. Carbon atoms are coloured green and mauve. Hydrogen atoms are blue and pink.

Carother's research team at the Du Pont chemical company produced synthetic rubber, neoprene, by addition polymerisation. Then, in 1935, the team synthesised nylon, the first completely synthetic condensation polymer.

The 1930s were probably the most important years in the development of polymers and the plastics industry. During this period, polythene, pvc, polystyrene and Perspex all came on the market. Since then research and development in the second half of the twentieth century has led to the production of many new and specialised polymers, including Teflon (ptfe), the polyamide Kevlar and biodegradable polymers.

Figure 14.15▲
A false colour electron micrograph of Gore-tex. The pink outer layers are nylon. The yellow and white layers consist of Teflon (ptfe). Magnification is ×170. Gore-tex is used to line outdoor wear such as anoraks and hiking boots. It is waterproof, yet it also allows perspiration to evaporate

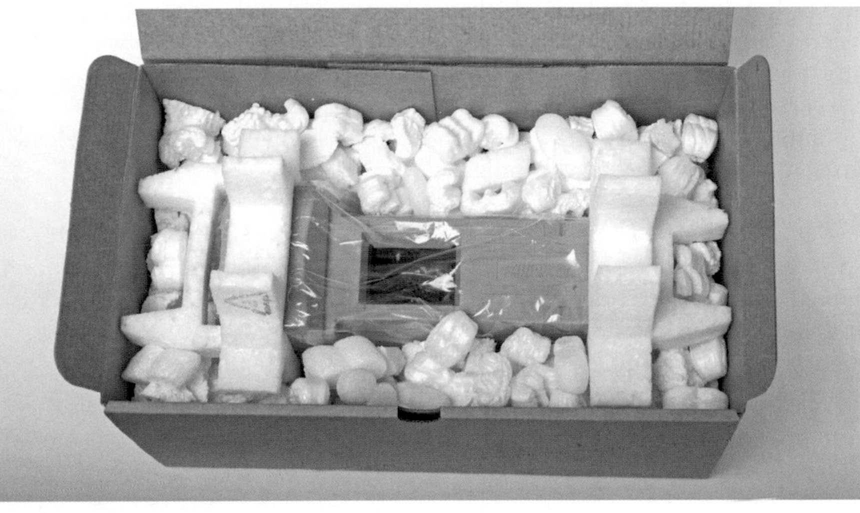

Figure 14.16▲
Expanded polystyrene has low density and is an excellent thermal insulator. It is used for packaging fragile goods because it absorbs shocks. Its correct name is poly(phenylethene).

14.6 Addition polymerisation

Addition polymerisation is a process for making polymers from compounds containing double bonds. The most important addition polymers are formed from compounds with the general formula $CH_2{=}CHX$ in which the nature of X determines the properties of the polymer.

Ethene for example, in which X = H, polymerises to form poly(ethene), commonly called polythene (Figure14.17).

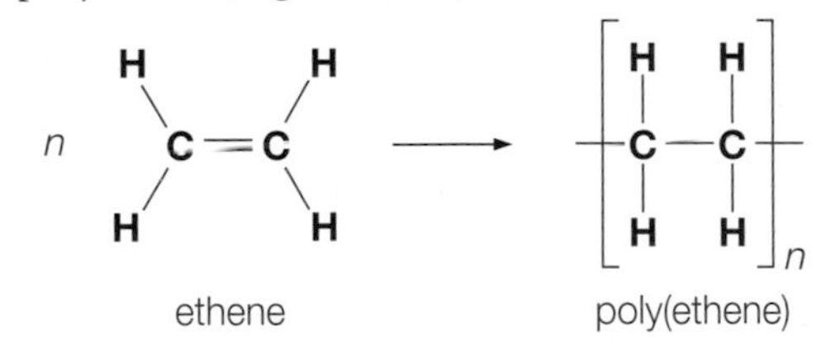

Figure 14.17▲
The formation of poly(ethene) from ethene.

Other widely used addition polymers include poly(propene), poly(phenylethene) – better known as polystyrene, poly(tetrafluoroethene) – often abbreviated to ptfe, and poly(chloroethene) – usually called pvc.

In these addition polymers, the repeat unit in the polymer chain has the general structure

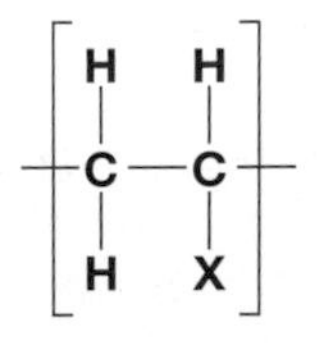

Definitions

Polymerisation is a process in which many small molecules (monomers) join up in long chains by addition or condensation reactions.

In **addition polymerisation**, the polymers form by addition reactions of monomers containing double bonds.

In **condensation polymerisation**, the polymers form by condensation reactions in which a small molecule such as water is split off between the functional groups of monomers (Section 14.7).

Figure 14.18▲
Inside the domes at the Eden Project in Cornwall, scientists have created the varying climatic conditions required by plants growing in different parts of the world. These domes consist of interconnecting steel pentagons and hexagons glazed with etfe. This is an addition polymer made by polymerising a mixture of two monomers, ethene and tetrafluoroethene. The polymer is lightweight and lets through the ideal spectrum of light for plants to photosynthesise.

Another important addition polymer, developed in recent years as a water-soluble plastic, is poly(ethenol) – sometimes called polyvinyl alcohol. 90–98% of poly(ethenol) is composed of the repeat unit

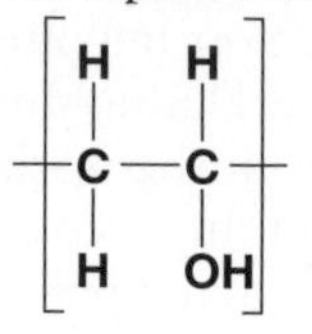

Poly(ethenol) is used to make plastic bags that dissolve in water. These are ideal as hospital laundry bags which can be handled without touching any infected contents. When washing begins, the bags dissolve in the water and the laundry is washed.

Poly(ethenol) is also used to make soluble capsules called 'liquitabs' containing liquid detergent. The soluble capsules slowly dissolve and release detergent as the washing cycle progresses.

Making addition polymers

One technique for making addition polymers uses an initiator and involves a free radical chain reaction at high temperature and pressure. The initiator is often a peroxo compound or an organic peroxide such as benzoyl peroxide. These peroxo compounds and peroxides act as a source of free radicals to initiate addition polymerisation (Figure 14.19).

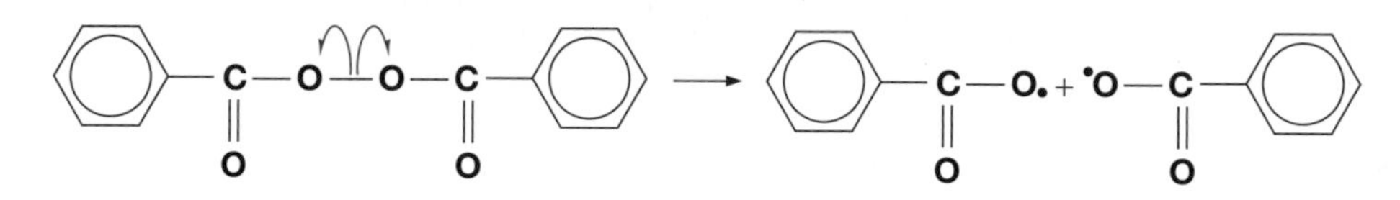

Figure 14.19▲
A molecule of benzoyl peroxide splitting to form two free radicals. The O–O bond is relatively weak.

Using the symbol RO• for a free radical, the addition polymerisation can be followed through the stages of initiation, propagation and termination as in Figure 14.20.

Figure 14.20►
The free radical chain reaction involved in the formation of an addition polymer.

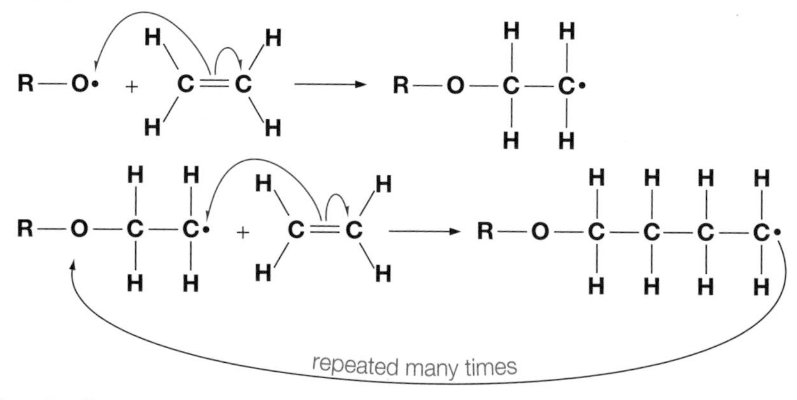

Figure 14.21◄
Extruding poly(ethene) to make plastic sheeting for the building industry. A high-pressure, high-temperature process with a peroxide initiator produces low-density poly(ethene) with branched chains. A low-pressure, low-temperature process with a special catalyst produces high-density poly(ethene) in which the polymer chains have very few branches and pack much closer.

An alternative method, which speeds up the polymerisation of alkenes such as ethene and propene, is to use special catalysts. These catalysts enable the production of addition polymers at relatively low temperatures and pressures.

Using the catalysts, there is very little chain branching in the polymers. So the poly(ethene) chains produced by this method can pack more closely, forming the high-density form of poly(ethene).

Test yourself

13 a) Draw two repeat units of the polymer poly(ethenol).
b) Why does poly(ethenol) dissolve in water, unlike poly(ethane) which is insoluble?

14 Draw two repeat units of the polymer chains formed from each of the monomers in Figure 14.22 and write the systematic IUPAC names of the polymers:

a) 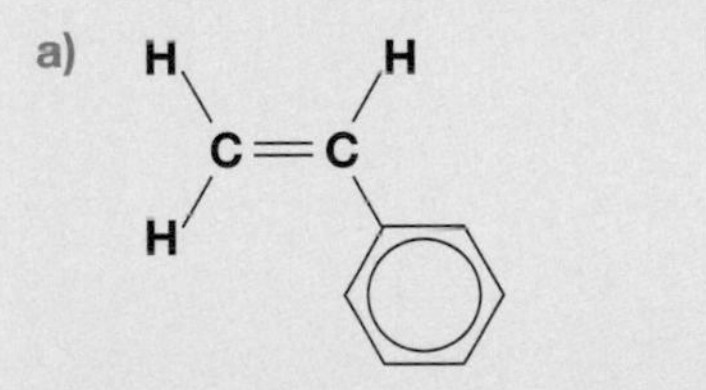

b) 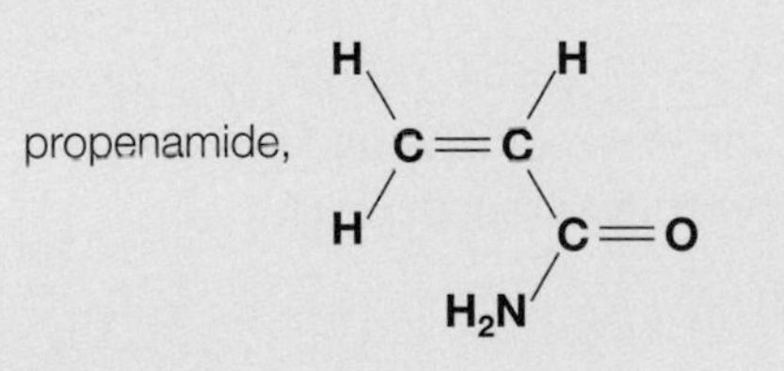

Figure 14.22▲

15 Identify the following polymers and write the names and displayed formulae of their monomers:

a) 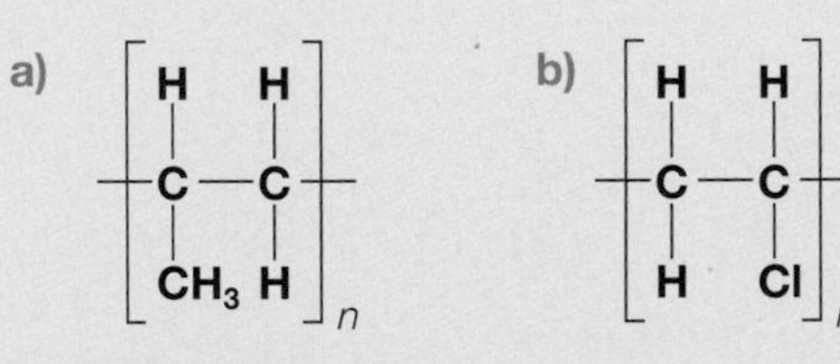

b) $\left[\begin{array}{cc} H & H \\ | & | \\ -C- & C- \\ | & | \\ H & Cl \end{array}\right]_n$

Figure 14.23▲

16 Explain the terms homolytic fission and free radical using the action of benzoyl peroxide as an initiator as your example.

17 The paste supplied with DIY wood fillers often contains phenylethene. The paste is supplied with a small tube of hardener.
a) Why is the hardener supplied in a separate tube and only mixed with the paste shortly before use?
b) Suggest a chemical for the hardener.
c) What factors will determine how fast the wood filler sets?

Activity

Covering the O_2

Figure 14.24▲
The O_2 at Greenwich, London.

The O_2 Dome at Greenwich, sometimes called 'the O_2', provided an important focus for celebrations of the millennium in 2000. The huge structure is a triumph for science and engineering which would have been impossible without the availability of a tough, lightweight, inert and non-flammable material for the roof.

The roof of the O_2 has an area of about 150 000 square metres. It is covered by 144 panels of ptfe-coated fibreglass supported by a network of steel cables suspended from 12 masts that reach to a height of 100 metres. The correct systematic name for ptfe is poly(tetrafluoroethene). It is, however, better known by its tradename, Teflon, and is used to coat non-stick saucepans and skis. Teflon is manufactured by polymerising tetrafluoroethene with a small amount of ammonium peroxodisulfate.

1 a) Draw the displayed formula of tetrafluoroethene and a short section of the ptfe polymer showing two repeat units.

b) Why is the polymer called ptfe?

2 Explain fully why ammonium peroxodisulfate is used in the synthesis of ptfe.

3 Predict the conditions used to synthesise ptfe.

4 Describe the general shape of ptfe molecules and the forces that hold these molecules together.

5 Poly(tetrafluoroethene) is an ideal material to cover the roof of the O_2 because its strong C–F bonds are resistant to chemical attack. What other properties does ptfe have that make it ideal for the O_2?

6 Suggest two advantages that ptfe-coated fibreglass has over ordinary glass as the roof covering of the O_2.

7 Suggest two disadvantages that ptfe-coated fibreglass has compared to ordinary glass as the roof covering of the O_2.

8 The roof of the O_2 is expected to be self-cleaning. Why is this?

14.7 Condensation polymerisation

Condensation polymers are produced by a series of condensation reactions in which small molecules such as water or hydrogen chloride are split off between the functional groups of the monomers.

Condensation reactions are sometimes described as 'addition plus elimination' reactions because the monomers undergo addition, but this occurs only by elimination of a small molecule between each repeating unit.

There are two important classes of condensation polymers: polyesters and polyamides.

Definitions

Polyesters are polymers with ester links between monomer units.

Polyamides are polymers with amide links between monomer units.

Polyesters

Polyesters are polymers formed by condensation polymerisation:

- between acids with two carboxylic acid groups and alcohols with at least two –OH groups
- or between monomers that have both a carboxylic acid group and an –OH group.

The units in the polyester chains are linked by a series of ester bonds.

The most common polyester is Terylene, used widely in fabrics. It is usually referred to simply as 'polyester'. Terylene is made by condensation reactions between benzene-1,4-dicarboxylic acid and ethane-1,2-diol (Figure 14.25). The traditional names for these two compounds are **ter**ephthalic acid and eth**ylene** glycol, hence the commercial name, Terylene. An alternative name for the polymer is poly(ethylene) terephthalate, which gives rise to the name PET when the same polymer is used to make plastic bottles for drinks.

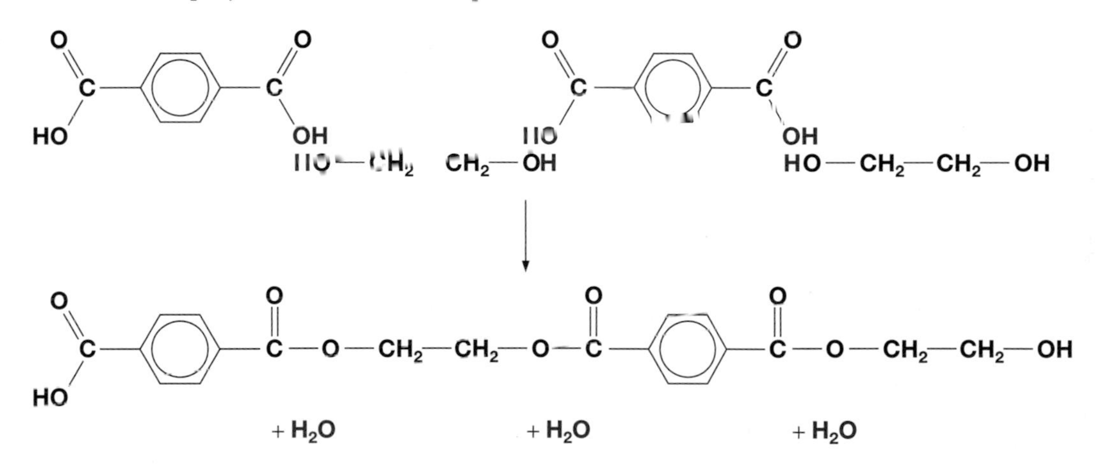

Figure 14.25▲
Condensation polymerisation to produce the polyester Terylene.

The condensation reactions shown in Figure 14.25 can be repeated again and again to produce a polymer with the repeat unit shown in Figure 14.26.

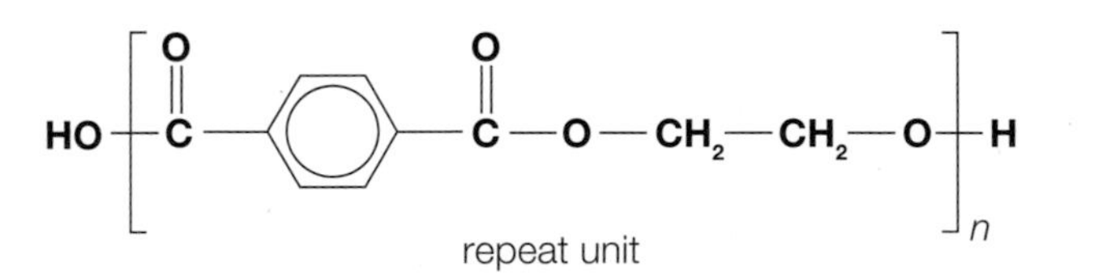

Figure 14.26◀
The repeat unit and structure of Terylene.

Polyesters have high tensile strength and, because of this, they are widely used as fibres in clothing and as the bonding resin in glass-fibre plastics.

Perhaps the most important development in polyester chemistry in recent years concerns poly(2-hydroxypropanoic acid), commonly called poly(lactic acid) or pla. Poly(lactic acid) is possibly the most useful and most versatile of the new biodegradable plastics. It is already used in such diverse goods as plant pots, disposable nappies and absorbable surgical sutures (stitches).

Figure 14.27▲
The blazer, tie, shirt and trousers that this schoolboy is wearing may all contain polyester (Terylene). The fabrics are hard wearing, washable and relatively cheap.

Poly(lactic acid) is manufactured by the condensation polymerisation of lactic acid, a single monomer that contains both a carboxylic acid group, –COOH, and an alcohol group, –OH (Figure 14.28).

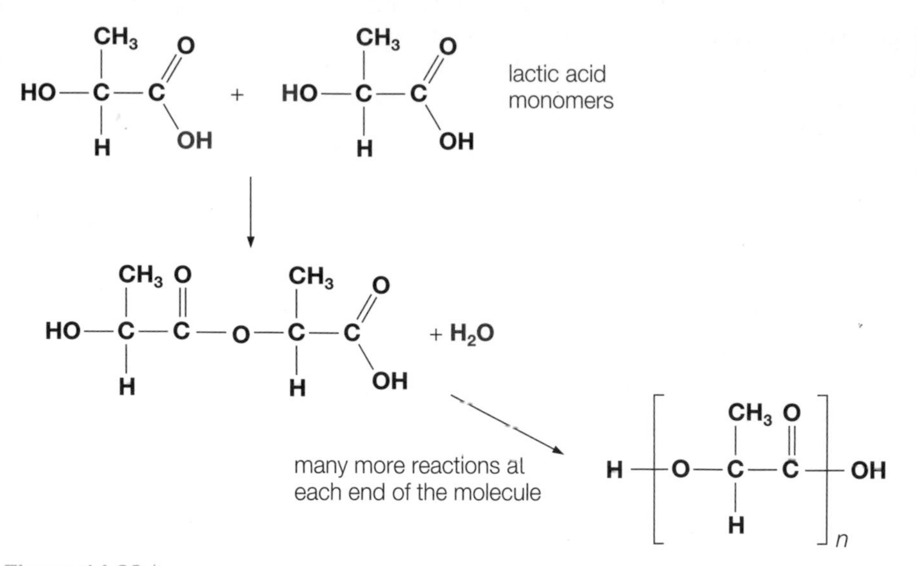

Figure 14.28▲
The synthesis of poly(lactic acid) by condensation polymerisation.

Polyamides

Polyamides are polymers in which the monomers are linked by an amide bond. This is exactly the same as the amide bond in proteins in which it is usually called the peptide bond (Figure14.6). So, proteins and polypeptides are naturally occurring polyamides.

From your studies earlier in this topic, you will know that polypeptides and proteins are synthesised in living things by condensation reactions between amino acids. In these reactions, the amine group, $-NH_2$, of one amino acid reacts with the carboxylic acid group, –COOH, of another amino acid to split out water and form an amide link (Figure14.6). This process is then repeated time after time to produce a polymer (protein) with tens, hundreds or, in some cases, thousands of units.

The first synthetic and commercially important polyamides were various forms of nylon. These were not, however, produced from amino acids. Instead, they were formed by condensation polymerisation between diamines and dicarboxylic acids. One of the commonest forms of nylon is nylon-6,6. This is made by a condensation reaction between 1,6-diaminohexane and hexanedioic acid (Figure 14.29). The product is named nylon-6,6 because both monomers contain six carbon atoms.

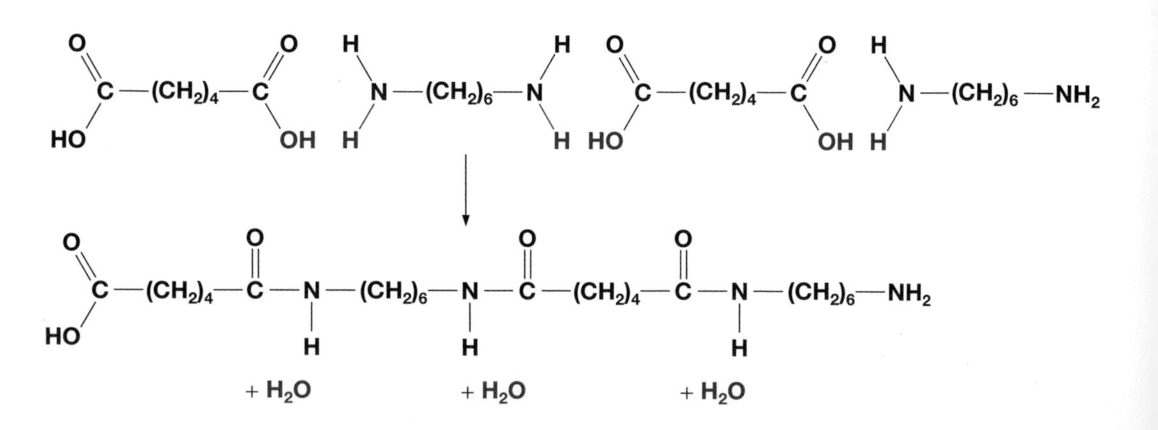

Figure 14.29►
Condensation polymerisation to make nylon-6,6.

Nylon-6,6 can be produced more readily in the laboratory using hexanedioyl dichloride in place of the less reactive hexanedioic acid. Hexanedioyl dichloride reacts readily with 1,6-diaminohexane at room temperature to produce nylon-6,6. In this case, hydrogen chloride molecules are eliminated in the condensation reaction (Figure 14.30).

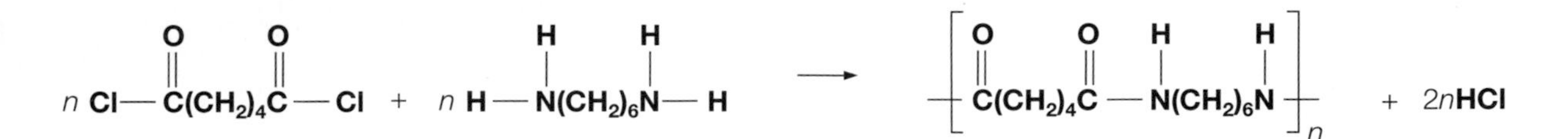

Figure 14.30▲
The reaction used to make nylon-6,6 in the laboratory.

Although nylon is similar in structure to wool and silk, it does not have the softness of the natural fibres. It is, however, much harder wearing and one of its earliest uses was as a substitute for silk in the manufacture of ladies' stockings.

Apart from their obvious use in stockings and tights, nylon fibres are used in various forms of clothing. In fact, about 75% of the UK nylon consumption goes on clothing, but its uses are many and varied. Nylon is used to make nylon ropes that don't rot, machine bearings that don't wear out, and it is mixed with wool to make durable carpets.

Nylon is the collective name for polymers with aliphatic hydrocarbon sections linked by amide bonds. They are aliphatic polyamides in which the polar amide bonds are fixed and inflexible, but the non-polar hydrocarbon sections are free to flex, rotate and twist. So, as the hydrocarbon sections become longer, we would expect the nylon polymers to become more flexible with weaker bonding between the molecules.

This suggests that the properties of polyamides can be modified by changing the length and nature of the hydrocarbon sections. Chemists have followed up these ideas to develop polyamides in which the hydrocarbon sections are aromatic rather than aliphatic. These polymeric **ar**omatic **amid**es are described as aramids. Aramids, such as Kevlar (Figure14.32), are extremely strong, rigid, fire-resistant and lightweight.

Figure 14.31▲
The American firm Du Pont patented nylon in February 1938. The first nylon stockings went on sale in the USA on 15 May 1940. In New York alone, four million pairs were sold in a few hours.

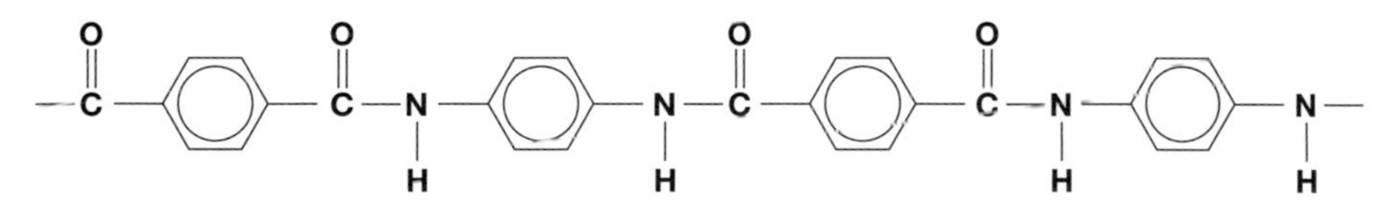

Figure 14.32▲
A section of the polymer chain in Kevlar.

Test yourself

18 a) What type of polymerisation would produce the polymer with a repeat unit like that in Figure 14.33?

b) Draw the structure of the monomer or monomers that would be used to prepare the polymer.

19 The compound in Figure 14.34 can form a polymer.

a) Identify the functional groups involved in forming the polymer.

b) What type of polymerisation will the monomer undergo?

c) What other product forms during polymerisation?

d) Draw a short length of the polymer chain showing two repeat units.

20 State two similarities and two differences between the structure of nylon-6,6 and the structure of a protein.

21 Identify the types of intermolecular forces that act between the polymer chains in:

a) poly(ethene)

b) nylon.

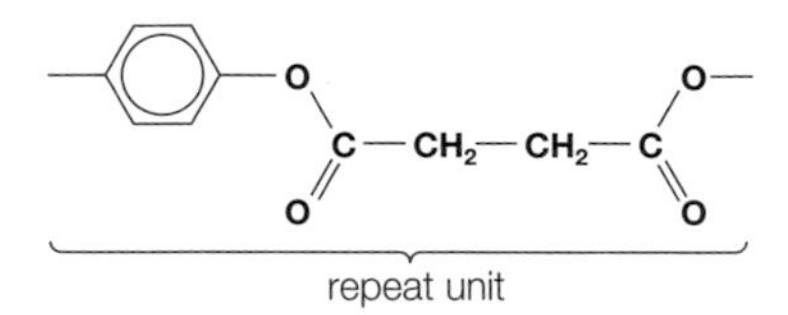

Figure 14.33▲

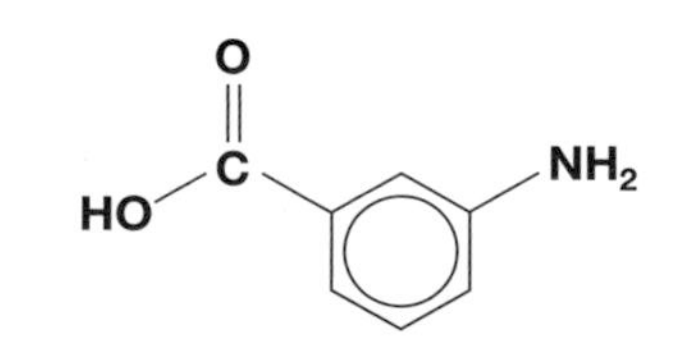

Figure 14.34▲

Activity

Modelling and synthesising polyamides

Experiments show that nylon polymers with longer hydrocarbon sections to their chains are more flexible than those with shorter sections. Kevlar is similar to nylon-6,6 but with benzene rings rather than aliphatic chains linked by the amide group. The repetition of benzene rings in its structure makes Kevlar exceptionally strong and very inflexible compared with nylon-6,6. Because of this, it is used extensively in tyres, brakes and clutch fittings, in ropes and cables and in protective clothing (Figure14.35).

Figure 14.35▲
This policeman is wearing a bulletproof jacket made from Kevlar.

1 Look closely at the structure of one chain of Kevlar in Figure 14.32.

- **a)** Explain how Kevlar is a condensation polymer of benzene-1,4-dicarboxylic acid and benzene-1,4-diamine.
- **b)** Weight for weight, Kevlar is five times stronger than steel. This exceptional strength of Kevlar is due to hydrogen bonding between the separate chains. Use Figure 14.32 to explain why interchain hydrogen bonding is so strong in Kevlar.
- **c)** Suggest a reason why Kevlar is made from monomers with functional groups in the 1,4 positions and not from isomers with functional groups in the 1,2 or 1,3 positions.

2 Using a molecular model kit, make one repeat unit for the structure of Kevlar and explore the flexibility of the structure. (*Hint*: Use the Kekulé structure with alternating double and single bonds for the benzene ring.)

Repeat the model making and flexibility testing with one repeat unit for the structure of nylon-6,6. Why is nylon-6,6 flexible whereas Kevlar is inflexible?

3 A condensation polymer can be prepared by mixing equal amounts of the monomers in Figure 14.36 at room temperature.

Figure 14.36▲

- **a)** Draw the structure of one repeat unit of the polymer formed from the two monomers.
- **b)** The polymer forms even more rapidly if the reaction mixture contains sodium carbonate. Why is this?
- **c)** The polymer molecules obtained at room temperature can be linked to one another (cross-linked) by a second reaction. Explain how this cross-linking can be achieved and state the conditions needed for it to happen.
- **d)** Explain how the choice of reaction conditions can control the extent of polymerisation and the extent of cross-linking.

14.8 Comparing addition and condensation polymerisation

Although both addition and condensation polymerisation result in the formation of long-chain organic molecules, known as polymers, from relatively small organic molecules, known as monomers, there are some clear differences between the two processes.

- The first difference concerns the **type of reaction** involved. As its name suggests, addition polymerisation involves only addition reactions, whereas condensation polymerisation involves addition plus elimination. As monomer units join together, a small molecule, usually water or hydrogen chloride, is eliminated and split off.
- The second difference between addition and condensation polymerisation involves the **type of links** along the polymer chain. In addition polymers, the central chain consists of carbon atoms linked by carbon–carbon single bonds. In condensation polymers, the central chain consists of short aliphatic or aryl sections linked by ester groups, $-\underset{\underset{O}{\|}}{C}-O-$, or amide groups, $-\underset{\underset{O}{\|}}{C}-\underset{\underset{H}{|}}{N}-$.
- The third difference concerns the **type of monomer** involved. In addition polymerisation, the monomers have molecules with carbon–carbon double bonds. In condensation polymerisation, the monomers have molecules with at least two functional groups which may be the same or different.
- A fourth difference concerns the **conditions for preparation** of the polymers. In general, addition polymerisations require an initiator together with high temperature and high pressure, unless a catalyst is involved. In contrast, condensation polymerisations do not require initiators and usually occur at a much lower temperature and atmospheric pressure.

Polymer properties

These differences between addition and condensation polymerisation lead to considerable variations in the properties of polymers. Polymeric materials include plastics, fibres and elastomers. As polymer science has grown, chemists and materials scientists have learnt how to develop new materials with particular properties.

Some of the ways of modifying the properties of polymers include:

- altering the average length of polymer chains
- changing the structure of the monomer to one with different side groups and different intermolecular forces
- varying the extent of cross-linking between chains
- selecting a monomer that produces a biodegradable polymer
- producing a co-polymer, such as etfe (Figure 14.18), which is made from two or more monomers, each of which could produce a polymer
- adding fillers and pigments
- making composites.

Definitions

Plastics are materials made of long-chain molecules which at some stage can be moulded into shapes that are retained.

Elastomers are materials made of long-chain molecules that can be moulded into new shapes but which spring back to their original shape when the pressure is removed.

Biodegradable materials break down due to the action of microorganisms.

Co-polymers are polymers made from two or more monomers, each of which could produce a polymer.

Composites are materials made up of two or more recognisable constituents, each of which contributes to the properties of the composite.

Figure 14.37 ◀ An electromicrograph of a glass fibre composite showing rods of glass fibre embedded in a polyester matrix. Magnification is ×660.

REVIEW QUESTIONS

Extension questions

1 An incomplete structure of the dipeptide threonylisoleucine is shown in Figure 14.38.

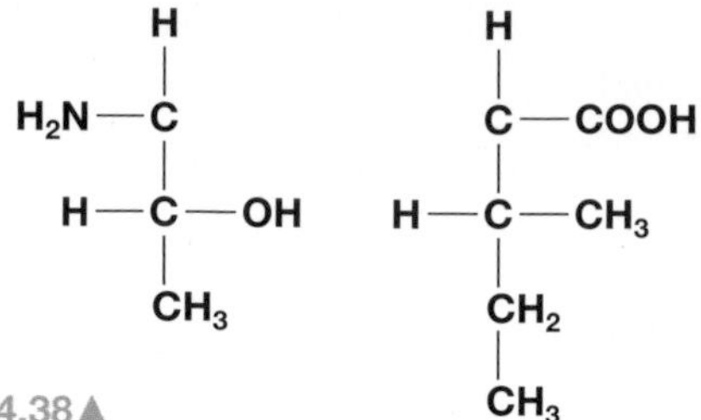

Figure 14.38▲

a) i) Redraw the structure of the dipeptide inserting the missing peptide link. (2)
ii) On your structure, circle all the chiral centres. (3)

b) What does the presence of a chiral centre tell you about a compound? (1)

c) Draw the structures of the products obtained when the dipeptide is refluxed with excess concentrated hydrochloric acid. (2)

2 A naturally occurring dipeptide, A, has the molecular formula $C_7H_{14}O_3N_2$. The dipeptide is hydrolysed forming two amino acids, B and C, on heating with concentrated hydrochloric acid. The two amino acids can be separated by paper chromatography using a solvent in which B has an R_f value of 0.60 and C has an R_f value of 0.26.

a) Draw a labelled diagram to scale showing the original and final spots and the solvent front on the chromatogram. (4)

b) Amino acid B is chiral, but C is non-chiral.
i) Draw the displayed formula of C. (1)
ii) What is the molecular formula of amino acid B? (1)
iii) Draw a possible structural formula for B. (2)

c) Draw a possible structural formula for the dipeptide A. (2)

d) What procedure would you use to make the 'spots' of amino acids visible on the chromatogram? (2)

e) How would you show that a sample of amino acid B was chiral? (2)

3 Describe three differences between addition polymerisation and condensation polymerisation using poly(ethene) and nylon-6,6 as your examples. (12)

4 Short sections of the molecular structures of two polymers are shown in Figure 14.39 below.

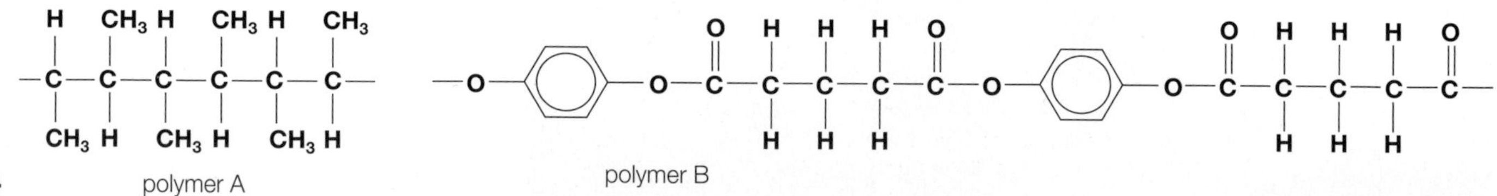

Figure 14.39▲

a) Draw the simplest repeat unit for each polymer. (2)

b) Draw and name the structural formula of the monomer used to prepare polymer A. (2)

c) i) Draw the structural formulae of the two monomers that could be used to prepare polymer B. (2)
ii) Name one of the two monomers. (1)

d) During the last decade, degradable polymers have been developed to reduce the quantity of plastic waste that is dumped in landfill sites.
State and explain why polymer B, which is a polyester, is more likely to be degradable than polymer A. (4)

5 An elastic tape consists of 60% polyester and 40% neoprene. Neoprene is a polymer similar to synthetic rubber. It is manufactured by polymerising 2-chlorobuta-1,3-diene as in the partially completed equation below.

$$n\,C{=}C{-}C{=}C \longrightarrow \left[\!-C{-}C{=}C{-}C-\!\right]_n$$

a) Copy and complete the equation above, showing the atoms or groups attached to the carbon skeletons. (2)

b) What type of polymerisation is illustrated by the manufacture of neoprene? (1)

c) What general name is given to polymers like neoprene and natural rubber? (1)

d) Polyester is manufactured by a reaction between benzene-1,4-dicarboxylic acid and ethane-1,2-diol.
i) Draw the structural formula of these two monomers. (2)
ii) Draw the structural formula of the molecule that forms when one molecule of each of these monomers reacts to produce an ester. (1)

e) Why do you think the manufacturers use a mixture of polyester and neoprene in the elastic tape? (2)

f) i) What reagent could you use to show the presence of neoprene in the elastic tape? (1)
ii) State and explain what you would observe if you used the reagent with some of the tape. (2)

g) Suggest two other properties, not covered earlier in this question, which should be considered in choosing a polymer for use in clothing. (2)

15 The synthesis of organic compounds

A lot of the purpose and pleasure of chemistry comes from making new materials such as pigments, perfumes, drugs and dyes. This making of new materials is called synthesis. Synthesis is at the heart of much of the chemical research that goes on today. We depend on synthesis for processed foods, for our fuels, for the clothes we wear and for many of the modern materials we use everyday. Synthesis is also important to our understanding of reactions and molecular structure, particularly those of organic molecules. It is not until someone has synthesised a molecule that chemists can be confident that they have determined its structure precisely.

As this is the final topic in the book, we will also use it to revise some of the key ideas in organic chemistry from previous topics.

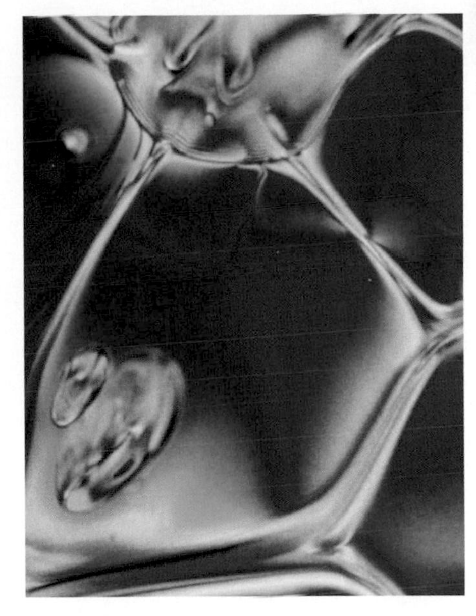

Figure 15.1▲
Liquid crystals photographed through a microscope with polarised light. Liquid crystals are used in the flat screens of modern computers and in the displays of calculators, digital cameras and digital watches.

15.1 Organic synthesis

The synthesis of organic compounds is very important in the research and production of new and useful products. Many features of modern life depend on the skills of chemists and their ability to synthesise new and complex materials. New colours, dyes and fabrics for the fashion industry are synthetic organic molecules. So also are the liquid crystals used in the flat screens of lightweight computers such as laptops. These organic compounds in the computer screen have been tailor-made by chemists to respond to an electric field and affect light.

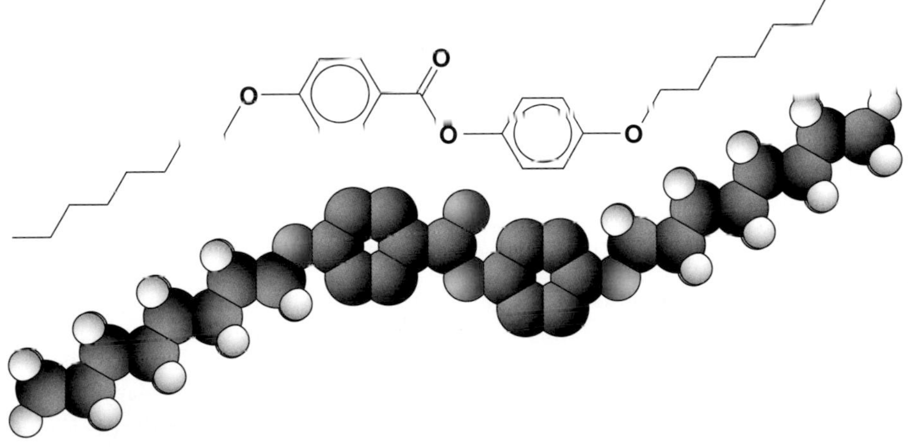

Figure 15.2▲
The structure of a liquid crystal molecule.

One of the major areas of chemical research today involves the synthesis of drugs and medicines. Every day, large numbers of compounds are synthesised for testing in pharmaceutical laboratories as potential drugs to cure or alleviate a particular disease. Medicines that have been synthesised by chemists include aspirin and paracetamol to relieve pain (Section 13.3), salbutamol to prevent asthma, chloramphenicol to treat typhoid and levodopa to alleviate Parkinson's disease (Figure 15.3).

Three other important areas of synthetic chemical research involve catalysts (Section 11.6), antiseptics (Section 12.12) and polymers (Sections 14.5 to 14.8).

The essential job of synthetic organic chemists is to consider the proposed structure for a target molecule and then devise a way of making it from simpler, readily available starting materials. The scale of work involved and the difficulties encountered in a complex organic synthesis are illustrated by the painstaking and ingenious synthesis of chlorophyll by a team of 17 scientists led by Robert Woodward at Havard University in 1959.

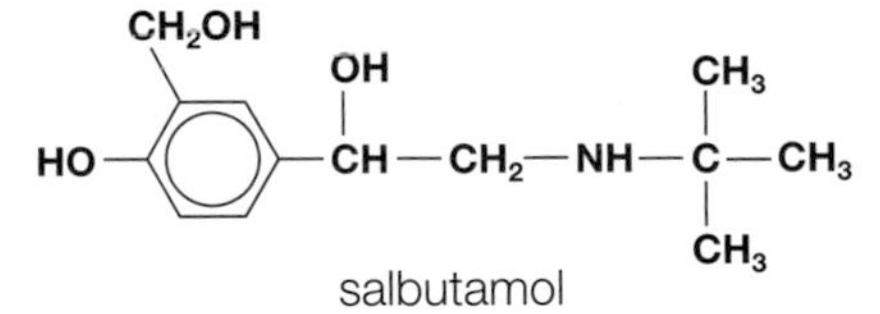

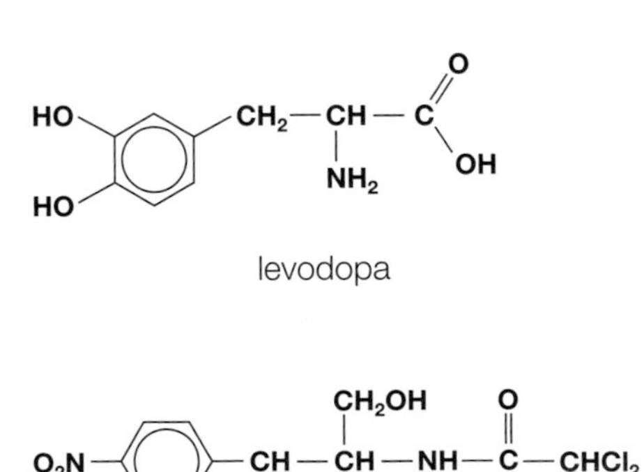

Figure 15.3▲
Three important drugs that have been synthesised by chemists – salbutamol, levodopa and chloramphenicol.

The synthesis of chlorophyll

Chlorophyll (Figure 15.4) is the green pigment in plants. Its structure was first proposed in 1940 by the famous chemist, Hans Fischer. When Woodward and his team started work in 1956, they could not be sure that Fischer's proposed structure was correct.

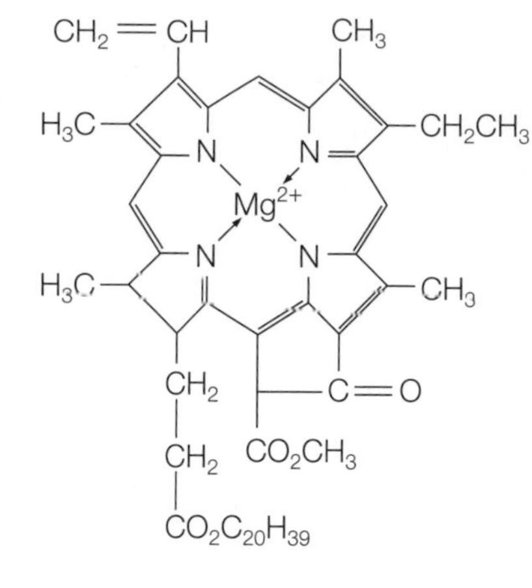

Figure 15.4▲
Chlorophyll is the green pigment in plants.

From the start, the project was planned in great detail. The chemists read all the papers concerning previous studies of chlorophyll to ensure that no clues to a successful synthesis were missed. They drew on their understanding of the mechanisms of organic reactions to predict the likely products at each stage and suggest routes to their target molecule.

The synthesis of chlorophyll would have been impossible without the newer methods of separation, purification and identification that had become available. The variety of spectroscopy techniques was also crucial to success. Woodward's team published their paper describing the successful synthesis of chlorophyll in 1960, opening up a new stage of research into the part that chlorophyll plays in photosynthesis.

Organic analysis

When complex molecules, such as chlorophyll, have been synthesised, chemists must use a variety of methods to analyse them and identify their precise composition and structure.

Traditionally, chemical tests were used to identify functional groups in organic molecules, together with combustion and quantitative analysis. Nowadays, however, modern laboratories rely on a range of highly sensitive, automated and instrumental techniques to identify the products of synthesis. These include chromatography, mass spectrometry and various kinds of spectroscopy.

Sensitive methods of analysis are very important in monitoring organic syntheses for several reasons.

- Sensitive methods of analysis will determine the degree of purity of a synthetic product.
- Sensitive methods of analysis will also identify any impurities, some of which may be toxic and in very small concentration.
- If analysis reveals an impurity in the product, it may be possible to limit its formation by changing the operating conditions for the reaction. Changes in the temperature, the pressure, the solvent used or the choice of catalyst may promote the formation of a desired product while reducing the formation of impurities.
- Many pharmaceutical laboratories that specialise in the development of new drugs produce thousands of compounds every year for further testing (Section

15.9). Some of their products are obtained in very small concentrations and particularly sensitive techniques are needed to analyse and identify them.

The food and drugs industries operate very high standards of purity in their products. Impurities, depending on their toxicity, may interfere with the health and well-being of consumers. For example, traces of sodium chloride in a medicine would probably not be considered a problem, but the slightest trace of sodium cyanide would be cause for alarm.

15.2 The formulae of organic molecules

Empirical formulae

From Section 1.6 in *Edexcel Chemistry for AS* you should know that the empirical formula of a compound shows the simplest ratio for the number of atoms of each element in it.

Finding the empirical formula of an organic compound is the first step towards understanding its chemistry. The usual way of doing this is to oxidise a weighed sample of the organic compound completely by burning it in pure dry oxygen, and then determine the masses of carbon dioxide and water produced. This process of combustion analysis then enables you to calculate the masses of carbon and hydrogen in a sample of the compound and hence its empirical formula.

Figure 15.5 shows a simplified diagram of the method and apparatus used. Modern methods of combustion analysis include refinements to ensure that the organic compound is completely oxidised and that all the carbon dioxide and water are absorbed and weighed.

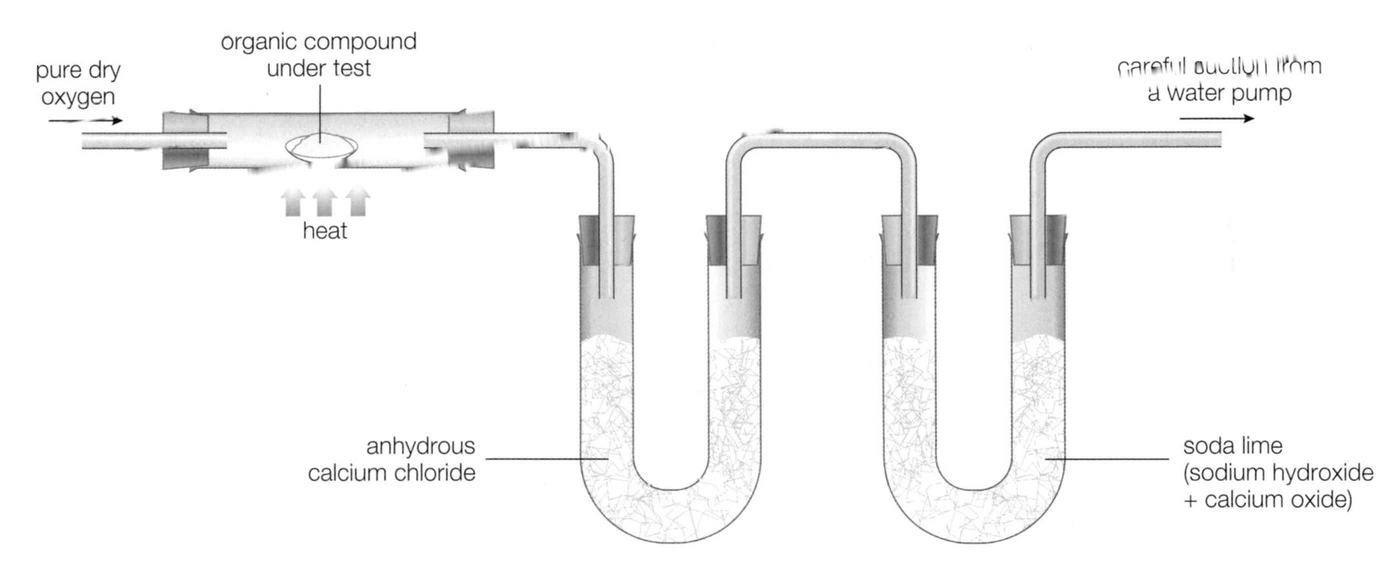

Figure 15.5▲.
Using careful suction, draw pure dry oxygen over a heated sample of the solid organic compound. (Liquid organic compounds can be burnt from a wick.) Pass the product gases through anhydrous calcium chloride (or anhydrous copper(II) sulfate) to absorb any water produced, and then through anhydrous soda lime (sodium hydroxide + calcium oxide) to absorb the carbon dioxide produced.

Molecular formulae

Molecular formulae are usually more helpful than empirical formulae because they show the actual number of atoms of each element in one molecule of a compound. All that is needed to find the molecular formula from the empirical formula is the molar mass of the compound. This can be determined from the mass spectrum of the compound (Section 9.3).

A molecular formula is always a simple multiple of the empirical formula. Methane, for example, has the empirical formula CH_4 and the molecular formula CH_4, benzene has the empirical formula CH and the molecular formula C_6H_6 and ethanoic acid has the empirical formula CH_2O and the molecular formula $C_2H_4O_2$.

C_3H_5
empirical formula

C_6H_{10}
molecular formula

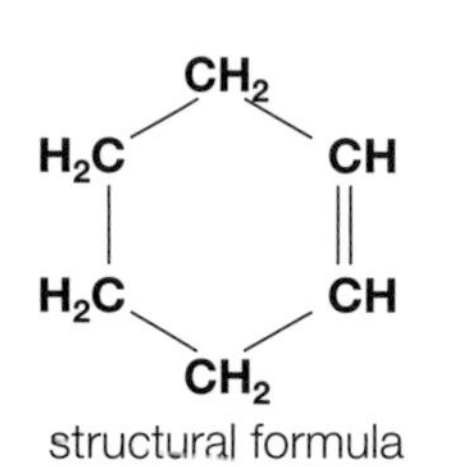

structural formula

Figure 15.6▲
The empirical, molecular and structural formulae of cyclohexene.

Structural formulae

The structural formula of a compound shows how the atoms link together in one molecule of the compound (Section 5.3 in *Edexcel Chemistry for AS*). Given the molecular formula of an organic compound and knowing something about its characteristic reactions, it is often possible to predict its structure by assuming that:

- carbon atoms form four covalent bonds
- nitrogen atoms form three covalent bonds
- oxygen atoms form two covalent bonds
- hydrogen and halogen atoms form one covalent bond.

Although the structural formulae of organic compounds can often be determined by combustion analysis to find their percentage composition followed by a study of their chemical reactions, structural formulae can be obtained more definitively and more precisely by spectrometry and spectroscopy.

Using mass spectrometry (Section 9.3) it is possible to identify the fragments of an organic molecule and then piece the whole molecule together.

Spectroscopic methods such as infrared spectroscopy (Section 9.3) and nuclear magnetic resonance (Section 9.2) provide information about the various bonds and functional groups in organic molecules in order to confirm their structural formulae.

Sometimes it is enough to show structural formulae in a condensed form, such as $CH_3CH_2CH_2CH{=}CHCH_3$ for hex-2-ene. At other times it is more helpful to write the full structural formula showing all the atoms and all the bonds. This type of formula is called a displayed formula.

Chemists have also devised a useful shorthand for showing the formulae of more complex molecules as skeletal structures. These skeletal formulae need careful study because they represent the hydrocarbon part of the molecule simply as lines for the bonds between carbon atoms, leaving out the symbols for carbon and hydrogen atoms (Figure 15.7).

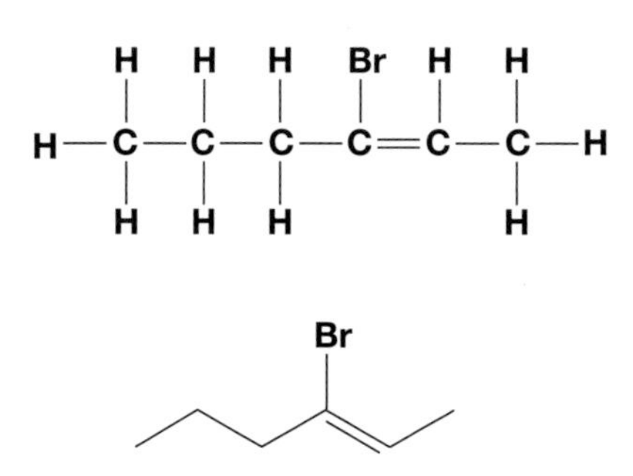

Figure 15.7▲
The displayed and skeletal formulae of 3-bromohex-2-ene.

Although the structural, displayed and skeletal formulae of an organic compound show how its atoms link together, they do not show its true shape in three dimensions. Sometimes, it is important to know and understand what the three-dimensional shape of a molecule is like and chemists use various models to do this. These include ball-and-stick models, space-filling models and various types of computer models.

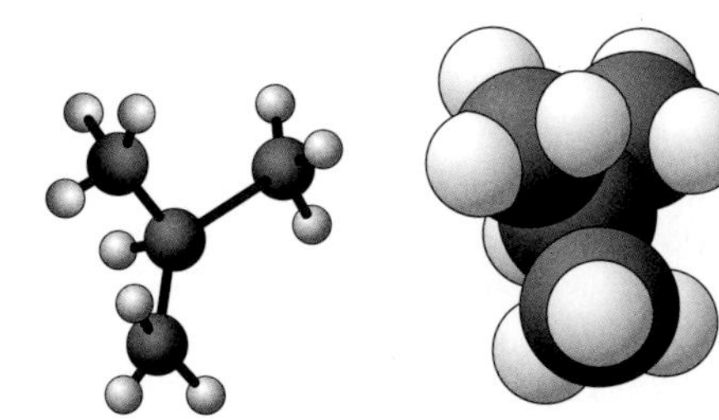

Figure 15.8▲
A ball-and-stick model and a space-filling model of 2-methylpropane.

Test yourself

1 Write out the empirical, molecular, structural, displayed and skeletal formulae of 2-methylpropane in Figure 15.8.

2 Describe briefly how you would determine the percentages of carbon and hydrogen in a solid organic compound.

3 A compound containing only carbon, hydrogen and fluorine was burnt in excess oxygen. 0.32 g of the compound produced 0.44 g of carbon dioxide and 0.09 g of water.
   a) What is the empirical formula of the compound?
   b) The relative molecular mass of the compound is 64. What is its molecular formula?
   c) Write all the possible displayed formulae of the compound and give the systematic name for each formula.

4 A compound, containing only carbon, hydrogen and oxygen, has prominent peaks in its mass spectrum which correspond to masses of 60, 43, 31, 29 and 17 relative to the standard $^{12}_{6}C = 12$.
   a) What is the relative molecular mass of the compound?
   b) Suggest possible fragments for the peaks corresponding to masses of 17, 29 and 31.
   c) Suggest a possible structure for the compound.

Activity

Using combustion analysis and nmr spectroscopy to identify an organic compound

An organic compound, X, containing carbon, hydrogen and oxygen only, was found to have a relative molecular mass of about 70. When 0.36 g of the compound was burnt in excess oxygen, 0.88 g of carbon dioxide and 0.36 g of water were formed.

1 Calculate the empirical formula of X.

2 What is the molecular formula of X?

3 X undergoes a condensation reaction with 2,4-dinitrophenylhydrazine to produce an orange solid. What can you conclude from this?

4 Write all the possible non-cyclic structural formulae for X and give the systematic name for each formula.

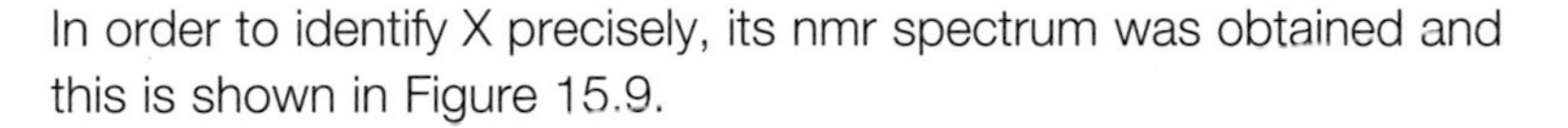

In order to identify X precisely, its nmr spectrum was obtained and this is shown in Figure 15.9.

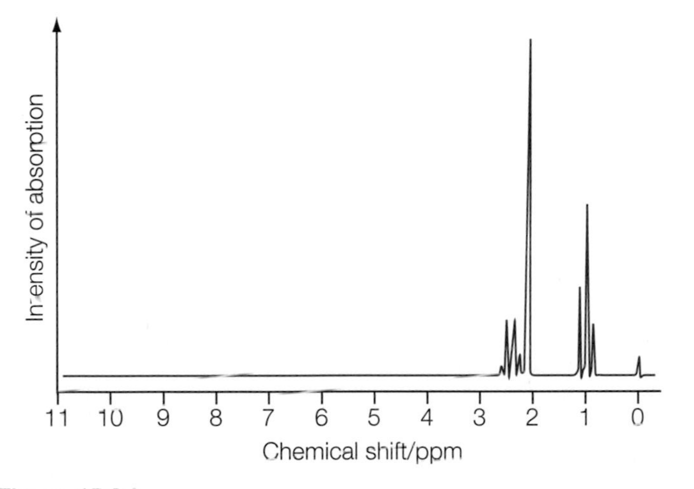

Figure 15.9▲
The nmr spectrum of compound X.

5 Using your Edexcel Data booklet, interpret the nmr spectrum and suggest the precise structural formula of X, indicating the evidence you have used from the spectrum.

15.3 Functional groups – the keys to organic molecules

Functional groups provide the key to organic molecules. A knowledge of the properties and reactions of a limited number of functional groups has opened up our understanding of most organic compounds.

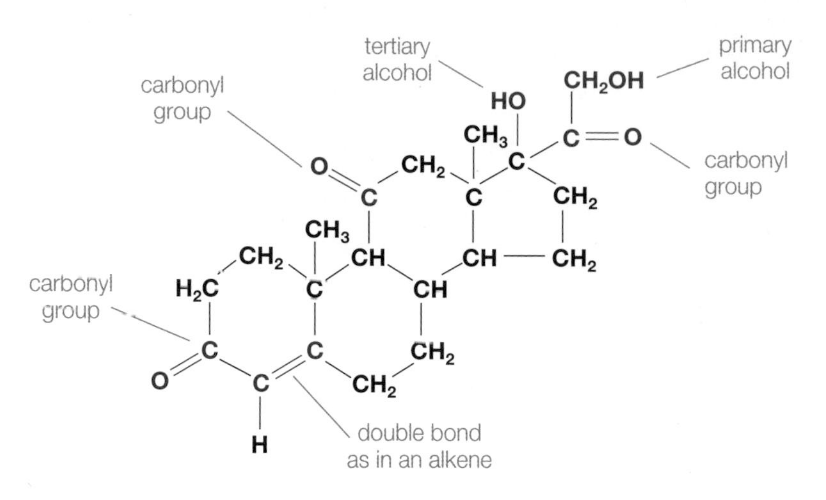

Figure 15.10▶
The structure of the steroid cortisone, labelled to show the reactive functional groups and the hydrocarbon skeleton.

A functional group is the atom or atoms that give a series of organic compounds their characteristic properties and reactions. Chemists often think of an organic molecule as a relatively unreactive hydrocarbon skeleton with one or more functional groups in place of hydrogen atoms. The functional group in a molecule is responsible for most of its reactions. In contrast, the carbon–carbon single bonds and carbon–hydrogen bonds are relatively unreactive, partly because they are both strong and non-polar.

Table 15.1 shows the major functional groups that you have met during your A-level studies, together with an example of one compound containing each group.

Definition

A **functional group** is the atom or group of atoms that give an organic compound its characteristic properties.

Table 15.1▼
The major functional groups.

Functional group		Example
Alcohol	—OH	propan-1-ol $CH_3CH_2CH_2OH$
Alkene	$>C=C<$	propene $CH_3CH=CH_2$
Halogenoalkane Hal = F, Cl, Br, I	—Hal	1-chloropropane $CH_3CH_2CH_2Cl$
Ether	$>C-O-C<$	methoxyethane $CH_3OCH_2CH_3$
Aldehyde	$-C(=O)H$	propanal CH_3CH_2CHO
Ketone	$>C=O$	propanone CH_3COCH_3
Carboxylic acid	$-C(=O)OH$	propanoic acid CH_3CH_2COOH

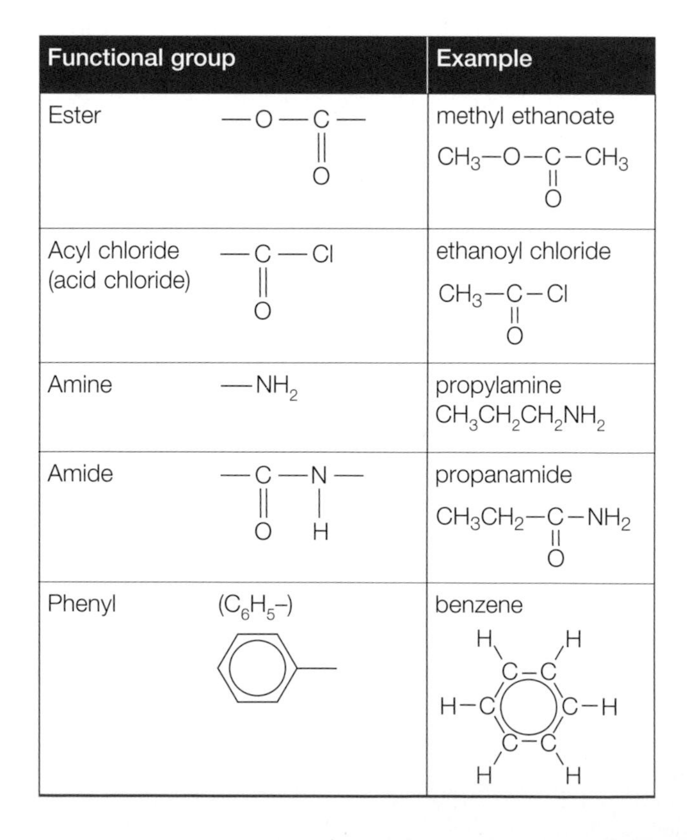

Functional group		Example
Ester	$-O-C(=O)-$	methyl ethanoate $CH_3-O-C(=O)-CH_3$
Acyl chloride (acid chloride)	$-C(=O)-Cl$	ethanoyl chloride $CH_3-C(=O)-Cl$
Amine	$-NH_2$	propylamine $CH_3CH_2CH_2NH_2$
Amide	$-C(=O)-N(H)-$	propanamide $CH_3CH_2-C(=O)-NH_2$
Phenyl	(C_6H_5-)	benzene C_6H_6 (H–C ring structure)

The characteristic properties and tests for most of these functional groups are shown on the Data sheets headed: 'Tests and observations on organic compounds' on the Dynamic Learning Student website.

Test yourself

5 Anaerobic respiration in muscle cells breaks down glucose to simpler compounds including the following two molecules. Identify the functional groups in these molecules:
a) $CH_2OH–CHOH–CHO$
b) $CH_3–CO–COOH$

6 Pheromones are messenger molecules produced by insects to attract mates or to give an alarm signal. Identify the functional groups in the pheromone in Figure 15.11 produced by queen bees.

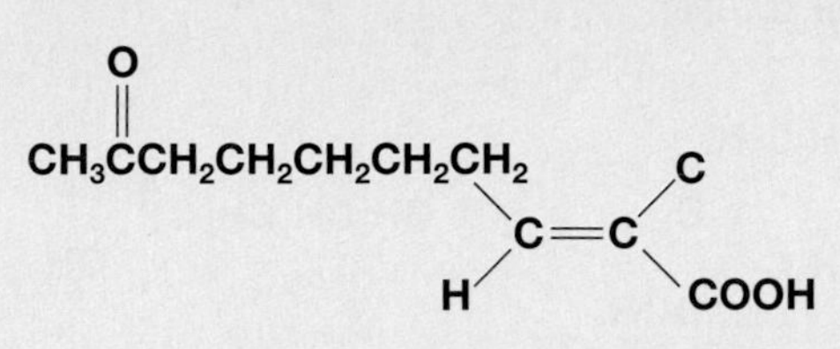

Figure 15.11▲

7 Use the Data sheets on the Dynamic Learning Student website headed 'Tests and observations on organic compounds' and the Reaction flow charts from previous topics to predict six important properties or reactions of the painkiller dextropropoxyphene and its mirror image which is an ingredient of cough mixtures (Figure 15.12).

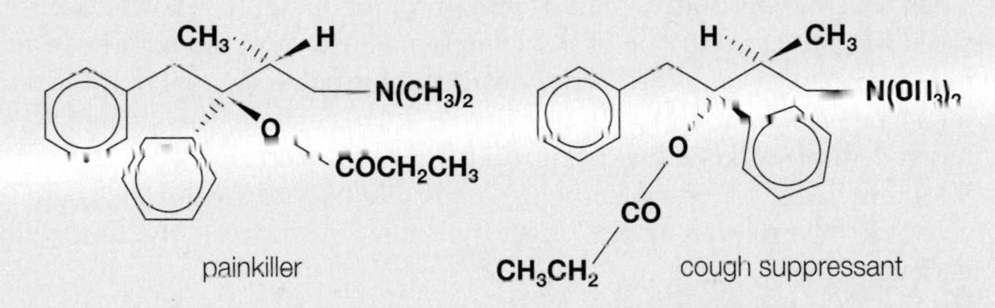

Figure 15.12▲
Dextropropoxyphene and its mirror image.

8 For each of parts a) to g) below only one of the compounds labelled A to D is correct.
A $CH_3CH_2NH_2$ B $C_6H_5NO_2$
C $C_6H_5NH_2$ D $C_6H_5N_2Cl$
a) Which is a strong electrolyte?
b) Which dissolves in dilute hydrochloric acid but not in water?
c) Which is insoluble in water, dilute acid and dilute alkali?
d) Which is explosive when pure?
e) Which has the highest vapour pressure at room temperature?
f) Which combines most readily with H^+ ions?
g) Which reacts with nitrous acid to produce nitrogen at 5 °C?

15.4 Organic routes

Organic chemists synthesise new molecules using their knowledge of functional groups, reaction mechanisms and molecular shapes, as well as the factors that control the rate and extent of chemical change.

A synthetic pathway leads from the reactants to the required product in one step or several steps. Organic chemists often start by examining the 'target molecule'. Then, they work backwards through a series of steps to find suitable starting chemicals that are cheap enough and available.

Figure 15.13 shows an example of the systematic way in which working back can be used in synthesising one 'target molecule' from a 'starting molecule'. In this case, the 'target molecule' is butanoic acid and the 'starting molecule' is 1-bromobutane.

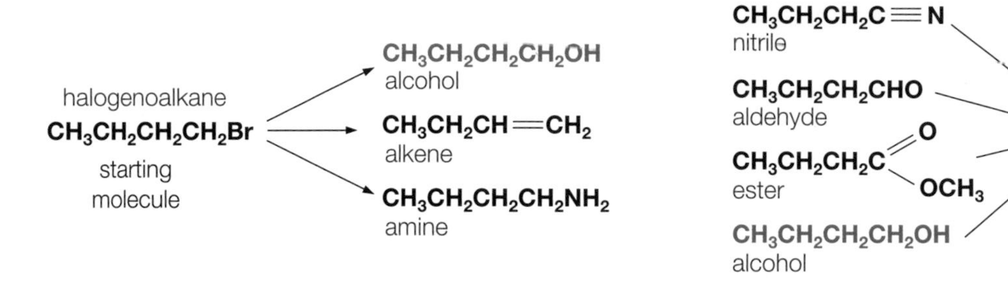

Figure 15.13▲ Working back from the target molecule to find a two-step synthesis of butanoic acid from 1-bromobutane.

- Begin by writing down the formulae of those compounds that could be readily converted to butanoic acid, the target molecule. These include the nitrile butanenitrile, the aldehyde butanal, the ester methyl butanoate and the alcohol butan-1-ol.
- Then look at your starting molecule, 1-bromobutane, to see whether it could be converted to one of the compounds which would readily form butanoic acid. If necessary, write down the formulae of compounds that might be produced from 1-bromobutane. These include the alcohol butan-1-ol, the alkene but-1-ene, and the amine butylamine.
- With any luck, you should now see a possible two-step synthetic route from your starting molecule to the target molecule. In this case, the route can go via butan-1-ol.
- If a two-step route is not clear at this point, then you might need to consider a three-step route involving the conversion of one of the products from the starting material to one of the reactants which will readily form the target material.

Chemists normally seek a synthetic route that has the least number of steps and produces a high yield of the product. The larger the scale of production, the more important it is to keep the yield high so as to avoid producing large quantities of wasteful by-products.

Changing the functional groups

All the reactions in organic chemistry convert one compound to another, but there are some reactions that are particularly useful for developing synthetic routes. These useful reactions include:

- the addition of hydrogen halides to alkenes
- substitution reactions that replace halogen atoms with other functional groups such as –OH or $–NH_2$
- substitution of a chlorine atom for the –OH group in an alcohol or a carboxylic acid
- elimination of a hydrogen halide from a halogenoalkane to introduce a carbon–carbon double bond
- oxidation of primary alcohols to aldehydes and then carboxylic acids
- reduction of carbonyl compounds to alcohols.

Activity

Converting one functional group to another

Make a copy of the flow chart in Figure 15.14. Beside each arrow, write the reagents and conditions needed for the conversion. You may need to refer to the Reaction flow charts from previous topics to do this.

1 Using your completed copy of Figure 15.14, suggest two-step syntheses, showing the reagents and conditions for each of the following conversions:

- **a)** ethene to ethylamine
- **b)** ethanol to ethyl ethanoate (using ethanol as the only carbon compound)
- **c)** propanoic acid to propanamide.

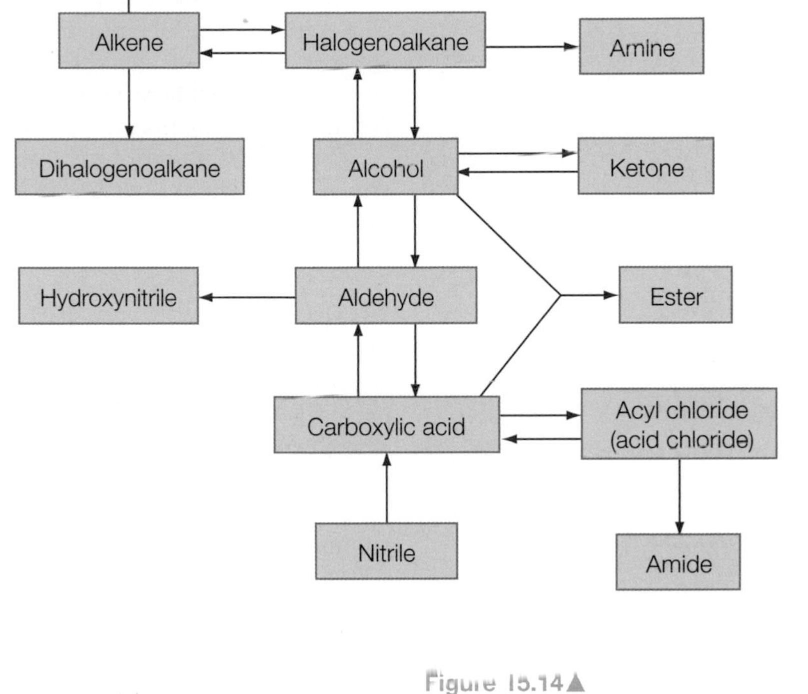

Figure 15.14▲ A flow diagram summarising the methods for converting one functional group to another.

2 Using your completed copy of Figure 15.14, suggest three-step syntheses, showing the reagents and conditions for each of the following conversions:

- **a)** ethene to ethanoic acid
- **b)** propan-2-ol to propane.

Test yourself

9 Draw the structural formula of the main organic product in each of the following reactions. Classify each reaction as addition, substitution or elimination and classify the reagent on the arrow as a free radical, nucleophile, electrophile or base.

a) $CH_2{=}CH_2(g) \xrightarrow{HBr(g)}$ b) $CH_3CH_2CH_2Br(l) \xrightarrow{KOH(aq)}$ c) C_6H_6 benzene $\xrightarrow[+ \text{ conc. } H_2SO_4]{\text{conc. } HNO_3}$

10 Identify substances A to H in the flow diagrams in Figure 15.15.

a) $CH_3COOH \xrightarrow[\text{then water}]{\text{A in ether}}$ B $\xrightarrow{PCl_5(s)} CH_3CH_2Cl \xrightarrow[\text{in ethanol}]{\text{conc. } NH_3}$ C

b) benzene $\xrightarrow[\text{conc. } H_2SO_4]{\text{D +}}$ nitrobenzene (NO_2) $\xrightarrow[\text{heat}]{\text{E + conc. F}}$ phenylamine (NH_2) $\xrightarrow[\text{(ii) H + dil. NaOH at 5°C}]{\text{(i) G + dil. HCl at 5°C}}$ C_6H_5–N=N–C_6H_4–OH

Figure 15.15▲

11 Give the reagents and conditions for converting:

- a) propanone to propene in three steps
- b) bromoethane to ethane-1,2-diol in three steps.

15.5 Synthetic techniques

www Tutorial

Chemists have developed a range of practical techniques and procedures for the synthesis of solid and liquid organic compounds. These methods allow for the fact that reactions involving molecules with covalent bonds are often slow and that it is difficult to avoid side reactions that produce by-products. There are five key stages in the preparation of an organic compound.

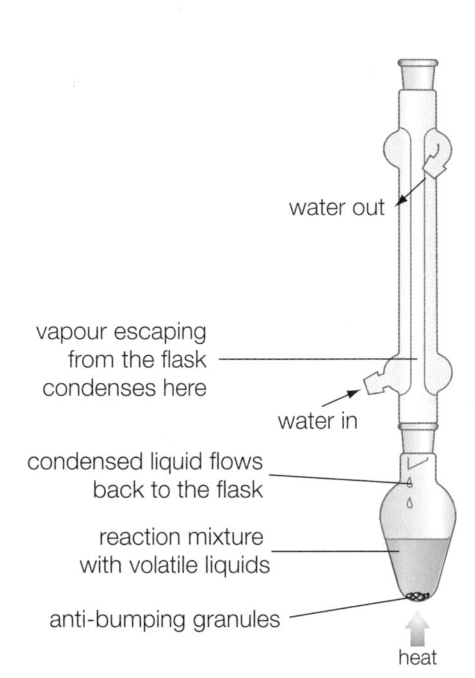

Figure 15.16▲ Heating in a flask with a reflux condenser prevents vapours escaping while the reaction is happening. Vapours from the reaction mixture condense and flow back (reflux) into the flask.

Stage 1: Planning

The starting point of any synthesis is to choose an appropriate reaction or series of reactions as described in the previous section. The next thing to do is to work out suitable reacting quantities from the equation and decide on the conditions for reaction.

An important part of the planning stage is a risk assessment. This should ensure that hazards have been identified and that appropriate safety precautions and control measures are used in order to reduce the risk during any synthesis.

Stage 2: Carrying out the reaction

During this stage, the reactants are measured out and mixed in suitable apparatus. Most organic reactions are slow at room temperature so it is usually necessary to heat the reactants using a flame, heating mantle or hotplate. One of the commonest techniques is to heat the reaction mixture in a flask fitted with a reflux condenser (Figure 15.16).

Organic reagents do not usually mix with aqueous reagents. So another common technique is to shake the immiscible reactants in a stoppered container.

Stage 3: Separating the product from the reaction mixture

Chemists talk of 'working up' the reaction mixture to obtain their crude product. If the product is a solid, it can be separated ('worked up') by filtration using a Buchner or Hirsch funnel with suction from a water pump. This is illustrated in Figure 15.20.

Liquids can often be separated by simple distillation, fractional distillation or steam distillation. Distillation with steam at 100 °C allows the separation of compounds that decompose if heated at their boiling temperatures. The technique works only with compounds that do not mix with water. When used to separate the products of organic preparations, steam distillation leaves behind those reagents and products that are soluble in water.

Figure 15.17► Setting up distillation apparatus to separate chemicals synthesised during research to develop new anti-cancer drugs.

Stage 4: Purifying the product

The 'crude' product separated from the reaction mixture is usually contaminated with by-products and unused reactants. The methods of purifying this 'crude' product depend on whether it is a solid or a liquid.

Purifying organic solids

The usual technique for purifying solids is recrystallisation, which is illustrated in part of Figure 7.18. The procedure for recrystallisation is based on using a solvent that dissolves the product when hot, but not when cold. The choice of solvent is usually made by trial and error. Use of a Buchner or Hirsch funnel and suction filtration speeds up filtering and facilitates recovery of the purified solid from the filter paper. The procedure is as follows.

- Dissolve the impure solid in the minimum volume of hot solvent.
- If the solution is not clear, filter the hot mixture through a heated funnel to remove insoluble impurities.
- Cool the filtrate so that the product recrystallises, leaving the smaller amounts of soluble impurities in solution.
- Filter to recover the purified product.
- Wash the purified solid with small amounts of pure solvent to wash away any solution containing impurities.
- Allow the solvent to evaporate from the purified solid in the air.

Purifying organic liquids

Chemists often begin to purify organic liquids that are insoluble in water by shaking with aqueous reagents in a separating funnel to extract impurities. This is followed by washing with pure water, drying and finally fractional distillation.

Fractional distillation separates mixtures of liquids with different boiling temperatures. On a laboratory scale, the process takes place in the distillation apparatus that has been fitted with a fractionating column between the flask and the still-head (Figure 15.18). Separation is improved if the column is packed with inert glass beads or rings to increase the surface area where rising

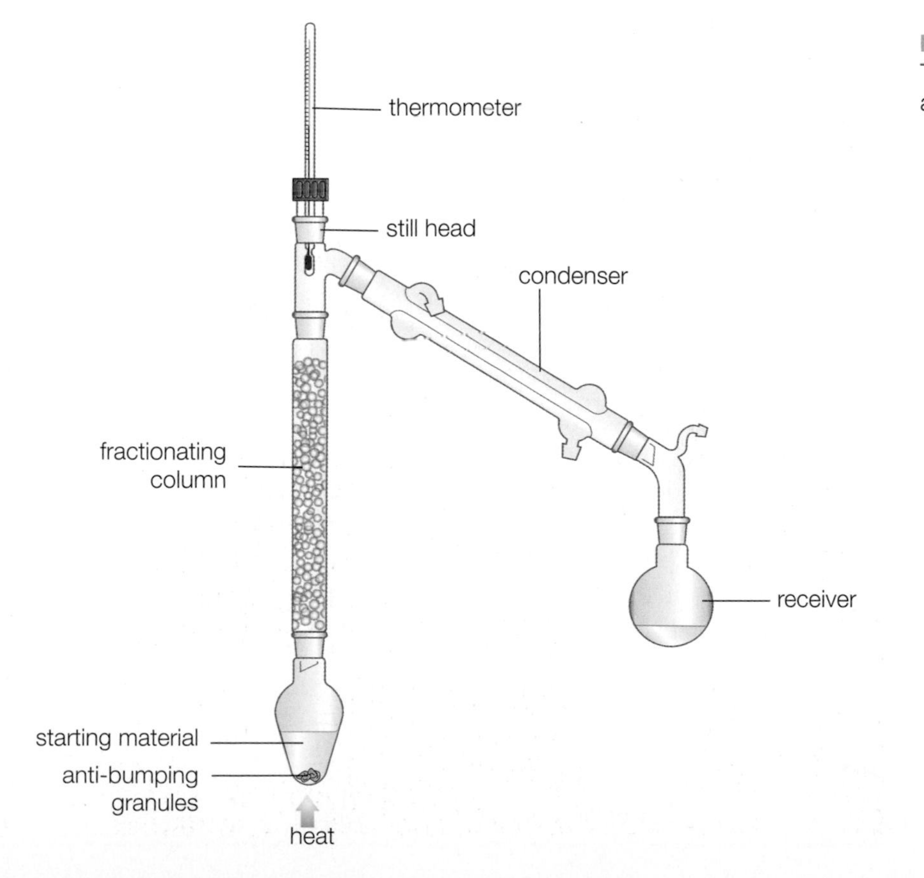

Figure 15.18 ◀
The apparatus for fractional distillation of a mixture of liquids.

Definitions

A **volatile liquid** evaporates easily, turning to a vapour.

Vapours are gases formed by evaporation of substances that are usually liquids or solids at room temperatures.

Chemists talk about 'hydrogen gas' but 'water vapour'. Vapours are easily condensed by cooling or increased pressure because of their relatively strong intermolecular forces.

vapour can mix with condensed liquid running back to the flask. The column is hotter at the bottom and cooler at the top. The thermometer reads the boiling temperature of the compound passing over into the condenser.

If the flask contains a mixture of liquids, the boiling liquid in the flask produces a vapour that is richer in the most volatile of the liquids present (the one with the lowest boiling temperature).

Most of the vapour condenses in the column and runs back. As it does so, it meets more of the rising vapour. Some of the vapour condenses. Some of the liquid evaporates. In this way, the mixture evaporates and condenses repeatedly as it rises up the column. But, every time it does so, the vapour becomes richer in the most volatile liquid present. At the top of the column, the vapour contains 100% of the most volatile liquid. So, during fractional distillation, the most volatile liquid with the lowest boiling temperature distils over first, then the liquid with the next lowest boiling temperature and so on.

Stage 5: Measuring the yield, identifying the product and checking its purity

Measuring the yield

Comparing the actual yield with the yield expected from the chemical equation is a good measure of the efficiency of a process.

The yield expected from the equation, assuming that the reaction is 100% efficient, is normally called the theoretical yield.

The efficiency of a synthesis, like that of other reactions, is normally calculated as a percentage yield. This is given by the relationship:

$$\text{percentage yield} = \frac{\text{actual yield of product}}{\text{theoretical yield of product}} \times 100\%$$

Worked example

a) What is the theoretical yield of glycine (2-aminoethanoic acid) from 15.5 g of chloroethanoic acid?

b) What is the percentage yield if the actual yield of glycine is 7.9 g?

Notes on the method

Start by writing an equation for the reaction. This need not be a full balanced equation so long as it includes the limiting reactant, the product and the molar amounts of reactant and product in the equation.

In this case we must assume that any other reactants are in excess and the limiting reactant is chloroethanoic acid.

Answer

a) *The equation:* $ClCH_2COOH \longrightarrow H_2NCH_2COOH$

The molar mass of chloroethanoic acid, $ClCH_2COOH = 94.5\ g\ mol^{-1}$

The molar mass of glycine, $H_2NCH_2COOH = 75\ g\ mol^{-1}$

According to the equation:

1 mol of chloroethanoic acid produces 1 mol of glycine

∴ 94.5 g of chloroethanoic acid produces 75 g of glycine

So, 15.5 g of chloroethanoic acid produces $\frac{75}{94.5} \times 15.5$ g of glycine

= 12.3 g of glycine

Theoretical yield of glycine = 12.3 g

b) $\text{Percentage yield} = \frac{\text{actual yield}}{\text{theoretical yield}} \times 100\%$

$= \frac{7.9}{12.3} \times 100\%$

$= 64\%$

Identifying the product and checking its purity

- **Qualitative tests**
  Simple chemical tests for functional groups can help to confirm the identity of the product. These tests for functional groups are shown in the Data sheets headed 'Tests and observations on organic compounds' on the Dynamic Learning Student website.

- **Measuring melting temperatures and boiling temperatures**
  Pure solids have sharp melting temperatures, but impure solids soften and melt over a range of temperatures. So watching a solid melt can often show whether or not it is pure. As databases now include the melting temperatures of all known compounds, it is possible to check the identity and purity of a product by checking that it melts sharply at the expected temperature (Figure 15.19).

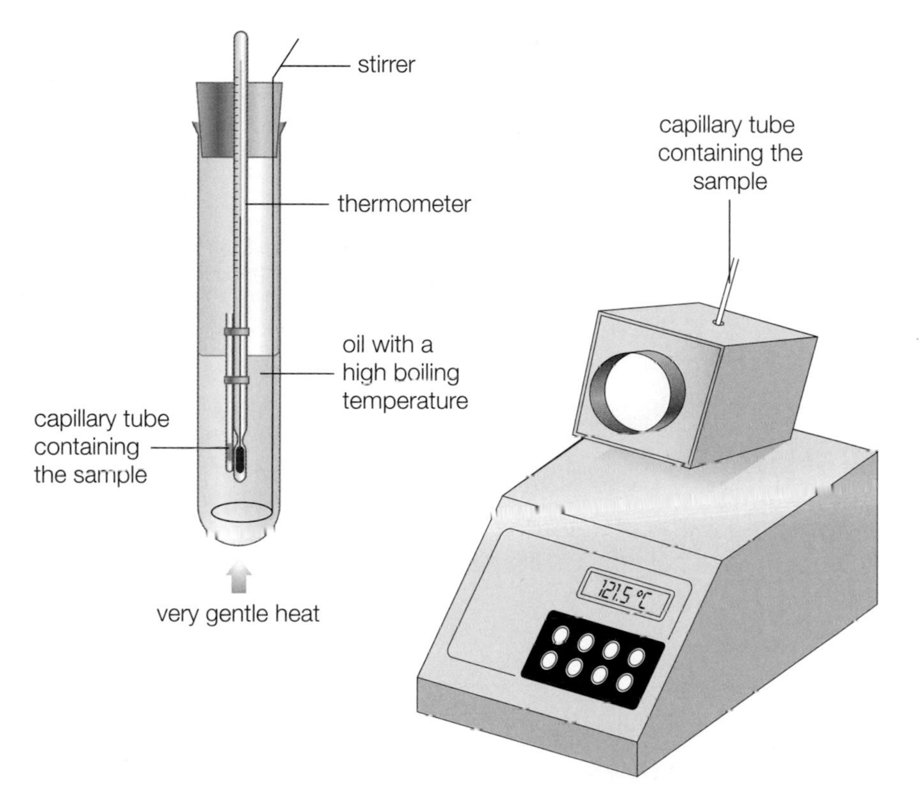

Figure 15.19▲
Two methods of measuring the melting temperature of a solid.

Like melting temperatures for solids, boiling temperatures can be used to check the purity and identity of liquids. If a liquid is pure, it should all distil over a narrow range at the expected boiling temperature. The boiling temperature can be measured as the liquid distils over during fractional distillation.

- **Chromatography and spectroscopy**
  The use of chromatography in identifying amino acids is described in Section 14.4, but the more general application of chromatography and spectroscopy in identifying compounds and checking their purity is covered in Topic 9.

Test yourself

12 A possible two-step synthesis of 1,2-diaminoethane first converts an alkene to a dihalogenoalkane and then reacts this with ammonia.
  a) Write out a reaction scheme for the synthesis, giving reagents and conditions.
  b) Calculate the mass of the alkene needed to make 2 g of the 1,2-diaminoethane, assuming a 60% yield in step 1 and a 40% yield in step 2. Which chemicals should be used in excess?
  c) What hazards does the synthesis pose and what safety precautions should be taken?

13 Give three reasons why the actual yield in an organic synthesis is always less than the theoretical yield.

14 A two-step synthesis converts 18 g of benzene first to 22 g of nitrobenzene and then to 12 g of phenylamine.
  a) State the reagents and conditions for each step.
  b) Calculate the theoretical yield and the percentage yield for each step.
  c) What is the overall percentage yield?

15 Read the sub-section headed 'Purifying organic solids' in Section 15.5 again and then answer the following questions.
  a) Why should the impure solid be dissolved in the *minimum* volume of *hot* solvent?
  b) Why is the solution sometimes cooled in ice when the pure product is being recrystallised?
  c) How could you improve the evaporation of excess solvent from the purified solid in the final stage?

Activity

The preparation of methyl 3-nitrobenzoate

⚠ Wear goggles

Pale yellow crystals of methyl 3-nitrobenzoate, $CH_3OCO–C_6H_4–NO_2$, can be prepared by the nitration of methyl benzoate as follows.

A Weigh out 2.7 g of methyl benzoate into a small flask and then dissolve it in 5 cm^3 of concentrated sulfuric acid. When the solid has dissolved, cool the solution in ice.

Prepare the nitrating mixture by carefully adding 2 cm^3 of concentrated sulfuric acid to 2 cm^3 of concentrated nitric acid and then cool this mixture in ice also.

B Add the nitrating mixture dropwise to the solution of methyl benzoate. Stir the mixture and keep the temperature below 10 °C.

C Allow the mixture to stand for about 15 minutes and then pour the reaction mixture onto crushed ice. Stir until all the ice has melted and crystalline methyl 3-nitrobenzoate has formed.

D Filter the crystals (Figure 15.20), wash them with cold water and then recrystallise them from the minimum volume of hot ethanol. Finally, dry the crystals between filter papers.

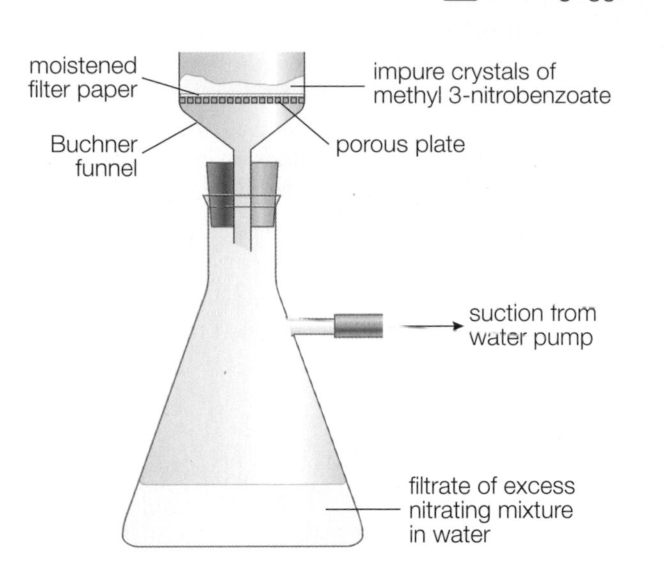

Figure 15.20▲ Filtering the impure crystals of methyl 3-nitrobenzoate using a Buchner funnel by suction from a water pump.

1 Draw the displayed formula for methyl benzoate.

2 Write an equation for the reaction which occurs in stages **B** and **C**.

3 State two procedures during the preparation that help to prevent further nitration of the product.

4 Suggest the names of two nitro-compounds that are likely to contaminate the impure crystals of methyl 3-nitrobenzoate.

5 Why were the crystals washed with water before recrystallisation?

6 Explain how the loss of product is kept to a minimum during recrystallisation.

7 How many moles of methyl benzoate and nitric acid were used in the preparation? (Assume that concentrated nitric acid is pure HNO_3 and that its density is 1.5 g cm^{-3}. H = 1, C = 12, N = 14, O = 16)

8 What is the theoretical yield of methyl 3-nitrobenzoate?

9 What are the hazards posed by the preparation and how is their risk reduced?

10 How might the purity of the product be checked?

15.6 Stereochemical synthesis

When organic compounds are prepared synthetically in the laboratory, the product is sometimes a mixture of two optical isomers.

One process in which this may happen is the reaction of carbonyl compounds with H^- ions from lithium tetrahydridoaluminate(III), $LiAlH_4$, or with cyanide ions, CN^-, from potassium cyanide (Section 7.4). Both of these reactions with H^- and CN^- involve nucleophilic addition. The initial product in each case is an intermediate anion which is converted to the final product containing an –OH group by adding dilute acid (Figure 15.21).

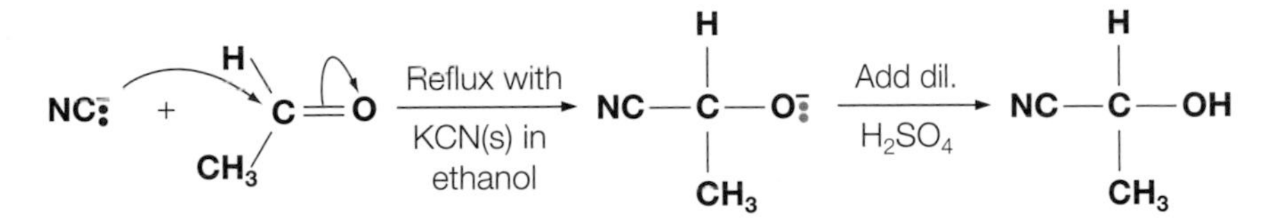

Figure 15.21 ◀ The nucleophilic addition of hydrogen cyanide to ethanal.

The reaction of ethanal with cyanide ions followed by dilute H_2SO_4 or dilute HCl increases the number of groups attached to the carbon atom in the original carbonyl group from three to four. In this case, there are now four different groups attached to the carbon atom. This means that the final product will be a mixture of two optical isomers. As the CN^- ions can attack the flat ethanal molecule equally well from either side, the product will be a racemic mixture containing equal amounts of the two optical isomers (Section 6.2).

The reaction that we have just considered is used as the first step in the two-step laboratory synthesis of lactic acid (2-hydroxypropanoic acid) from ethanal. In the second step, the product shown in Figure 15.21 is refluxed with the dilute acid and this converts the –CN group to a carboxylic acid group, –COOH (Figure 15.22).

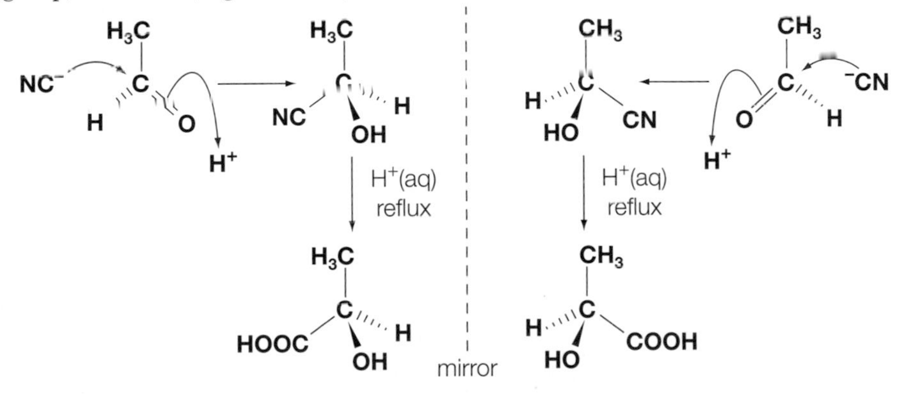

Figure 15.22 ◀ The laboratory synthesis of lactic acid (2-hydroxypropanoic acid) from ethanal produces a racemic mixture.

In contrast to the lactic acid produced synthetically in the laboratory, that formed in our muscles during excessive activity is the single (+) isomer. The reason for this is that the lactic acid formed in our muscles is produced naturally through processes catalysed by enzymes.

In fact, most optically active compounds in living systems are present as just one optical isomer. For example, all the optically active glucose in living systems is the (+) isomer.

Almost all the chemical reactions in living organisms are controlled by enzymes, which are natural catalysts consisting of proteins (Section 14.3). When enzymes act as catalysts, the reactions take place at an active site somewhere on the enzyme molecule. The shape of the active site and the functional groups around it are very stereospecific and give the enzyme its catalytic properties.

The stereospecific nature of the active site means that enzymes can usually accommodate only one particular molecule at their active site. This molecule is sometimes only one of the optical isomers and, after the reaction is completed at the active site, the product released is also just one of the optical isomers.

Figure 15.23 ▲ Chris Hoy winning his third gold medal at the Beijing Olympics. All the lactic acid in his weary muscles is the (+) isomer.

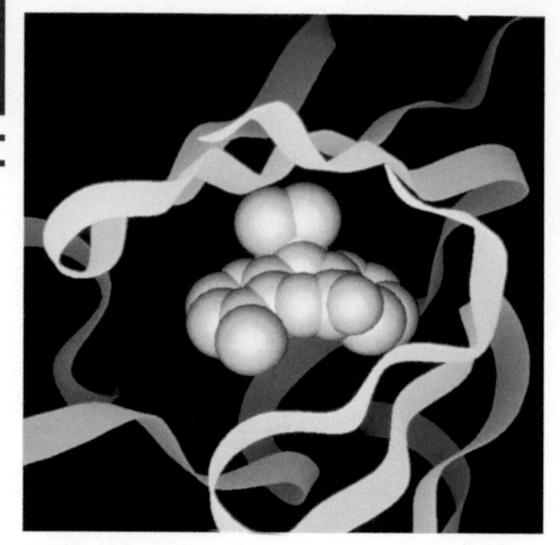

Figure 15.24▲
A computer graphic of an anti-HIV drug (yellow) blocking the active site of the enzyme reverse transcriptase (green). The drug stops the HI virus from reproducing.

15.7 Stereochemistry and drugs

Pharmaceutical chemists seek to discover and synthesise drugs (pharmaceuticals) that will prevent diseases, cure them or at least alleviate the symptoms. Drugs act on the molecules within the cells in our bodies and most drugs act on proteins either in enzymes or in sensitive receptors on the surface of cells. Receptors make cells responsive to the chemicals from nerve endings and hormones.

One of the ways in which drugs act is to target the active site of a specific enzyme. Many drugs that kill harmful bacteria or stop their reproduction work by targeting enzymes. The sulfonamide drugs and penicillin antibiotics work in this way. What is crucial is that the drug affects an enzyme vital to the biochemistry of the bacteria, but does not damage any of the enzymes in humans.

Drugs that target proteins in receptors are also very important and these include the active ingredients in medicines to treat pain, heart failure, asthma and Parkinson's disease.

Chirality in drugs

Many molecules in our bodies interact selectively with the active sites in the protein structures of enzymes and receptors. These molecules are all chiral but our body chemistry works with only one of the mirror image forms. This means that most drugs are also chiral and often the two optical isomers act on the body in different ways.

Often, one isomer is active while the other is inactive, but this is not always the case (Figure 15.25).

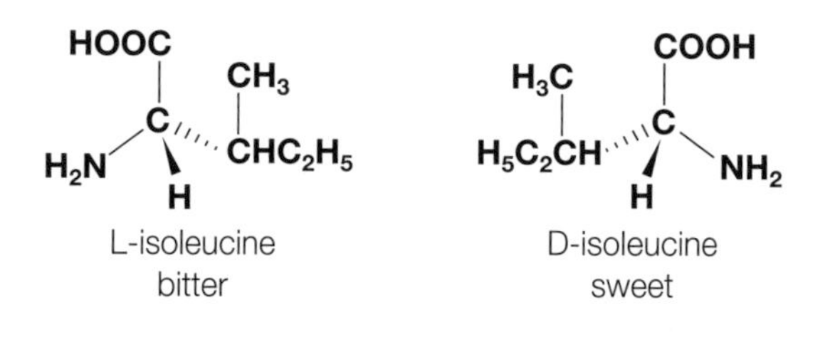

Figure 15.25▶
Optical isomers of isoleucine produce completely different tastes.

The pharmaceutical industry was alerted to the crucial importance of chirality early in the 1960s. A new drug, called thalidomide, was introduced and used to treat morning sickness in the first few months of pregnancy. Soon it was realised that thalidomide was responsible for serious malformations in babies, who were born with stunted limbs.

Test yourself

16 a) Write an equation for the nucleophilic attack of ethanal by H^- ions from $LiAlH_4$ followed by the addition of dilute sulfuric acid similar to that in Figure 15.21.

b) In what particular way is the product different from that in Figure 15.21?

c) Give the name and structure of the simplest carbonyl compound that will give a chiral product with H^- ions followed by dilute acid.

17 Copy the structure shown in Figure 15.26 and then identify its chiral centres with asterisks.

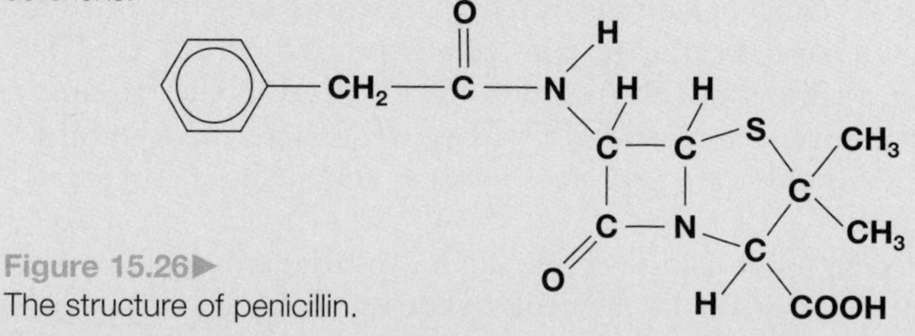

Figure 15.26▶
The structure of penicillin.

18 Explain in your own words with the help of Figure 15.22 why a laboratory synthesis of a chiral compound usually produces a racemic mixture.

Activity

Thalidomide

Thalidomide first appeared in Germany in October 1957. It was marketed as a sedative with very few side-effects. The pharmaceutical company that developed thalidomide thought it was so safe that it could be prescribed to women in their first few months of pregnancy to alleviate morning sickness. No one could have imagined what would follow.

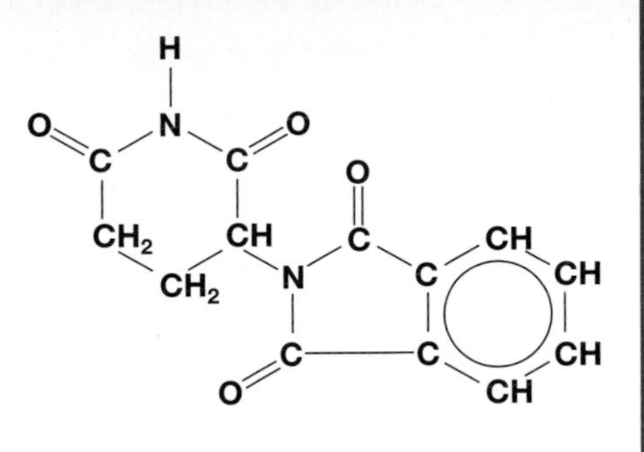

Figure 15.27▲
The structure of thalidomide.

At the beginning of the 1960s, babies began to be born with shortened, flipper-like limbs. The disabilities were traced to thalidomide and the drug was withdrawn, but not before an estimated 20 000 babies had been affected.

Drug testing was much less thorough in the 1950s. Tests had been conducted on rats and mice, but not on primates or humans. Years later, research showed that rodents metabolised thalidomide in a different way to humans.

Thalidomide is a chiral molecule (Figure 15.27) and the drug was a racemic mixture of the two isomers. Pharmacologists have since discovered that the (+) isomer is an effective and harmless sedative, while the (–) isomer is harmful to unborn babies. The adjective for this is 'teratogenic'.

A vast amount of time and money has gone into researching the properties of thalidomide and its devastating effects on the development of limbs in the uterus. Research has shown that:

- the (–) isomer blocks the development of blood vessels
- the body can convert one optical isomer of thalidomide to the other
- a few of the women with thalidomide deformities have given birth to babies with limb defects.

Today, the testing of drugs and other pharmaceutical products has become much more rigorous. However, there is still the risk that a new drug may have unforeseen side-effects on some patients.

In recent years, suggestions have been made that thalidomide would be effective in the treatment of certain cancers and in curtailing weight loss in HIV patients. This has led to rigorous clinical research trials.

Figure 15.28▲
These technicians are preparing bottles of pills for a clinical trial.

1 What functional groups are present in thalidomide?

2 a) What is the molecular formula of thalidomide?

b) What is its empirical formula?

3 Use a model-building kit to construct a molecular model of one of the thalidomide isomers.

a) Identify the chiral centre.

b) Is the character of thalidomide acidic, basic or neutral? Explain your answer.

4 Thalidomide is moderately soluble in water and in non-polar solvents. Which groups in thalidomide help to make it:

a) soluble in water

b) soluble in non-polar solvents?

5 Why were tests of thalidomide on rodents flawed?

6 Why would the harmful effects of thalidomide still have happened even if the pure (+) isomer had been prescribed to pregnant women?

7 What conclusions can be drawn from the fact that a few children born to the first victims of thalidomide have limb deformities even though their mothers have not taken the drug?

8 Suggest a reason why thalidomide may be effective in the treatment of certain cancers. (*Hint:* Cancer cells grow faster than normal cells.)

9 Suggest three precautions that should be taken in the present clinical research trials with thalidomide.

15.8 Synthesising drugs

The testing of drugs and other pharmaceuticals is now much more thorough than it was in the 1950s and 1960s. When stereospecific chiral molecules are involved, licensing authorities require pharmaceutical companies to carry out research with both isomers and identify their individual properties. Very often, this leads to a requirement to produce a single optical isomer.

This requirement to synthesise particular isomers of stereospecific drugs means that it is important to understand the mechanisms of the reactions used to produce them. By understanding the mechanisms of these reactions, we can make the synthesis of stereospecific drugs much more efficient.

This synthesis and production of drugs that contain a single optical isomer has clear benefits both for patients and for drug companies.

- It reduces the possible side-effects of the product.
- It reduces the risk of companies being taken to court for negligence.
- It improves pharmacological activity and therefore reduces the quantity needed for each dose.
- If the synthesis can be planned so that only the desired optical isomer is produced at each stage, the overall process is much more economical and there is no need to discard half the product.

Figure 15.29▶ A molecular model of ibuprofen in which some of the atoms have non-standard colours. Ibuprofen, $CH_3CH(COOH)–C_6H_4–CH_2CH(CH_3)_2$, is a chiral drug.

There is, however, one very significant disadvantage. This is the increased cost of production when it is necessary to separate optical isomers with almost identical properties.

In order to overcome the problems of producing a pure single optical isomer of a particular drug, chemists have employed some very innovative techniques. These include the use of:

- enzymes or bacteria that metabolise only one of the optical isomers and introduce the required stereospecific structure
- natural chiral molecules, such as L-amino acids or D-glucose, as starting molecules
- chiral catalysts that lead to the synthesis of stereospecific drug molecules in a similar way to enzymes.

15.9 Combinatorial chemistry in drug research

Large pharmaceutical companies synthesise and test thousands and thousands of different compounds for possible use as medicines each year. Yet only about one in ten thousand will prove suitable and safe for eventual use.

In order to synthesise such large numbers of different compounds each year, the companies use a technique known as combinatorial chemistry (Figure 15.30). Compounds can be produced in quantities of a few milligrams using computer-controlled equipment which will add reagents, stir mixtures and heat to different temperatures in a programmed sequence.

Figure 15.30▲
This photo shows a combinatorial chemistry laboratory. Using the computer-controlled techniques of combinatorial chemistry, vast numbers of new compounds can be made and tested in a short time.

Once synthesised, the compounds can be quickly tested for possible use as drugs. Those that show some suitability and effectiveness can then be prepared in larger quantities using more conventional equipment for further tests and trials.

One of the common techniques of combinatorial chemistry uses machines with 96 small reaction tubes arranged in eight rows of 12. So, using this computer-controlled equipment, 96 different esters could be produced simultaneously using 8 different carboxylic acids and 12 different alcohols.

With small-scale combinatorial chemistry techniques such as this, pharmaceutical companies can produce 10 000 compounds in one day. In comparison, a chemist using traditional methods might produce just one compound in the same time.

Another innovative technique of combinatorial chemistry involves the use of insoluble polymer beads or polymer supports containing functional groups to which reactant molecules can be bonded. A product can then be synthesised in several steps using one reactant after another, with excess reactant being washed away after each step (Figure 15.31). Throughout all the steps, the product remains bonded to the polymer support, without any being lost. At the end of the reaction sequence, the product can be removed from the polymer support using an appropriate reactant.

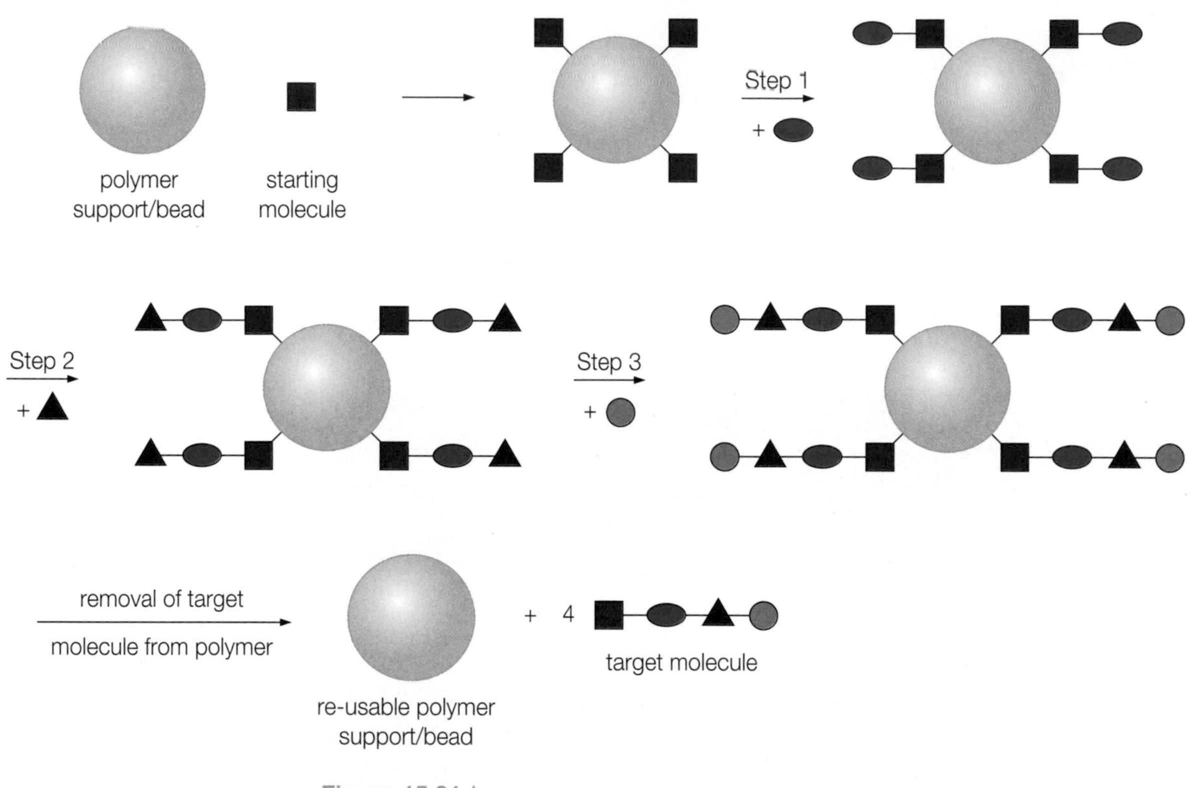

Figure 15.31▲
Using polymer supports or beads (grey spheres) to which the starting molecule is attached, target molecules (■–⬬–▲–●) can be synthesised in several steps without needing to isolate the product after each step.

REVIEW QUESTIONS

Extension questions

1 A series of tests were carried out on three organic compounds, A, B and C. The results of the tests are described below.

State the deductions which you could make from the tests on each of A, B and C. You are not expected to identify compounds A, B and C.

a) i) A is a colourless liquid which does not mix with water. (1)

ii) After warming a few drops of A with aqueous sodium hydroxide, the resulting solution was acidified with nitric acid. Silver nitrate solution was then added and a cream-coloured precipitate formed. (2)

b) i) B is a white solid that chars on heating and gives off a vapour which condenses to a liquid that turns cobalt chloride paper from blue to pink. (2)

ii) A solution of B turns universal indicator red. (1)

iii) A solution of B reacts with aqueous sodium carbonate to produce a colourless gas that turns limewater milky. (2)

iv) When a little of B is warmed with ethanol and one drop of concentrated sulfuric acid, a sweet-smelling product can be detected on pouring the reaction mixture into cold water. (2)

c) i) C is a liquid which burns with a very smoky, yellow flame. (1)

ii) C does not react with sodium carbonate solution. (1)

iii) C fizzes with sodium and gives off a gas that produces a 'pop' with a burning splint. (2)

2 The following steps were used in one method of synthesising ethyl ethanoate (boiling temperature 77 °C).

A Heat ethanol and ethanoic acid under reflux for about 45 minutes with a little concentrated sulfuric acid.

B Then, distil the reaction mixture, collecting all the liquid that distils below 84 °C.

C Shake the distillate with aqueous sodium carbonate solution.

D Add two spatula measures of anhydrous sodium sulfate or anhydrous calcium chloride to the organic product.

E Finally, redistill the organic product, collecting the liquid that boils between 75 and 79 °C.

a) Draw a diagram of the apparatus for heating under reflux. (3)

b) State the reasons for each of the procedures in steps **A** to **E**. (8)

3 Salicylic acid has been used as a painkiller. Its displayed formula is shown in Figure 15.32.

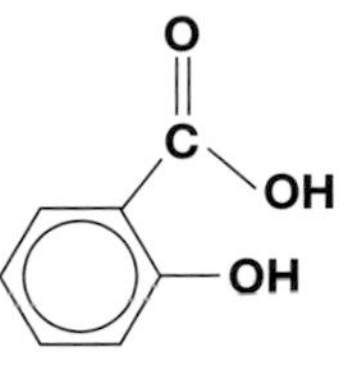

Figure 15.32▲

a) Identify the functional groups in salicylic acid. (2)

b) Write the molecular formula of salicylic acid. (1)

c) Draw the displayed formula of the organic product that forms when salicylic acid:

i) is heated under reflux with ethanol and concentrated sulfuric acid (1)

ii) reacts with bromine water (1)

iii) is warmed with aqueous sodium hydroxide. (2)

4 Salbutamol is used in inhalers to relieve the symptoms of asthma. Its displayed formula is shown in Figure 15.33.

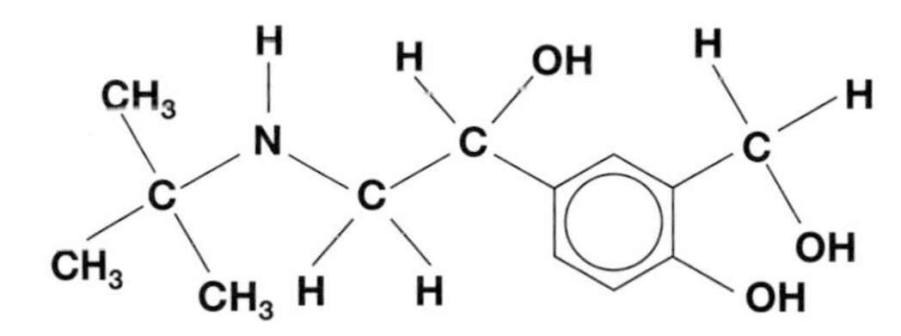

Figure 15.33▲

a) Identify the chiral centre in salbutamol. (1)

b) List three reasons why salbutamol is used as a single optical isomer in pharmaceutical products. (3)

c) Draw a displayed formula of the organic products that form when salbutamol is refluxed for some time with acidified potassium dichromate(VI). (2)

5 The skeletal formula of compound D is shown in Figure 15.34. D is a constituent of jasmine oil and it is partly responsible for the taste and smell of black tea.

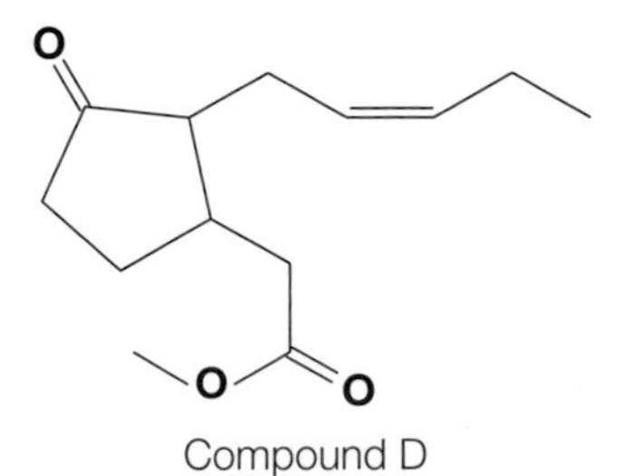

Figure 15.34▲

a) What is the molecular formula of D? (1)

b) Name the functional groups in D. (3)

c) Compound D is a stereoisomer.

i) Draw the structure of D and identify each stereochemical component with an asterisk. **(2)**

ii) Label each of these components with the type of stereoisomerism involved. **(1)**

iii) How many stereoisomers are there with the skeletal formula shown? Explain your answer. **(3)**

d) Point out four important factors that pharmaceutical companies must consider in their production of chiral compounds that are to be used in medicines. **(4)**

6 Mass spectrometry of the hydrocarbon E shows three very prominent peaks at relative atomic masses of 29, 77 and 106, which is the heaviest mass identified.

Laboratory tests on E show that it burns with a smoky flame.

a) What is the relative molecular mass of E? **(1)**

b) Suggest structures for the fragments of masses 29 and 77 in the mass spectrum of E. **(2)**

c) What further evidence, besides mass spectrometry, is there for the fragment of mass 77? **(1)**

d) Predict the structural formula of E. **(1)**

Figure 15.35 shows the simplified nmr spectrum of E.

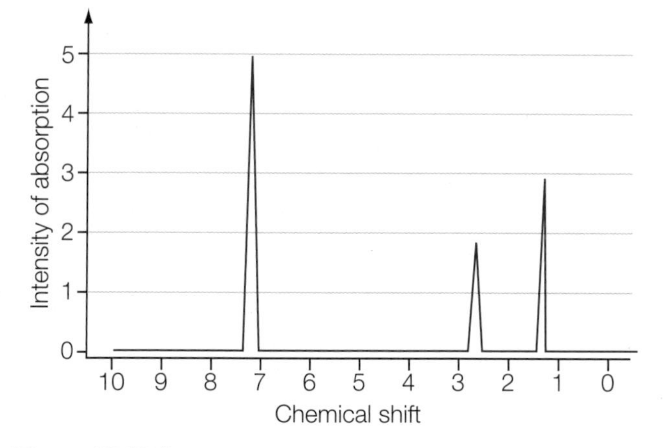

Figure 15.35▲
The proton nmr spectrum of hydrocarbon, E.

e) i) Why is the type of nmr spectrum in Figure 15.35 described as a proton nmr spectrum? **(1)**

ii) What does the size of each peak in an nmr spectrum correspond to? **(1)**

f) Identify fragments in the structure of E which relate to the three peaks in its nmr spectrum. **(3)**

7 a) Two isomeric compounds, F and G, with the molecular formula C_3H_8O can be oxidised to H and J respectively.

H reacts with Fehling's solution to produce a red-brown precipitate of copper(I) oxide.

J has no reaction with Fehling's solution, but gives a yellow crystalline product with 2,4-dinitrophenylhydrazine.

Give the structural formulae and names of F, G, H and J. **(8)**

b) i) Describe what you would observe and name all the products formed when H is warmed with sodium dichromate(VI) solution acidified with dilute sulfuric acid. **(4)**

ii) Write an equation or equations for the reactions taking place. **(3)**

8 Figure 15.36 shows a simplified infrared spectrum for ethanol in which the bonds associated with the different troughs are shown and also labelled A, B, C, D and E.

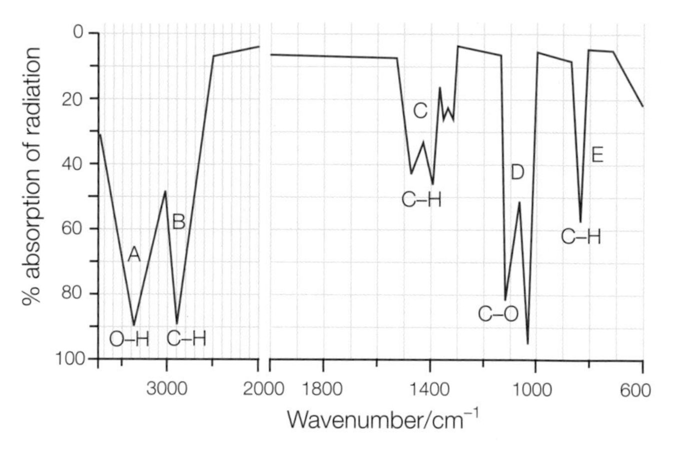

Figure 15.36▲
A simplified infrared spectrum for ethanol.

a) Why are the units on the horizontal axis cm^{-1}? **(2)**

b) Which bonds in ethanol are not identified in its infrared spectrum? **(1)**

c) Why are there troughs in the infrared spectra of organic compounds? **(2)**

d) i) Draw the displayed formula of propane. **(1)**

ii) Use the troughs labelled A to E in Figure 15.36 to explain what the infrared spectrum of propane will look like. **(2)**

e) i) Draw the displayed formula of propan-1-ol. **(1)**

ii) How will the infrared spectrum of propan-1-ol compare with that of ethanol? **(1)**

Index

Page numbers in **bold** refer to illustrations.